Springer Remote Sensing/Photogrammetry

More information about this series at http://www.springer.com/series/10182

Constantin Andronache
Editor

Remote Sensing of Clouds and Precipitation

Springer

Editor
Constantin Andronache
Boston College
Chestnut Hill, MA, USA

ISSN 2198-0721 ISSN 2198-073X (electronic)
Springer Remote Sensing/Photogrammetry
ISBN 978-3-030-10231-9 ISBN 978-3-319-72583-3 (eBook)
https://doi.org/10.1007/978-3-319-72583-3

Softcover re-print of the Hardcover 1st edition 2018

Printed on acid-free paper

This Springer imprint is published by the registered company Springer International Publishing AG part of Springer Nature.
The registered company address is: Gewerbestrasse 11, 6330 Cham, Switzerland

Preface

The objective of this book is to present a series of advanced research topics on remote sensing of clouds and precipitation. The motivation of this project comes from the recognized importance of the role that clouds and precipitation play in the weather and climate and their applications. Clouds influence the atmospheric radiative balance and hydrological cycle of the Earth. Reductions in the current uncertainties in weather forecast and climate projections require obtaining accurate cloud observations and improving their representation in numerical models. In this effort to better understand the role of cloud systems, the remote sensing of clouds and precipitation presents particular challenges and opportunities, many of them illustrated in this book.

The book has two parts, covering a wide range of topics. The first part, "Ground-Based Remote Sensing," contains articles on weather radar and its use in meteorological applications of detection, characterization, and forecasting of tornadoes, storms, and thunderstorms. The material also illustrates recent developments of radar systems for meteorological and hydrological use. One chapter presents the use of cloud radar and its applications. The second part, "Spaceborne Remote Sensing," covers satellite remote sensing of precipitation, illustrating major programs such as Tropical Rainfall Measuring Mission (TRMM) and Global Precipitation Measurement (GPM), as well as recent progress in cloud detection and characterization using cloud radar in satellite missions. It presents observations of cloud structure from CloudSat and Cloud-Aerosol Lidar and Infrared Pathfinder Satellite Observation (CALIPSO) and their implications for future satellite missions. Since water vapor plays a major role in the climate system and in cloud and precipitation processes, the book includes a chapter on water vapor detection using satellite microwave sensors. Another chapter discusses advances in cloud modeling in numerical models supported by remote sensing observations. Recent research shows that using remote sensing data can improve the performance of numerical weather prediction systems and quantitative precipitation estimation.

While the topic of remote sensing of clouds and precipitation is very broad, this book brings evidence of the enormous power and potential of increasing refined remote sensing observation techniques at multiple scales. These methods have significant societal benefits, and will contribute substantially to solve complex problems of the role cloud systems play in Earth sciences. The book is addressed to upper-level undergraduates, graduate students, researchers in atmospheric sciences, meteorology, environmental sciences, hydrology, and remote sensing, and engineers involved in weather and cloud radar technology and applications.

Boston, MA, USA
October 2017

Constantin Andronache

Acknowledgments

The publication of this book would not have been possible without the help, interest, and enthusiasm of the contributing authors. I would like to thank all authors and their supporting institutions for making this project possible. For many suggestions during the review process, I am particularly grateful to Dr. Scott Collis (Argonne National Laboratory, USA), Dr. Susanne Crewell (Institute of Geophysics and Meteorology, University of Cologne, Germany), Dr. Raquel Evaristo (Meteorological Institute, University of Bonn, Germany), Dr. Patrick N. Gatlin (National Aeronautics and Space Administration, USA), Dr. Karen Kosiba (Center for Severe Weather Research, USA), Dr. Hirohiko Masunaga (Institute for Space-Earth Research Institute, Nagoya University, Japan), and Dr. Angela Rowe (Department of Atmospheric Sciences, University of Washington, USA). I would like to thank the American Geophysical Union, the European Geosciences Union, and the American Meteorological Society for a series of recent stimulating conferences on remote sensing and their applications. I am grateful to John Wiley & Sons, Inc., for permission to use a set of figures published in their journals. Finally, it is a pleasure to acknowledge Zachary Romano, Susan Westendorf, Aaron Schiller, Kalaiselvi Ramalingam, and John Ram Kumar from Springer for their willingness and enthusiastic assistance during the project.

Contents

Contributors

Elisa Adirosi National Research Council of Italy (CNR), Institute of Atmospheric Sciences and Climate ISAC, Rome, Italy

Filipe Aires LERMA, Observatoire de Paris, Paris, France

Estellus, Paris, France

Constantin Andronache Boston College, Chestnut Hill, MA, USA

Luca Baldini National Research Council of Italy (CNR), Institute of Atmospheric Sciences and Climate ISAC, Rome, Italy

Wesley Berg Department of Atmospheric Science, Colorado State University, Fort Collins, CO, USA

David J. Bodine University of Oklahoma, Norman, OK, USA

David T. Bolvin Science Systems and Applications, Inc., Lanham, MD, USA

NASA Goddard Space Flight Center, Greenbelt, MD, USA

Mircea Grecu Goddard Earth Sciences Technology and Research, Morgan State University, Baltimore, MD, USA

NASA Goddard Space Flight Center, Greenbelt, MD, USA

George J. Huffman NASA Goddard Space Flight Center, Greenbelt, MD, USA

Takamichi Iguchi Mesoscale Atmospheric Processes Laboratory, NASA Goddard Space Flight Center, Greenbelt, MD, USA

Earth System Science Interdisciplinary Center, University of Maryland, College Park, MD, USA

Chris Kidd University of Maryland, College Park, MD, USA

NASA Goddard Space Flight Center, Greenbelt, MD, USA

Dalia B. Kirschbaum NASA Goddard Space Flight Center, Greenbelt, MD, USA

Matthew R. Kumjian Department of Meteorology and Atmospheric Science, The Pennsylvania State University, University Park, PA, USA

James M. Kurdzo MIT Lincoln Laboratory, Lexington, MA, USA

Takeshi Maesaka National Research Institute for Earth Science and Disaster Resilience, Tsukuba, Ibaraki, Japan

Toshihisa Matsui Mesoscale Atmospheric Processes Laboratory, NASA Goddard Space Flight Center, Greenbelt, MD, USA

Earth System Science Interdisciplinary Center, University of Maryland, College Park, MD, USA

Mario Montopoli National Research Council of Italy (CNR), Institute of Atmospheric Sciences and Climate ISAC, Rome, Italy

Hajime Okamoto Research Institute for Applied Mechanics, Kyushu University, Kasuga, Fukuoka, Japan

Walter A. Petersen NASA Marshall Space Flight Center, Huntsville, AL, USA

Nicoletta Roberto National Research Council of Italy (CNR), Institute of Atmospheric Sciences and Climate ISAC, Rome, Italy

Kaori Sato Research Institute for Applied Mechanics, Kyushu University, Kasuga, Fukuoka, Japan

Gail Skofronick-Jackson NASA Goddard Space Flight Center, Greenbelt, MD, USA

Yukari N. Takayabu Center for Climate System Research, The University of Tokyo, Kashiwa, Chiba, Japan

Acronyms

AIR	Atmospheric Imaging Radar
ARM	Atmospheric Radiation Measurement
ARRC	Advanced Radar Research Center
AVSET	Automated Volume Scan Evaluation and Termination
CALIOP	Cloud-Aerosol Lidar with Orthogonal Polarization
CALIPSO	Cloud-Aerosol Lidar and Infrared Pathfinder Satellite Observation
CaPE	Convection and Precipitation/Electrification project
CASA	Collaborative Adaptive Sensing of the Atmosphere
CRM	Cloud Resolving Model
CSWR	Center for Severe Weather Research
DMSP	US Defense Meteorological Satellite Program
DOW	Doppler on Wheels
ECMWF	European Centre for Medium-Range Weather Forecasts
EOS	Earth Observing System
EOV	Effective Field of View
GBVTD	Ground-Based Velocity Tracking Display
GMI	GPM Microwave Imager
GPM	Global Precipitation Measurement mission
GPM-CO	GPM Core Observatory
GPROF	Goddard profiling algorithm
G-SDSU	Goddard Satellite Data Simulator Unit
HCS	Hydrometeor classification scheme
ISCCP	International Satellite Cloud Climatology Project
JAXA	Japan Aerospace Exploration Agency
JMANHM	Japan Meteorological Agency Nonhydrostatic Model
LANL	Los Alamos National Laboratory
LEO	Low Earth Orbit
LES	Large Eddy Simulations
LIS	Lightning Imaging Sensor
LPVEx	Light Precipitation Validation Experiment
MDA	Mesocyclone Detection Algorithm

MJO	Madden-Julian oscillation
ML	Melting layer
MODIS	Moderate-resolution imaging spectroradiometer
MPAR	Multi-function Phased Array Radar sensing experiments
NASA	National Aeronautics and Space Administration
NCAR	National Center for Atmospheric Research
NEXRAD	Next-Generation Radar
NIC	Non-inductive charging theory
NPOESS	National Polar-Orbiting Operational Environmental Satellite System
NSF	National Science Foundation
NSSL	National Severe Storms Laboratory
NWP	Numerical Weather Prediction
PAIR	Polarimetric Atmospheric Imaging Radar
PIA	Path-integrated attenuation
PR	Precipitation Radar
PSD	Particle Size Distribution
QPE	Quantitative precipitation estimation
Radar	Radio detection and ranging
RaXPol	Rapid X-band Polarimetric Radar
RFGF	Rear-flank gust front
RHI	Range height indicator mode
ROC	Radar Operations Center
ROTATE	Radar Observations of Tornadoes and Thunderstorms Experiment
SAILS	Supplemental Adaptive Intra-Volume Low-Level Scans
SMART-R	Shared Mobile Atmospheric Research and Training Radar
SRT	Surface return technique
SSM/I	Special sensor microwave/imager
STAR	Simultaneous transmit and receive mode
TDA	Tornado detection algorithm
TDS	Tornado debris signature
TDWR	Terminal Doppler Weather Radar
TIP	Tornado Intercept Project
TKE	Turbulence Kinetic Energy
TMI	TRMM Microwave Imager
TMPA	TRMM Multi-satellite Precipitation Analysis
TOA	Time of arrival
TRMM	Tropical Rainfall Measuring Mission
TVS	Tornado vortex signature
TWIRL	Tornadic Winds: In situ and Radar observations at Low levels
VAD	Velocity azimuth display
VCP	Volume coverage pattern
VIIRS	Visible Infrared Imaging Radiometer Suite
VORTEX	Verification of the Origins of Rotation in Tornadoes
WSR-88D	Weather Surveillance Radar 1988 Doppler

Part I
Ground Based Remote Sensing

Introduction

Constantin Andronache

1 Overview

Clouds have a significant influence on the atmospheric radiation balance and hydrological cycle. They impact the energy budget of the Earth by interacting with incoming shortwave radiation and outgoing longwave radiation. Clouds also have an important role in the Earth's hydrological cycle by affecting water transport and precipitation [15]. Observation, measurement, and prediction of precipitation using remote sensing methods are research and operational activities with significant economic and societal impact [23, 24, 26, 55]. Since the development of weather radar and its use in meteorology, there is a continuous progress in remote sensing capabilities involving multiple disciplines. The growth of applications and the ability to address challenges in weather and climate monitoring and prediction, are reflected in many recent publications [2, 3, 10, 14, 33, 38, 39, 45, 46, 57, 59].

This volume presents a series of topics on advanced remote sensing of clouds and precipitation. Its motivation comes from the recognition that cloud systems are important in weather, hydrological and climate research and applications. Traditionally, the precipitating clouds have been the primary focus, due to the direct interest in precipitation and its impact on various economic activities and daily life. Research in Numerical Weather Prediction (NWP) and climate modeling revealed a need to characterize and understand the non-precipitating clouds as well. Thus, cloud radar was developed and used extensively in fixed sites, field campaigns, and satellites [25, 54, 60]. Together with other remote sensing methods, the cloud radar provided the means to advance the characterization of cloud structure. There is ongoing effort toward a comprehensive measurement system, capable to characterize all clouds

C. Andronache (✉)
Boston College, Chestnut Hill, MA, USA
e-mail: andronac@bc.edu

C. Andronache (ed.), *Remote Sensing of Clouds and Precipitation*,
Springer Remote Sensing/Photogrammetry,
https://doi.org/10.1007/978-3-319-72583-3_1

(non-precipitating and precipitating) on a continuous basis, at the global scale. Such goal involves the integration of measurements from ground-based and satellite sensors, and can provide much needed data sets for global atmospheric models. In practice, many measurements systems have been designed for specific applications and have particular limitations, including range, coverage, resolution, precision, and continuity. Using only ground-based measurements results in a lack of coverage over the ocean or polar regions. Satellites provide valuable measurements, covering large regions of the earth, as illustrated in this volume. A synergistic approach, using ground-based and satellite measurements, aided by numerical modeling, emerges as a promising venue to address current challenges.

The first part of this volume contains chapters on ground-based remote sensing, with focus on weather radars used in surveillance of storms, thunderstorms, and tornadoes. It also introduces the cloud radar and its applications. The second part treats the use of passive and active remote sensing from satellites. It includes the Tropical Rainfall Measuring Mission (TRMM) and the Global Precipitation Measurement (GPM) programs, to illustrate the measurements of precipitation from space. Other topics include the cloud observations from CloudSat and Cloud-Aerosol Lidar and Infrared Pathfinder Satellite Observation (CALIPSO), and the progress made in the global characterization of cloud structure and microphysics. One chapter discusses the remote sensing of water vapor using microwave radiometers, to address the importance of water substance in the atmosphere. A distinct chapter discusses the use of remote sensing data in numerical models. It concentrates on problems such as cloud microphysics schemes, comparisons between model results and remote sensing measurements, and discusses techniques used to improve numerical simulations of cloud systems. Before a brief presentation of the chapters in this book, we review a few outstanding developments in the characterization of cloud systems, not treated in the present volume.

A series of missions by the National Aeronautics and Space Administration (NASA) in the context of the Earth Observing System (EOS) employed a multitude of sensors that contributed to the understanding of Earth-atmosphere system. While EOS is designed to obtain comprehensive observations of land, ocean and atmosphere, this volume concentrates only on aspects related to clouds and precipitation. One of the most used satellite remote sensing platforms for earth science investigations is the Moderate-Resolution Imaging Spectroradiometer (MODIS). It was launched on NASA's Terra satellite on December 18, 1999, followed by Aqua on May 4, 2002. The instrument was designed to provide global observations of the earth's atmosphere, land, and oceans. MODIS uses spectral channels ranging from the visible (VIS) to the infrared (IR) to measure the reflected solar and emitted thermal radiation in 36 spectral channels. MODIS provides spectral and spatial capabilities for retrieving cloud top, optical, and microphysical properties. Cloud top properties measured are pressure, temperature, and height during both day and night. Cloud optical and microphysical properties are: cloud optical thickness, effective particle radius, and derived water path for both liquid water and ice cloud thermodynamic phases during daytime only [44]. A follow-on instrument to the MODIS with similar capabilities is the Visible and Infrared Imaging Radiometer

Suite (VIIRS). These sensors are part of the following operational missions named the National Polar-Orbiting Environmental Satellite Series (NPOESS) and the NPOESS Preparatory Project (NPP). The NPP was designed to bridge the gap between the Terra and Aqua missions and the NPOESS series [45, 46].

A significant contribution to remote sensing of clouds and precipitation was provided by the U.S. Department of Energy (DOE) Atmospheric Radiation Measurement (ARM) Program with its focus on the role of clouds in the climate system [37, 56]. ARM established a set of permanent ground-based observatories and facilitated many field missions [51, 58]. Advances in ground-based remote sensing capabilities developed by the ARM program, aided by numerous field campaigns, resulted in accurate methods to observe atmospheric processes related to water vapor, aerosol, clouds, and radiation [50]. The ability to detect and characterize cloud systems at ARM sites provided the basis for the development of additional observation stations in other parts of the world. ARM contributed substantially to the validation of satellite cloud detection schemes, as well as satellite retrievals of cloud properties. These ARM activities were possible through its set of advanced ground-based instruments such as the millimeter-wavelength cloud radar, the Raman lidar, as well as the passive microwave radiometers and infrared interferometers [36, 50].

One significant development in Europe was the Cloudnet program, which established a standard set of ground-based remote sensing instruments capable of providing cloud parameters that can be compared with current operational NWP models [9, 21]. Further developments following the Cloudnet program and the expansion of ARM capabilities and collaborations have resulted in a more comprehensive approach to monitor cloud systems, at a variety of sites, enabling the evaluation and improvement of high-resolution numerical models [17].

2 Ground-Based Remote Sensing

The first part of the volume shows examples of weather radar applications to detect, monitor severe weather phenomena including storms, tornadoes, and thunderstorms. One chapter is dedicated to the use of cloud radar and its applications.

Chapter 2 provides a conceptual overview of the principles of dual-polarization Doppler weather radar as one of the essential tools in remote sensing of clouds and precipitation [10, 11, 14, 27, 28]. It starts with a perspective on the fundamentals of electromagnetic scattering by individual nonspherical particles. It follows with the treatment of scattering by a population of particles and the concept of equivalent reflectivity factor, the Doppler principle and Doppler velocity estimation, and discusses the essential overview of the radar system. The chapter introduces the dual-polarization radar variables and their physical interpretation. Dual-polarization Doppler weather radar observations have facilitated new insights into cloud and precipitation physics, as well as operationally significant advances in the detection and monitoring the hazardous weather. Such weather cases include severe convective storms (hail, tornadoes, damaging winds, floods) and winter storms. The chapter concludes with a survey of the numerous polarimetric Doppler weather radar applications, highlighting novel research avenues for improved understanding of clouds and precipitation. Algorithms for classification of radar echoes and quantitative precipitation estimation have also made substantial improvements in recent decades. Future advances will improve our understanding of fundamental physical processes in clouds and precipitation and the blending of numerical weather prediction models and polarimetric Doppler weather radar observations through data assimilation techniques.

Chapter 3 starts with a discussion of the role of radar for tornado studies. Tornadoes cause many fatalities and significant damage during major outbreaks [5]. Each year, numerous tornadoes are reported in the United States and other regions of the world. The chapter follows with a description of theory and simulations of tornadoes. Conceptual models for the flow structure of the tornado were developed based on theory, vortex chamber experiments, and numerical Large Eddy Simulation (LES). The next section presents the history of radars for tornadoes, from the early radar to current mobile Doppler radar systems. One outstanding development was WSR-88D, also commonly referred to as NEXRAD ("Next-Generation Radar"), widespread, ground-based, fixed-site, Doppler radar coverage to the vast majority of the continental United States during the late 1980s and early 1990s. From a historical perspective, a significant advancement was the realization of major tornado field projects using radar. The chapter continues with a section on achievements from ground-based radar observations since the advent of Doppler radar in the 1970s. The accomplishments include tornado detection, radar analysis techniques, tornado genesis and dynamics, and dual-polarization and rapid-scan radar observations of tornadoes [6–8, 31, 32]. Finally, the chapter presents several unsolved tornado research areas and needed improvements in radar analysis techniques.

Chapter 4 examines the applications of radar for detection and characterization of thunderstorms. Weather radars are essential tools to identify and analyze thunderstorms, often associated with intense precipitation, lightning, and strong winds [12, 13, 41, 47]. Lightning in convective thunderstorms can impact aviation and management of electric infrastructures. Various features of a thunderstorm can be revealed such its severity, trajectory, and particle type distribution within the storm cloud system. This chapter describes how different radar measurements are utilized to identify essential characteristics of thunderstorms. The chapter presents the characteristics of radar signatures of convection. It follows with a discussion on cloud electrification processes inferred from dual-polarization radar observations. Dual-polarization weather radar allows the detection and measurement of graupel particles within a cumulonimbus that can be associated with the electrical activity of the thunderstorm. Finally, authors present applications of combined radar and lightning measurements.

Chapter 5 describes advances in cloud radar and its applications in ground-based networks to provide advanced warning of heavy rainfall events. The chapter presents the requirements of cloud radars and provides a typology of these instruments based on antenna scanning and capabilities. After an analysis of physical conditions affecting radar reflectivity and considering the Bragg scattering and transmissivity of the atmosphere, it is shown that Ka- and W-bands are suitable for cloud observations. Finally, observations using a dual-polarization scanning cloud radar system are illustrated and discussed, particularly from the perspective of their ability to provide cloud system details not available from a precipitation radar. Dual-polarization cloud radar can become widespread, and it is expected to make a great contribution to cloud microphysics studies [25, 35].

3 Spaceborne Remote Sensing

The second part of the volume presents satellite measurements of precipitation from the TRMM and GPM programs. One chapter is dedicated to cloud characterization from CloudSat and CALIPSO. Another chapter treats water vapor measurements with satellite microwave sensors. Finally, one chapter analyses the advances in numerical cloud modeling using remote sensing data.

Chapter 6 presents the applications of radar to detection and characterization of precipitation in the Tropical Rainfall Measuring Mission (TRMM). This joint mission between NASA and the Japan Aerospace Exploration Agency (JAXA) was launched in 1997 and operated until April 2015. TRMM was designed to study rainfall for weather and climate research and provided critical precipitation measurements in the tropical and subtropical regions [29, 30]. TRMM generated long-term rainfall records, contributed to tropical cyclone forecasting, numerical weather prediction, and precipitation climatologies, as well as a broad range of societal applications [24]. The TRMM most important instruments, i.e., the Precipitation Radar (PR) and the TRMM microwave Imager (TMI), as well as the algorithms used to derive precipitation estimates from their observations, are described. The PR provided new insights into tropical storm physics, and the TMI enabled measurements of water vapor, cloud water, and rainfall intensity in the atmosphere. The challenges in obtaining precipitation from TRMM observations and the strategies to overcome them are also examined. A section of the chapter is devoted to the TRMM Multi-satellite Precipitation Analysis (TMPA) precipitation product, developed within the project to mitigate temporal sampling errors in the precipitation estimates. The chapter also includes a brief presentation of directions and examples of usage of TRMM observations and products in scientific and societal applications [19, 34, 61].

Chapter 7 reviews the applications of radar to detection and characterization of precipitation in the Global Precipitation Measurement (GPM) program, a follow-on of the successful TRMM program. GPM is an international satellite mission initiated by the NASA and JAXA to expand global precipitation measurements from space [18]. The GPM Core Observatory (GPM-CO, launched February 2014) carries a Ka/Ku-band Dual-frequency Precipitation Radar (DPR) and a multifrequency (10–183 GHz) microwave radiometer, the GPM Microwave Imager (GMI) [16]. The GPM-CO serves as an observatory of precipitating systems and a calibration reference for precipitation estimates from a constellation of microwave radiometers [18].

The mission is designed to provide the next-generation, constellation-based global precipitation products for research and societal applications [53]. With a focus on the Earth's water and energy cycle, GPM delivers near-real-time (NRT) observations of precipitation for monitoring severe weather events, freshwater resources, and other societal applications. GPM NRT data availability enables the use of space-based precipitation observations in a variety of practical applications [24]. This chapter presents the GPM mission design, descriptions of sensor characteristics, inter-satellite calibration, retrieval methodologies, ground validation activities, and applications.

Chapter 8 presents a perspective on remote sensing of clouds using active sensors with a focus on CloudSat, CALIPSO, and EarthCARE. Cloudsat and CALIPSO are parts of the A-Train satellite constellation [55] which included the NASA Aqua and Aura satellites, and the French satellite PARASOL. The A-Train approach to Earth observation can advance our understanding of atmospheric hydrological processes. The chapter describes how the cloud profiling radar onboard CloudSat and CALIPSO lidar together can provide the vertical structure of cloud macroscale and its microphysical properties. To assess the role of clouds in the climate system, numerical models need ample observations, including the atmospheric temperature, pressure, water vapor, the 3D multi-layer structure of clouds, cloud phase, and microphysical parameters. The chapter presents the principles of cloud radar and lidar, algorithms used to determine physical parameters of interest in cloud characterization and discusses the challenges involved in detection of various types of clouds [42].

Chapter 9 presents the methodology of atmospheric water vapor profiling over ocean and land and for clear and cloudy situations using microwave observations. The focus is on a water vapor retrieval algorithm that uses satellite observations in the microwave region. A neural network scheme is used and includes a dedicated calibration system of the satellite observations. The water vapor is retrieved for clear and cloudy conditions, over ocean and land surfaces, in the absence of precipitation. The atmospheric relative humidity profile is retrieved together with the total column water vapor. The algorithm has been developed for the instruments AMSR-E/HSB (resp. AMSUA/MHS) onboard AQUA (resp. MetOp) platforms. The chapter describes the principles of the inversion method and the theoretical retrieval uncertainties. Then, the algorithm is tested using real observations from HSB/AMSRE (resp. MHS/AMSU-A) instruments on board AQUA (resp. MetOp) platforms. The results are compared to European Center for Medium range Weather Forecasting (ECMWF) analyses and data from radiosondes. Posteriori validation tests on the brightness temperatures reveal an overall positive impact of the retrievals relatively to the a priori ECMWF analyses [1, 4].

Chapter 10 provides an overview of recent efforts towards improving parameterizations for cloud and precipitation microphysics in atmospheric models, in comparison with remote sensing measurements. Significant research effort has been dedicated to the representation of cloud microphysical processes in numerical models at various scales [20, 22, 40, 43, 48, 49, 52]. The chapter describes the structure of typical cloud microphysics parameterization. It follows with the presentation of two conventional approaches used to compare the numerical model output with remote sensing measurements. One approach is to compare physical quantities estimated from measurements, against corresponding quantities calculated from model simulations. This method is a direct way to identify errors in target quantities but has potential drawbacks caused by the uncertainties in the quantities retrieved from remote sensing measurements. An alternative approach is to use signal simulators which enables signal-based comparison between the measurements and the atmospheric model simulations without inherent uncertainty within the retrieval algorithms. Advantages and challenges of these approaches are discussed.

4 Concluding Remarks

The recent research on remote sensing of clouds and precipitation presented in this volume, as well as the selected references for each chapter, provide an overview of current efforts to characterize cloud systems using remote sensing. Understanding the role of clouds and precipitation in the earth system is increasingly imperative for contemporary activities such as short-term weather forecast, accurate climate projections, and their multiple applications. In response to these challenges, there is a need to refine atmospheric observation techniques, numerical models, and use a synergistic approach to address complex problems. Remote sensing methods play a crucial role in this undertaking. These efforts are sustained by many evolving research programs and by a vigorous community of scientists and engineers. The book "Remote sensing of clouds and precipitation" provides the essential information to help readers access the current status of remote sensing of clouds systems, and their implications for weather and climate monitoring and prediction, and many practical applications.

Acknowledgements I want to express my sincere appreciation to all authors and reviewers who contributed to this volume. Many thanks to Springer for continuous assistance during this project.

References

1. Aires, F., F. Bernardo, and C. Prigent. 2013. Atmospheric water-vapour profiling from passive microwave sounders over ocean and land. Part I: methodology for theMegha-Tropiques mission. *Quarterly Journal of the Royal Meteorological Society* 139: 852–864. https://doi.org/10.1002/qj.1888.
2. Andronache, C., ed. 2017. *Mixed-Phase Clouds: Observations and Modeling*. Amsterdam: Elsevier.
3. Atlas, D., ed. 2015. *Radar in Meteorology: Battan Memorial and 40th Anniversary Radar Meteorology Conference*. New York: Springer.
4. Bernardo, F., F. Aires, and C. Prigent. 2013. Atmospheric water-vapour profiling from passive microwavesounders over ocean and land. Part II: validation using existing instruments. *Quarterly Journal of the Royal Meteorological Society* 139: 865–878. https://doi.org/10.1002/qj.1946.

5. Bluestein, H.B. 2013. *Severe Convective Storms and Tornadoes: Observations and Dynamics*. Berlin/Heidelberg: Springer.
6. Bluestein, H.B., M.M. French, I. PopStefanija, R.T. Bluth, and J.B. Knorr. 2010. A mobile, phased-array Doppler radar for the study of severe convective storms. *Bulletin of the American Meteorological Society* 91(5): 579–600.
7. Bodine, D.J., R.D. Palmer, and G. Zhang. 2014. Dual-wavelength polarimetric radar analyses of tornadic debris signatures. *Journal of Applied Meteorology and Climatology* 53: 242–261.
8. Bodine, D.J., R.D. Palmer, T. Maruyama, C.J. Fulton, Y. Zhu, and B.L. Cheong. 2016. Simulated frequency dependence of radar observations of tornadoes. *Journal of Atmospheric and Oceanic Technology* 33(9): 1825–1842.
9. Bouniol, D., et al. 2010. Using continuous ground-based radar and lidar measurements for evaluating the representation of clouds in four operational models. *Journal of Applied Meteorology and Climatology* 49: 1971–1991. https://doi.org/10.1175/2010JAMC2333.1.
10. Bringi, V.N., and V. Chandrasekar. 2001. *Polarimetric Doppler Weather Radar: Principles and Applications*. Cambridge: Cambridge University Press.
11. Fabry, F. 2015. *Radar Meteorology: Principles and Practice*. Cambridge: Cambridge University Press.
12. Federico, S., E. Avolio, M. Petracca, G. Panegrossi, P. Sano, D. Casella, and S. Dietrich. 2014 Simulating lightning into the RAMS model: implementation and preliminary results. *Natural Hazards and Earth System Sciences* 14: 2933–2950. https://doi.org/10.5194/nhess-14-2933-2014.
13. Formenton, M., G. Panegrossi, D. Casella, S. Dietrich, A. Mugnai, P. Sano, F. Di Paola, H.-D. Betz, C. Price, and Y. Yair. 2013. Using a cloud electrification model to study relationships between lightning activity and cloud microphysical structure. *Natural Hazards and Earth System Sciences* 13: 1085–1104. https://doi.org/10.5194/nhess-13-1085-2013.
14. Fukao, S., K. Hamazu, and R.J. Doviak. 2014. *Radar for Meteorological and Atmospheric Observations*. Tokyo: Springer.
15. Gettelman, A., and S.C. Sherwood. 2016. Processes responsible for cloud feedback. *Current Climate Change Reports* 2: 179–189. https://doi.org/10.1007/s40641-016-0052-8.
16. Grecu, M., W.S. Olson, S.J. Munchak, S. Ringerud, L. Liao, Z. Haddad, B.L. Kelley, and S.F. McLaughlin. 2016. The GPM combined algorithm. *Journal of Atmospheric and Oceanic Technology* 33: 2225–2245. https://doi.org/10.1175/JTECH-D-16-0019.1.
17. Haeffelin, M., et al. 2016. Parallel developments and formal collaboration between European atmospheric profiling observatories and the U.S. ARM research program. In *The Atmospheric Radiation Measurement (ARM) Program: The First 20 Years*. Meteorological Monographs. Vol. 57. Boston: American Meteor Society. https://doi.org/10.1175/AMSMONOGRAPHS-D-15-0045.1.
18. Hou, A.Y., et al. 2014. The global precipitation measurement mission. *Bulletin of the American Meteorological Society* 95: 701–722. https://doi.org/10.1175/BAMS-D-13-00164.1.
19. Huffman, G.J., R.F. Adler, D.T. Bolvin, G. Gu, E.J. Nelkin, K.P. Bowman, Y. Hong, E.F. Stocker, and D.B. Wolff. 2007. The TRMM multi-satellite precipitation analysis: quasi-global, multi-year, combined-sensor precipitation estimates at fine scale. *Journal of Hydrometeorology* 8(1): 38–55.
20. Iguchi, T., T. Matsui, J. Shi, W. Tao, A. Khain, A. Hou, R. Cifelli, A. Heymsfield, and A. Tokay. 2012. Numerical analysis using WRF-SBM for the cloud microphysical structures in the C3VP field campaign: impacts of supercooled droplets and resultant riming on snow microphysics. *Journal of Geophysical Research-Atmospheres* 117. https://doi.org/10.1029/2012JD018101.
21. Illingworth, A.J., et al. 2007. CloudNet: continuous evaluations of cloud profiles in seven operational models using ground-based observations. *Bulletin of the American Meteorological Society* 88: 883–898.
22. Khain, A.P., K.D. Beheng, A. Heymsfield, A. Korolev, S.O. Krichak, Z. Levin, M. Pinsky, V. Phillips, T. Prabhakaran, and A. Teller. 2015. Representation of microphysical processes in cloud-resolving models: spectral (bin) microphysics versus bulk parameterization. *Revista Geografica* 53(2): 247–322.

23. Kidd, C., and V. Levizzani. 2011. Status of satellite precipitation retrievals. *Hydrology and Earth System Sciences* 15: 1109–1116. https://doi.org/10.5194/hess-15-1109-2011.
24. Kirschbaum, D.B., et al. 2017. NASA's remotely sensed precipitation: a reservoir for applications users. *Bulletin of the American Meteorological Society* 98: 1169–1184. https://doi.org/10.1175/BAMS-D-15-00296.1.
25. Kollias, P., E.E. Clothiaux, M.A. Miller, B.A. Albrecht, G.L. Stephens, and T.P. Ackerman. 2007. Millimeter-wavelength radars: new frontier in atmospheric cloud and precipitation research. *Bulletin of the American Meteorological Society* 88: 1608–1624. https://doi.org/10.1175/BAMS-88-10-1608.
26. Kucera, P.A., E.E. Ebert, F.J. Turk, V. Levizzani, D. Kirschbaum, F.J. Tapiador, A. Loew, and M. Borsche. 2013. Precipitation from space: advancing earth system science. *Bulletin of the American Meteorological Society* 94: 365–375. https://doi.org/10.1175/BAMS-D-11-00171.1.
27. Kumjian, M.R., and W. Deierling. 2015. Analysis of thundersnow storms over northern Colorado. *Weather Forecast* 30: 1469–1490. https://doi.org/10.1175/WAF-D-15-0007.1.
28. Kumjian, M.R., and K.A. Lombardo. 2017. Insights into the evolving microphysical and kinematic structure of Northeastern U.S. winter storms from dual-polarization Doppler radar. *Monthly Weather Review* 145: 1033–1061.
29. Kummerow, C., W. Barnes, T. Kozu, J. Shiue, and J. Simpson. 1998. The tropical rainfall measuring mission (TRMM) sensor package. *Journal of Atmospheric and Oceanic Technology* 15: 809–817.
30. Kummerow, C., J. Simpson, O. Thiele, W. Barnes, A.T.C. Chang, E. Stocker, R.F. Adler, A. Hou, R. Kakar, F. Wentz, P. Ashcroft, T. Kozu, Y. Hong, K. Okamoto, T. Iguchi, K. Kuriowa, E. Im, Z. Haddad, G. Huffman, B. Ferrier, W.S. Olson, E. Zipser, E.A. Smith, T.T. Wilheit, G. North, T. Krishnamurti, and K. Nakamura. 2000. The status of the tropical rainfall measuring mission (TRMM) after two years in orbit. *Journal of Applied Meteorology, Part 1* 39(12): 1965–1982.
31. Kurdzo, J.M., and R.D. Palmer. 2012. Objective optimization of weather radar networks for low-level coverage using a genetic algorithm. *Journal of Atmospheric and Oceanic Technology* 29(6): 807–821.
32. Kurdzo, J.M., F. Nai, D.J. Bodine, T.A. Bonin, R.D. Palmer, B.L. Cheong, J. Lujan, A. Mahre, and A.D. Byrd. 2017. Observations of severe local storms and tornadoes with the atmospheric imaging radar. *Bulletin of the American Meteorological Society* 98(5): 915–935.
33. Levizzani, V., P. Bauer, and F.J. Turk. 2007. *Measuring Precipitation from Space: EURAINSAT and the Future*. Vol. 28. Dordrecht: Springer Science and Business Media.
34. Liu, Z., D. Ostrenga, W. Teng, and S. Kempler. 2012. Tropical rainfall measuring mission (TRMM) precipitation data and services for research and applications. *Bulletin of the American Meteorological Society* 93: 1317–1325. https://doi.org/10.1175/BAMS-D-11-00152.1.
35. Maesaka, T., K. Iwanami, S.-I. Suzuki, Y. Shusse, and N. Sakurai. 2015. Cloud radar network in Tokyo metropolitan area for early detection of cumulonimbus Generation. Paper Presented at 37th Conference on Radar Meteorology. Norman: American Meteorological Society. https://ams.confex.com/ams/37RADAR/webprogram/Paper275910.html.
36. Marchand, R. 2016. ARM and satellite cloud validation. *Meteorological Monographs* 57: 30.1–30.11. https://doi.org/10.1175/AMSMONOGRAPHS-D-15-0038.1.
37. Mather, J.H., D.D. Turner, and T.P. Ackerman. 2016. Scientific maturation of the ARM program. In *The Atmospheric Radiation Measurement (ARM) Program: The First 20 Years*. Meteorological Monographs. Vol. 57. Boston: American Meteorological Society. https://doi.org/10.1175/AMSMONOGRAPHS-D-15-0053.1.
38. Meischner, P. 2003. Weather radar-principle and advanced applications. In *Weather Radar – Principle and Advanced Applications Physics of Earth and Space Environments*, 337. Berlin: Springer. ISBN: 3-540-000328-2.
39. Michaelides, S.C., ed. 2008. *Precipitation: Advances in Measurement, Estimation and Prediction*. Berlin: Springer Science and Business Media.

40. Morrison, H., J.A. Curry, and V.I. Khvorostyanov. 2005. A new double-moment microphysics parameterization for application in cloud and climate models. Part I: description. *Journal of the Atmospheric Sciences* 62: 1665–1677. https://doi.org/10.1175/JAS3446.1.
41. Mosier, R.M., C. Schumacher, R.E. Orville, and L.D. Carey. 2011. Radar nowcasting of cloud-to-ground lightning over Houston, Texas. *Weather Forecast* 26: 199–212. https://doi.org/10.1175/2010WAF2222431.1.
42. Okamoto, H., K. Sato, and Y. Hagihara. 2010. Global analysis of ice microphysics from CloudSat and CALIPSO: incorporation of specular reflection in lidar signals. *Journal of Geophysical Research* 115: D22209. https://doi.org/10.1029/2009JD013383.
43. Phillips, V.T.J., P.J. DeMott, C. Andronache, K. Pratt, K.A. Prather, R. Subramanian, and C. Twohy. 2013. Improvements to an empirical parameterization of heterogeneous ice nucleation and its comparison with observations. *Journal of the Atmospheric Sciences* 70: 378–409.
44. Platnick, S., K.G. Meyer, M.D. King, G. Wind, N. Amarasinghe, B. Marchant, G.T. Arnold, Z. Zhang, P.A. Hubanks, R.E. Holz, and P. Yang. 2017. The MODIS cloud optical and microphysical products: collection 6 updates and examples from Terra and Aqua. *IEEE Transactions on Geoscience and Remote Sensing* 55(1): 502–525.
45. Qu, J.J., W. Gao, M. Kafatos, R.E. Murphy, and V.V. Salomonson, eds. 2006. *Earth Science Satellite Remote Sensing: vol. 1: Science and Instruments*. Beijing/Berlin, Heidelberg: Tsinghua University Press/Springer.
46. Qu, J.J., W. Gao, M. Kafatos, R.E. Murphy, and V.V. Salomonson, eds. 2006. *Earth Science Satellite Remote Sensing: vol. 2: Data, Computational Processing, and Tools*. Beijing: Tsinghua University Press/Berlin, Heidelberg: Springer.
47. Roberto, N., E. Adirosi, L. Baldini, D. Casella, S. Dietrich, P. Gatlin, G. Panegrossi, M. Petracca, P. Sano, and A. Tokay. 2016. Multi-sensor analysis of convective activity in central Italy during the Hy-MeX SOP 1.1. *Atmospheric Measurement Techniques* 9: 535–552. https://doi.org/10.5194/amt-9-535-2016.
48. Seifert, A., and K.D. Beheng. 2001. A double-moment parameterization for simulating autoconversion, accretion and selfcollection. *Atmospheric Research* 59: 265–281.
49. Seifert, A., and K.D. Beheng. 2006. A two-moment cloud microphysics parameterization for mixed-phase clouds. Part 1: model description. *Meteorology and Atmospheric Physics* 92(1): 45–66.
50. Shupe, M.D., J.M. Comstock, D.D. Turner, and G.G. Mace. 2016. Cloud property retrievals in the ARM program. In *The Atmospheric Radiation Measurement (ARM) Program: The First 20 Years*. Meteorological Monographs. Vol. 57. Boston: American Meteor Society. https://doi.org/10.1175/AMSMONOGRAPHS-D-15-0030.1.
51. Sisterson, D., R. Peppler, T.S. Cress, P. Lamb, and D.D. Turner. 2016. The ARM southern great plains (SGP) site. In *The Atmospheric Radiation Measurement (ARM) Program: The First 20 Years*. Meteorological Monographs. Vol. 57. Boston: American Meteor Society. https://doi.org/10.1175/AMSMONOGRAPHS-D-16-0004.1.
52. Skamarock, W.C., J.B. Klemp, J. Dudhia, D.O. Gill, D.M. Barker, M.G. Duda, X.Y. Huang, W. Wang, and J.G. Powers. 2008. A description of the advanced research WRF version 3, NCAR technical note, Mesoscale and Microscale Meteorology Division, National Center for Atmospheric Research, Boulder, CO. Shupe, M., et al. 2008. A focus on mixed-phase clouds: the status of ground-based observational methods. *Bulletin of the American Meteorological Society* 87: 1549–1562.
53. Skofronick-Jackson, G., W.A. Petersen, W. Berg, C. Kidd, E.F. Stocker, D.B. Kirschbaum, R. Kakar, S.A. Braun, G.J. Huffman, T. Iguchi, P.E. Kirstetter, C. Kummerow, R. Meneghini, R. Oki, W.S. Olson, Y.N. Takayabu, K. Furukawa, and T. Wilheit. 2017. The global precipitation measurement (GPM) mission for science and society. *Bulletin of the American Meteorological Society* 98: 1679–1695. https://doi.org/10.1175/BAMS-D-15-00306.1.
54. Stephens, G.L., and C.D. Kummerow. 2007. The remote sensing of clouds and precipitation from space: a review. *Journal of the Atmospheric Sciences* 64: 3742–3765.
55. Stephens, G.L., et al. 2002. The CLOUDSAT mission and the A-train – a new dimension of space-based observations of clouds and precipitation. *Bulletin of the American Meteorological Society* 83: 1771–1790.

56. Stokes, G.M. 2016. Original ARM concept and launch. In *The Atmospheric Radiation Measurement (ARM) Program: The First 20 Years*. Meteorological Monographs. Vol. 57. Boston: American Meteor Society. https://doi.org/10.1175/AMSMONOGRAPHS-D-15-0054.1.
57. Tao, W.K., ed. 2015. *Cloud Systems, Hurricanes, and the Tropical Rainfall Measuring Mission (TRMM): A Tribute to Joanne Simpson*. Berlin: Springer.
58. Verlinde, J., B. Zak, M.D. Shupe, M. Ivey, and K. Stamnes. 2016. The ARM North Slope of Alaska (NSA) sites. In *The Atmospheric Radiation Measurement (ARM) Program: The First 20 Years*. Meteorological Monographs. Vol. 57. Boston: American Meteor Society. https://doi.org/10.1175/AMSMONOGRAPHS-D-15-0023.1.
59. Wakimoto, R.M., and R. Srivastava, eds. 2003. *Radar and Atmospheric Science: A Collection of Essays in Honor of David Atlas*. Vol. 30(52). Boston: American Meteorological Society.
60. Winker, D.M., M.A. Vaughan, A.H. Omar, Y. Hu, K.A. Powell, Z. Liu, W.H. Hunt, and S.A. Young. 2009. Overview of the CALIPSO mission and CALIOP data processing algorithms. *Journal of Atmospheric and Oceanic Technology* 26: 2310–2323. https://doi.org/10.1175/2009JTECHA1281.1.
61. Yong, B., D. Liu, J.J. Gourley, Y. Tian, G.J. Huffman, L. Ren, and Y. Hong. 2015. Global view of real-time TRMM multisatellite precipitation analysis: implications for its successor global precipitation measurement mission. *Bulletin of the American Meteorological Society* 96: 283–296. https://doi.org/10.1175/BAMS-D-14-00017.1.

Weather Radars

Matthew R. Kumjian

1 Introduction

Weather radar is the most indispensable tool for remote sensing of clouds and precipitation. Although initially developed for military purposes of detecting aircraft and ships, radar operators almost immediately noticed its ability to detect precipitation. Since then, advancements in radar engineering, signal processing, and meteorology have improved the accuracy, resolution, and information available from weather radars. This has led to earlier warnings, fewer false alarms, and improved understanding of hazardous weather, including severe convective storms and associated hail, tornadoes, floods, and damaging winds, as well as winter storms and associated heavy snow, precipitation transitions, and disruptive mixed-phase precipitation. The added information from new weather radar technology has led to the development and refinement of precipitation classification schemes that characterize radar pixels as one of a variety of precipitation types or non-precipitation signals, or *echoes*. This provides operational meteorologists a quick estimate of the types of radar echoes present in a given scan. These improvements have also benefited hydrometeorologists by providing valuable quantitative precipitation estimates. More recently, weather radar data are being assimilated into numerical models to improve model analyses and forecasts. Many of these topics will be treated in this and other chapters in this volume.

Radar is an acronym coined in 1940 by the U.S. Navy that stands for RAdio Detection And Ranging. Radars operate by transmitting powerful radio waves, electromagnetic radiation invisible to the human eye. At any point in space or

M. R. Kumjian (✉)
Department of Meteorology and Atmospheric Science, The Pennsylvania State University, University Park, PA, USA
e-mail: kumjian@psu.edu

C. Andronache (ed.), *Remote Sensing of Clouds and Precipitation*,
Springer Remote Sensing/Photogrammetry,
https://doi.org/10.1007/978-3-319-72583-3_2

time, one can characterize these electromagnetic waves by their *frequency* (rate of oscillation given in cycles per second or Hz), *amplitude* (intensity of the oscillating electric and magnetic fields), *phase* (point in the oscillation cycle of the electric or magnetic field), and *polarization* (direction in which the electric field vector oscillates, which is orthogonal to the direction of wave propagation). These radio waves propagate through the atmosphere and can interact with cloud and precipitation particles known as *hydrometeors*, as well as other nonmeteorological particles, biota, and ground targets. When illuminated by the radar signal, these particles undergo an electromagnetic excitation, extracting some of the radar wave's energy, radiating some of it in all directions in a process called *scattering*, and dissipating some of it as thermal energy in a process called absorption. Some of this radiation may be scattered back towards the radar (*backscattering*). The amount of the transmitted wave's power lost during propagation owing to scattering and absorption is called *attenuation*.

The backscattered signal provides important information for characterizing clouds and precipitation. For example, the amplitude of the received signal can provide information about the precipitation intensity. Because electromagnetic waves propagate through the atmosphere at a known speed (the speed of light), the time between when a signal is transmitted and when the backscattered signal is received provides the distance, or *range*, to the precipitation. The measured time rate of change of the received signal phase is the Doppler shift that tells of particles' motion towards or away from the radar. Radars that transmit and/or receive electromagnetic radiation at different polarizations or frequencies can provide additional insights into the shapes, orientations, and compositions of hydrometeors. These and other radar fundamentals will be reviewed from a conceptual framework in this chapter. More thorough discussions may be found in standard radar textbooks (e.g., [20, 34, 37, 126]).

Typically, radars are characterized by the frequency (or wavelength) of radiation they transmit. The radar's operating frequency has an important control on numerous aspects of the radar's ability to remotely sense clouds and precipitation, including the effective beam width for a given antenna size, the largest Doppler velocity that can be detected, sensitivity to precipitation, how hydrometeors respond to the incident radiation, etc. For example, for a parabolic reflector like those used by many operational weather radar networks, the effective beam width θ (in degrees) for which the radiation pattern drops to half its peak power is related to the diameter of the reflector D_R and the radar wavelength λ as $\theta \approx 70°\lambda/D_R$; thus, for a given antenna reflector diameter D_R, the beam width decreases with decreasing wavelength λ (increasing frequency). Alternatively, for a given radar wavelength λ, a larger reflector size leads to a smaller beam width. Further, radiation of different λ responds differently to cloud and precipitation particles. For example, smaller-λ radiation attenuates much more than longer-λ for a given amount of hydrometeors. The tradeoff between antenna reflector size (e.g., cost, portability), effective beam width (e.g., resolution), and measurement capabilities (e.g., attenuation) are important considerations in the design of weather radar systems.

Table 1 Naming conventions for different frequency bands and wavelengths

Naming convention	Nominal frequency	Nominal wavelength
LF	30–300 kHz	10–1 km
MF	0.3–3 MHz	1000–100 m
HF	3–30 MHz	100–10 m
VHF	30–300 MHz	10–1 m
UHF	300–3000 MHz	1–0.1 m
L	1–2 GHz	30–15 cm
S	2–4 GHz	15–8 cm
C	4–8 GHz	8–4 cm
X	8–12 GHz	4–2.5 cm
Ku	12–18 GHz	2.5–1.7 cm
K	18–27 GHz	1.7–1.2 cm
Ka	27–40 GHz	1.2–0.75 cm
W	75–110 GHz	4.0–2.73 mm
G	110–300 GHz	2.73–0.1 mm

Different electromagnetic radiation frequency bands are given letter names by convention, some of which were developed during World War II and intended to be confusing to enemy spies. Table 1 provides a list of these bands and their naming convention. Typically, LF through UHF bands have been used for upper atmospheric applications (e.g., [19, 54, 57, 58, 145, 167]), whereas higher-frequency radars are used for remote sensing of clouds and precipitation. Traditional weather radars can be classified into precipitation radars (S-X bands) and cloud radars (Ku-W). Recently, G-band radars also have been proposed for cloud studies [6]. Among operational radar networks, the United States National Weather Service's Weather Surveillance Radar-1988 Doppler (WSR-88D) radars operate at S band, whereas Canadian and several South American and European countries have C-band operational radar networks. Recently, some operational networks in the United States, Asia, and Europe have incorporated gap-filling X-band radars [9, 105, 108] for improved data coverage of low levels (e.g., [137]). Owing to their smaller antenna sizes, X-band radars can be transportable, making them attractive choices for mobile radars that are often used to study tornadoes and severe convective storms (e.g., [12, 13, 39, 118, 171, 174]). Higher-frequency cloud radars predominantly have been used for research purposes (e.g., [70, 73, 92–94, 107]) and spaceborne applications (e.g., [53]).

Given this volume's focus on remote sensing of clouds and precipitation, this chapter will be limited in scope to meteorological returns. The rest of the chapter proceeds as follows. Section 2 introduces electromagnetic scattering by a single particle. Section 3 extends this discussion to scattering of radiation at different polarizations by single nonspherical particles. Scattering by a population of particles and the equivalent reflectivity factor are introduced in Sect. 4. Section 5 describes the detection and ranging of weather echoes, and Sect. 6 introduces the Doppler effect and principles of Doppler velocity estimation. Section 7 is a basic introduction to the radar system. Section 8 introduces dual-polarization radar variables and their interpretation. Section 9 is a brief survey of the numerous weather radar applications and new avenues of research. The chapter concludes with a brief summary in Sect. 10.

2 Scattering by a Single Particle

How a hydrometeor scatters incident electromagnetic radiation is a function of its size, shape, orientation, composition (e.g., ice, liquid, mixed phase), temperature, and the frequency of the illuminating radiation. The particle's physical composition, temperature, and the radar wavelength are characterized by the *dielectric constant*, or *relative permittivity* (ϵ_r), which is a complex number that quantifies a material's response to electromagnetic radiation. For weather radar frequencies, a larger "real" part of ϵ_r leads to a greater amount of backscattered energy from a given particle size/shape and a larger phase shift imparted on a wave propagating through a collection of such particles. A larger "imaginary" part indicates greater absorption in the particles. For example, at S band (the wavelength of the U.S. WSR-88D radar network), the real part of ϵ_r for liquid water is much greater than the real part of ϵ_r for ice. Thus, liquid raindrops backscatter far greater energy than equal-sized ice particles. For precipitation radars, ϵ_r is related to the complex *refractive index* m as $\epsilon_r = m^2$. The real part of the refractive index can be thought of as the ratio of the wave's phase speed in a vacuum to that in the medium. Equivalently, the wavelength of radiation decreases in the medium compared to in a vacuum. Thus, a radar wave propagating through a dielectric medium will experience a smaller phase speed (or smaller wavelength) in the medium than in the surrounding air.

Consider an arbitrarily shaped particle composed of some dielectric material. We can decompose this particle into tiny *finite scattering elements* (e.g., [16]), represented as spheres in Fig. 1a. When an incident electric field is applied to the particle (Fig. 1b), the electrostatic force on the electrons and protons in the tiny spheres induces a net *dipole moment*, which is the vector sum of the charge magnitude times the charge separation distance for each tiny sphere summed over the particle. This can be thought of as dipoles being induced in each tiny sphere, all of which are aligned with the electric field vector (Fig. 1b). If the incident electric field is oscillating (as is the case for radar waves), then these induced dipoles oscillate at the same frequency as the incident radiation. Thus, each tiny sphere behaves as a dipole oscillator, transmitting electromagnetic wavelets in all directions. The sum of these wavelets at some point far from the particle is the total scattered wave.

The particle's size compared to the wavelength is important for determining the total scattering response. Often, the *size parameter* is used to assess the relationship between particle size and wavelength: $x = 2\pi a/\lambda$ where a is the particle characteristic length (e.g., particle equivalent spherical radius). This size

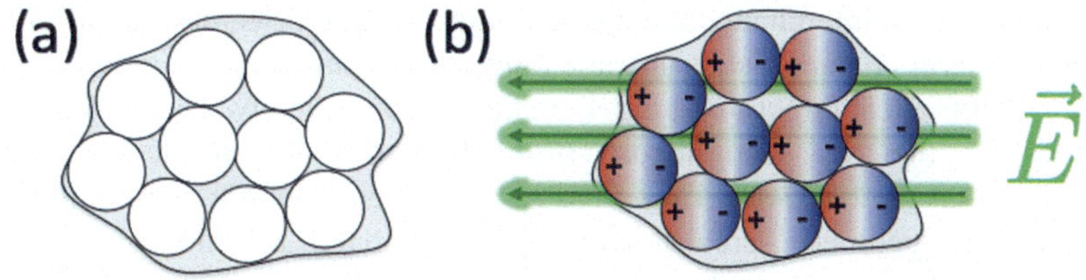

Fig. 1 (**a**) Schematic showing an arbitrary particle comprising tiny spherical finite scattering elements (white circles). (**b**) When an electric field (green vectors) is applied to the particle, a dipole moment is induced (red shading and plus signs indicate net positive charge; blue shading and minus signs indicate net negative charge), and dipoles align themselves in the direction of the electric field vector

Fig. 2 Schematic showing (**a**) a particle small compared the wavelength (traced out by the green line) in which the electric field (green vectors) is uniform throughout the particle, and (**b**) a particle large compared to the wavelength in which the electric field is nonuniform throughout the particle. In (**a**), the induced dipoles oscillate in phase with one another. In (**b**), the induced dipoles oscillate out of phase with one another

parameter can be thought of as the number of complete wave cycles of the radiation per characteristic particle length. When the particle is small compared to the wavelength of the radiation (i.e., $x \ll 1$), then there is little variation in the electric field across all the tiny spheres within the particle at some instant in time (Fig. 2a). Thus, the induced dipoles throughout the particle oscillate in phase with one another. For small spherical particles, the resulting scattered radiation pattern thus resembles that of a dipole antenna. Nonspherical particles are discussed in the following section. In contrast, for a particle large compared to the wavelength, large variations in the electric field can exist across the particle (Fig. 2b) at a given instant in time. Thus, the induced dipoles oscillate out of phase with one another, which can lead to constructive and destructive interference of the scattered wavelets. The likelihood of interference or "resonance" effects increases as the resonance parameter:

$$\Re = \frac{D\sqrt{|\epsilon_r|}}{\lambda} \quad (1)$$

approaches unity (e.g., [131]). In this expression, D is the particle's equivalent volume spherical diameter. The inclusion of relative permittivity in (1) captures the stronger scattering response (and thus greater likelihood of strong resonance effects) for particles with larger ϵ_r.

The amount of radiation scattered back towards the radar from a target is characterized by its *backscattering cross section* σ_b or *radar cross section* (traditionally, these terms are used interchangeably in radar meteorology). Because we never know the exact scatterer characteristics at some range, we assume that the received power is from a hypothetical target that scatters radiation isotropically. The backscattering cross section is the cross-sectional area of this hypothetical isotropically scattering target. Note that the backscattering cross section often does not correspond to the actual particle's geometric area. For example, at S band, a 1-mm raindrop has a larger backscattering cross section than a 1-mm ice pellet because of the difference in relative permittivities described above.

Mie [110] provides the exact solution to σ_b of a dielectric sphere. If the spherical particles are small compared to the wavelength, the so-called Rayleigh approximation to the Mie solution for the backscattering cross section of a dielectric sphere can be employed:

$$\sigma_b \approx \frac{\pi^5}{\lambda^4} D^6 |K|^2 \tag{2}$$

where λ is the radar wavelength, and the dielectric factor K is a function of the complex refractive index m:

$$| K |^2 \equiv \left| \frac{m^2 - 1}{m^2 + 2} \right|^2 \tag{3}$$

For operational weather radars that transmit at S and C bands, many cloud and precipitation particles can be considered small compared to the wavelength. Exceptions include the largest raindrops and large melting ice particles such as hail and some snow aggregates. Additionally, many hydrometeors are not spherical, an important fact exploited by dual-polarization radars that will be discussed in Sect. 3. Approximate formulae for the backscattering cross sections of nonspherical particles are obtained by modifying the Rayleigh approximation and may be found in, e.g., [16].

3 Polarization

Recall that electromagnetic radiation's *polarization* describes the orientation direction of its electric field vector. Because electromagnetic waves are transverse waves, the polarization direction is always perpendicular to the direction of wave propagation. To understand the importance of polarization and nonspherical particles, consider again an electromagnetically small particle that comprises a large number of tiny spheres. This time, consider a hexagonal plate crystal that is aligned with its major axis approximately in the horizontal (Fig. 3). The crystal is highly nonspherical, so its response to incident radiation depends on the polarization of that radiation. Consider an illuminating horizontally polarized wave. This wave excites

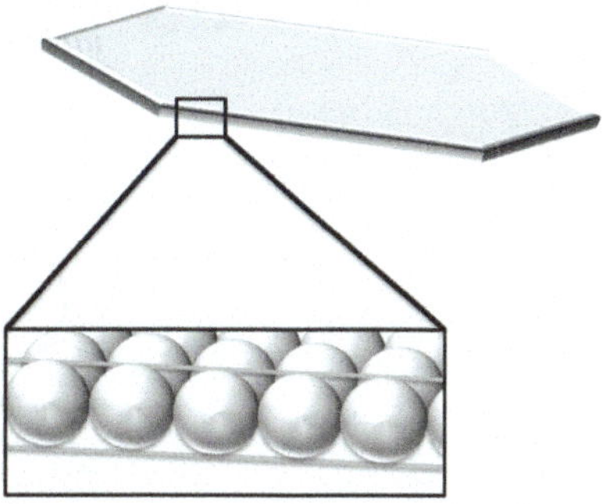

Fig. 3 Conceptual model for understanding scattering by nonspherical particles. Consider a hexagonal plate comprising a large number of tiny spheres (zoomed-in panel) that behave as dipole oscillators when illuminated with electromagnetic radiation

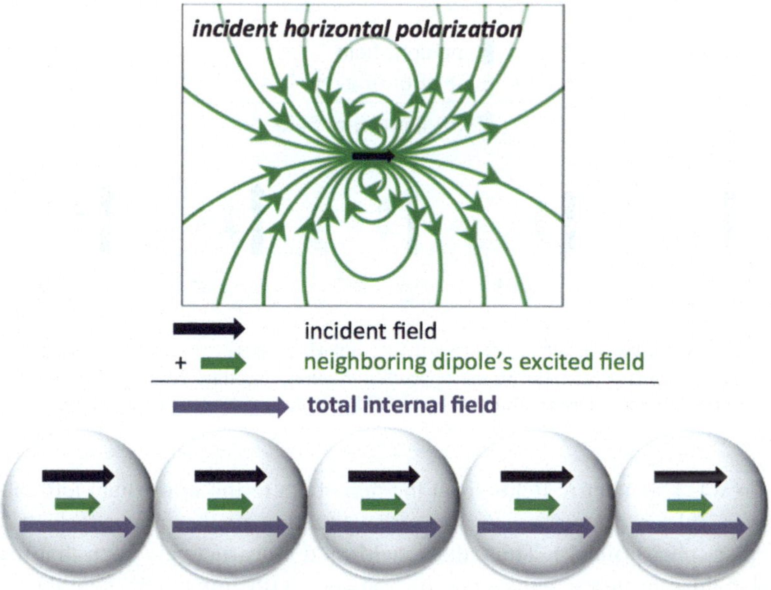

Fig. 4 Top panel: electric field lines (green) for a horizontally aligned dipole excited by an incident horizontally polarized electric field (black vector). Below: conceptual model of a horizontally oriented particle comprising a number of small spheres (cf. Fig. 3) being illuminated by an incident horizontally polarized wave (black arrow shows incident electric field). Each sphere produces a radiation pattern like the top panel. The electric field in each sphere induced by its neighbors is shown as the green vector. The resultant total internal electric field excited within each sphere is the summation of the incident and neighbors' electric fields (black+green), shown in purple. The neighboring sphere's electric field adds constructively to the incident electric field, exciting a larger total internal field. Adapted from Lu et al. [101]

the tiny spheres by inducing a dipole in each that is aligned with the electric field (in this case, horizontally). The radiation pattern from each tiny sphere is shown in Fig. 4, top panel.

Even though the induced dipoles are oscillating in phase with one another, each tiny sphere's excited field also influences its neighboring spheres (Fig. 4,

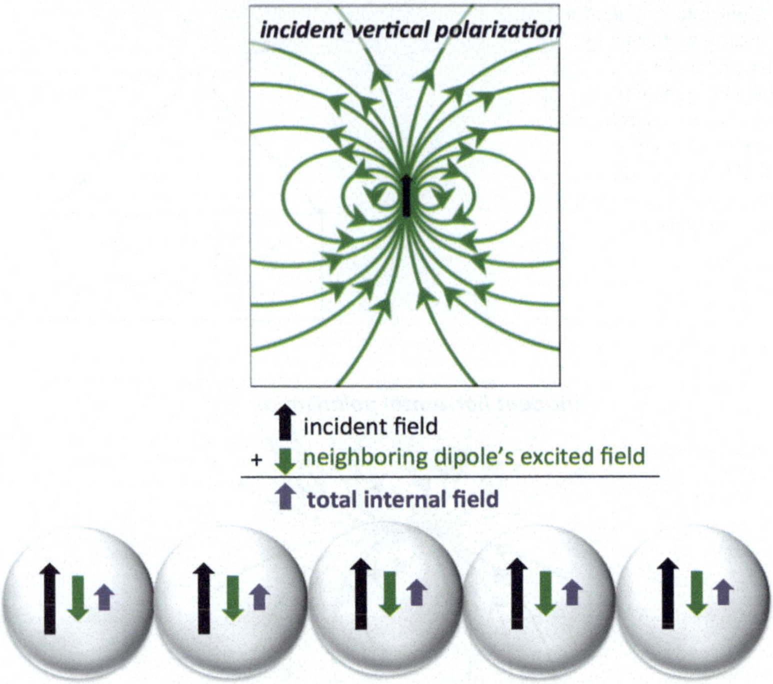

Fig. 5 As in Fig. 4, but for an incident vertically polarized wave. In this case, the neighboring spheres' electric fields (green) act in the opposite direction, and thus destructively interfere with the incident field (black). The resulting internal field (purple) is smaller in magnitude

bottom row). The electric fields from the neighbors to the immediate left and right of a given sphere (green) are in the same direction as the induced electric field within that sphere (black). Thus, these fields add constructively such that the total internal field (purple) is enhanced for a sphere with immediate neighbors to its left and right relative to that of an isolated sphere. Because the internal electric field is enhanced for such a sphere, the radiation scattered from it is also enhanced compared to an isolated sphere. Therefore, the crystal's total scattered radiation at horizontal polarization is also enhanced by these so-called *near-field interactions* (e.g., [16, 101]).

Now consider the same hexagonal crystal being illuminated by a vertically polarized wave. Again, each tiny sphere has an induced dipole, this time vertically oriented (Fig. 5, top row). However, this time the electric fields from neighbors to the immediate left and right of a given sphere (Fig. 5, bottom row, green) point in the opposite direction to the incident field (black). This leads to destructive interference, resulting in a total internal field (purple) reduced for a sphere with neighbors relative to an isolated sphere. Thus, the radiation scattered by such a sphere is reduced. Therefore, the near-field interactions reduce the crystal's total scattered radiation at vertical polarization.

By comparing Figs. 4 and 5, we see that the incident horizontally polarized wave excites a stronger internal electric field in the crystal compared to the vertically polarized wave, and thus the scattered wave from the hexagonal crystal has a larger amplitude at horizontal polarization than at vertical polarization. This is the principle behind radar polarimetry: nonspherical particles will scatter radiation differently at different polarizations, and this information may be exploited to better characterize clouds and precipitation.

Consider a hydrometeor oriented in such a way that its major axis no longer aligns perfectly with the direction of incident polarization (Fig. 6). For incident

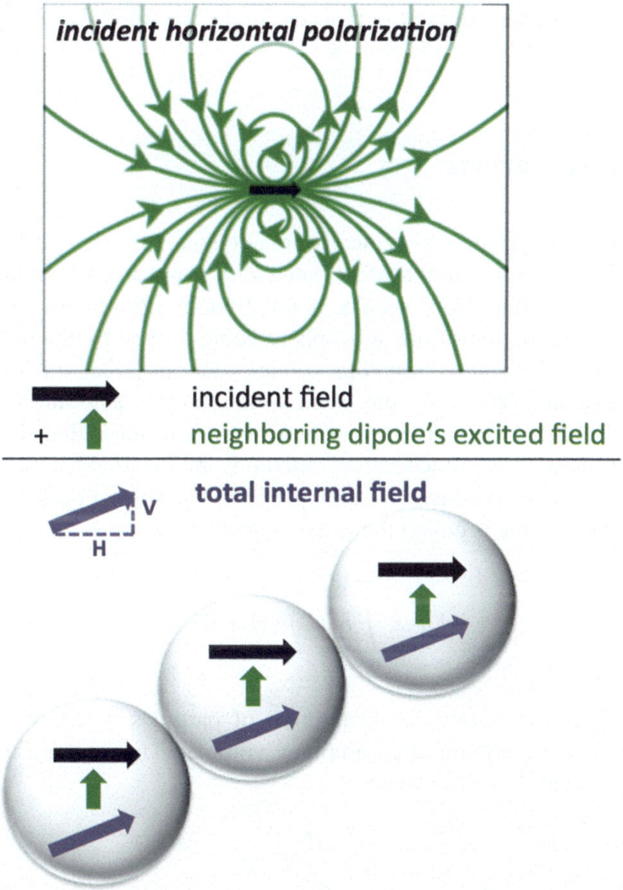

Fig. 6 As in Figs. 4 and 5, but for an incident horizontally polarized wave and a particle aligned at an angle with respect to the incident polarization direction. In this case, the electric field from neighboring spheres has a component orthogonal to the incident wave polarization direction. The resulting total internal field thus has components at both horizontal and vertical polarizations. The backscattered wave thus has components at both H and V polarizations, despite the incident wave being only horizontally polarized

horizontal polarization, each tiny sphere has an induced dipole radiation pattern with horizontal polarization, as before. This time, however, the spheres are oriented in such a way that they experience an electric field from their neighbors that has a vertical component. The total internal electric field excited in each sphere thus has components in both horizontal and vertical directions (dashed purple lines in Fig. 6). As a result, the scattered wave now has components that are both horizontally and vertically polarized, despite the incident wave being only horizontally polarized. This process, in which the particle changes the polarization of the radiation upon scattering, is known as *depolarization*. Some weather radars transmit radiation at one polarization and receive radiation at two orthogonal polarizations in order to measure the amount of depolarization that occurs in clouds and precipitation. This, too, can be used to characterize the physical properties of cloud and precipitation particles and will be discussed in Sect. 8.

4 Distributed Targets

In the atmosphere, hydrometeors typically are not isolated, single scatterers. Rather, they are found in large populations distributed within and around clouds. Therefore, when a radar wave intercepts such a population, it illuminates many particles simultaneously. The region of the atmosphere being probed by the radar at a given moment is called the *radar sampling volume*. The physical dimensions of the sampling volume are defined by the beam width (both in azimuth and elevation), and for radars that transmit pulses, the pulse length in the radial direction. Each particle in this sampling volume scatters radiation, and the total backscattered signal is the sum of the waves produced by each particle. The sum of backscattering cross sections per unit volume is called the *radar reflectivity*, η:

$$\eta \equiv \int_0^\infty \sigma_b(D)N(D)\,dD \tag{4}$$

where $N(D)dD$ is the number concentration of particles of equivalent spherical diameters D to $D + dD$ in the sampling volume. Thus, where the Rayleigh approximation is valid, we can write

$$\eta \approx \frac{\pi^5}{\lambda^4}|K|^2 Z \tag{5}$$

where we have defined the *radar reflectivity factor* (Z) as

$$Z \equiv \int_0^\infty N(D)D^6\,dD \tag{6}$$

Of course, these expressions are only useful in practice if the properties of the targets are known ahead of time. Generally, the dielectric properties of the targets (expressed through K) in the sampling volume are unknown. Further, it is generally unknown if the Rayleigh approximation is valid. Because of these uncertainties, another quantity is defined: the *equivalent radar reflectivity factor* (Z_e), which assumes a priori that the scatterers comprise small, spherical liquid drops whose backscattering cross sections can be described by the Rayleigh approximation. Thus,

$$\eta \approx \frac{\pi^5}{\lambda^4}|K_w|^2 Z_e \tag{7}$$

where the dielectric factor of liquid water K_w is used. For typical operational weather radar wavelengths, $|K_w|^2 = 0.93$. Note that $Z = Z_e$ only in the case where targets are small liquid drops. It is Z_e that is shown in operational weather radar displays, typically in base-10 logarithmic scale in units of decibels, or dBz:

$$Z_e[\text{dBz}] = 10\log_{10}\left[\frac{Z_e[\text{mm}^6\,\text{m}^{-3}]}{1\,\text{mm}^6\,\text{m}^{-3}}\right] \tag{8}$$

This logarithmic scale is used because Z_e values can span many orders of magnitude in clouds and precipitation, from $<10^{-2}$ mm^6 m^{-3} (<-20 dBz) in non-precipitating clouds to $>10^7$ mm^6 m^{-3} (>70 dBz) in hail-bearing severe convective storms.

Each particle's scattered wavelets can constructively or destructively interfere with those from other particles in the sampling volume. This results in fluctuating received signal power as the particles "reshuffle" their relative positions within the sampling volume as they fall or move with the air flow. Thus, multiple samples are needed to estimate the mean Z_e, with sufficient time between each sample such that the samples are statistically independent. For operational weather radars under normal atmospheric conditions, this time needed for sufficient resampling and sample independence is on the order of ≤ 0.01 s. Typically, on the order of 10^1 or 10^2 samples are used to obtain a good estimate of Z_e. This combination of an adequate number of samples and the time it takes the samples to become independent through reshuffling governs the *dwell time* required for a radar to sample a particular region of the atmosphere, and thus limits the temporal resolution of the scanning. However, some research radars employ advanced techniques to mitigate this issue, allowing for faster sampling. This includes transmitting pulses of slightly different frequencies that provide independent samples in a shorter amount of time in a technique sometimes called *frequency hopping* (e.g., [13, 34, 44, 55, 56, 118]). Other radars use multiple frequencies to generate multiple, nearly simultaneous beams at different fixed elevation angles (e.g., [172]), allowing the radar to scan a larger portion of the atmosphere at once. Still others use advanced signal post-processing techniques such as *digital beam forming* (e.g., [64]) to image a large region of the atmosphere instantaneously. These advancements have substantially improved the ability of weather radars to rapidly sample hazardous storms.

Despite being one of the most often-used and familiar quantities, Z_e is challenging to calibrate to the accuracy necessary for its quantitative use. For example, most operational weather radar networks require absolute Z_e calibration to within 1 dB for acceptable errors in quantitative precipitation estimation (e.g., [47, 130]). Atlas [4] summarizes numerous techniques used for absolute radar calibration. These include using targets of known backscattering cross sections like corner reflectors or spheres suspended from balloons or aircraft, and comparisons with ground-based disdrometer measurements. Additionally, self-consistency amongst the dual-polarization radar variables in rain has been exploited for absolute radar calibration [45–47, 63, 130]. In practice, there is no clear consensus yet on the optimal calibration method.

5 Detection and Ranging of Weather Echoes

For a radar to detect a cloud or precipitation echo at a given range, the received signal must be strong enough to discern from noise. The power received from a target depends on properties of the wave propagation path to the target and back, characteristics of the radar system such as the antenna, and the properties of the target itself. These are collectively described by the so-called *radar equation*; different forms of the radar equation and derivations thereof may be found in standard radar texts (e.g., [7, 20, 34, 37, 126]). The radar equation is important because it allows the equivalent reflectivity factor Z_e to be estimated from the returned power. For example, one may increase the strength of the returned signal by increasing the radar's transmitted power, increasing the duration of the transmitted pulse, and/or increasing the size of the antenna reflector (which helps focus the transmitted beam and collect the received signal). Alternatively, more power is received if the targets scatter more radiation back to the radar and/or they are located closer to the radar, because radiation intensity decreases as $1/r^2$, where r is the distance from the source of the radiation.

Many modern weather radars are pulsed-Doppler radar systems that typically transmit 10^2–10^3 pulses per second. The amount of time a pulse lasts is the *pulse length*, which is typically on the order of a microsecond. This results in a *pulse width* on order of a few hundred meters. The rate at which these pulses are transmitted is called the *pulse repetition frequency* (PRF), which can be inverted to obtain the *pulse repetition time* (PRT), or time between consecutive pulses. The transmitted waves propagate through the atmosphere at the speed of light c. The distance to a target, or its *range* (r), thus can be determined by the time Δt it takes for a pulse to travel out and return to the radar:

$$r = \frac{c\Delta t}{2} \tag{9}$$

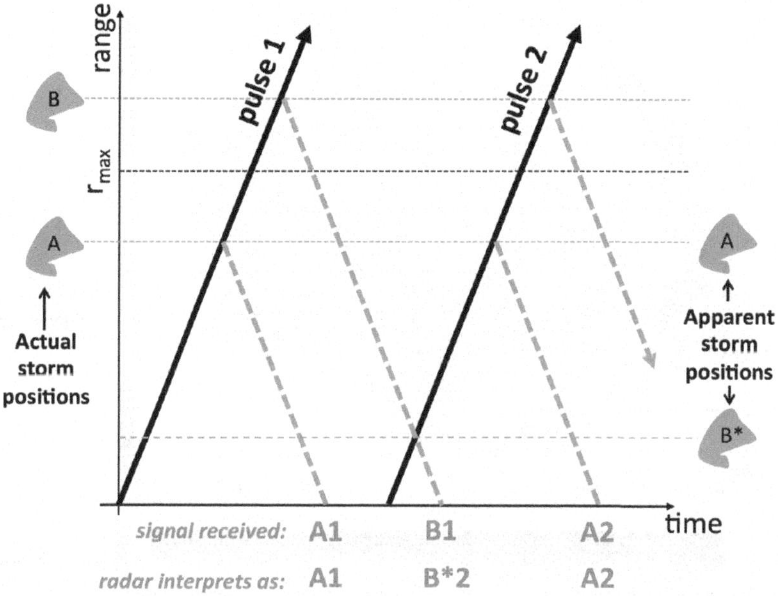

Fig. 7 Conceptual model of range aliasing. The radar fires pulse 1 (slanted black arrow) at the initial time. There are two storms, A and B, located at positions shown to the left of the graph. The maximum unambiguous range (r_{max}) is shown as the dotted black line. The backscattered signals from storms A and B are shown as dashed gray lines. The top row below the graph shows the times when these backscattered signals are actually received, where "A1" stands for the signal from storm A from pulse 1, etc. However, because storm B is located beyond r_{max}, the radar interprets its echo as coming from an apparent storm B* at a much closer range (equivalent to the actual range of storm B minus r_{max}) from pulse 2. Thus, storm B's range is said to be "aliased" or "folded"

The factor of two is because the pulse must travel out to the target and back. The fact that the radar transmits pulses every PRT limits the maximum range to which the pulse can travel to and back from before the next pulse is transmitted: $r_{\max} = c \times \mathrm{PRT}/2$. If a target is located beyond $r_{\max}$, then the signal scattered from that target is received after the next pulse has been transmitted. In this case, the radar interprets the echo as coming from the target scattering the second transmitted pulse. As a result, the target's true range is *aliased* or *folded* and incorrectly indicated as being from a range between 0 and $r_{\max}$ (Fig. 7). Modern weather radars employ more sophisticated techniques to mitigate this so-called range ambiguity (e.g., [10, 159, 160]).

Many weather radars operate in surveillance scanning modes in which the radar antenna is set at a fixed elevation and rotated in azimuth. Typically, the data are displayed from a "bird's eye" view in what has been historically called a *plan position indicator* or PPI. More modern terms synonymous with PPI include *sweep* and *surveillance scan*. Radar pixel locations in such a surveillance scan are specified by range and azimuth coordinates, which are sometimes interpolated onto a Cartesian grid (Fig. 8a). Operational scanning strategies typically comprise several

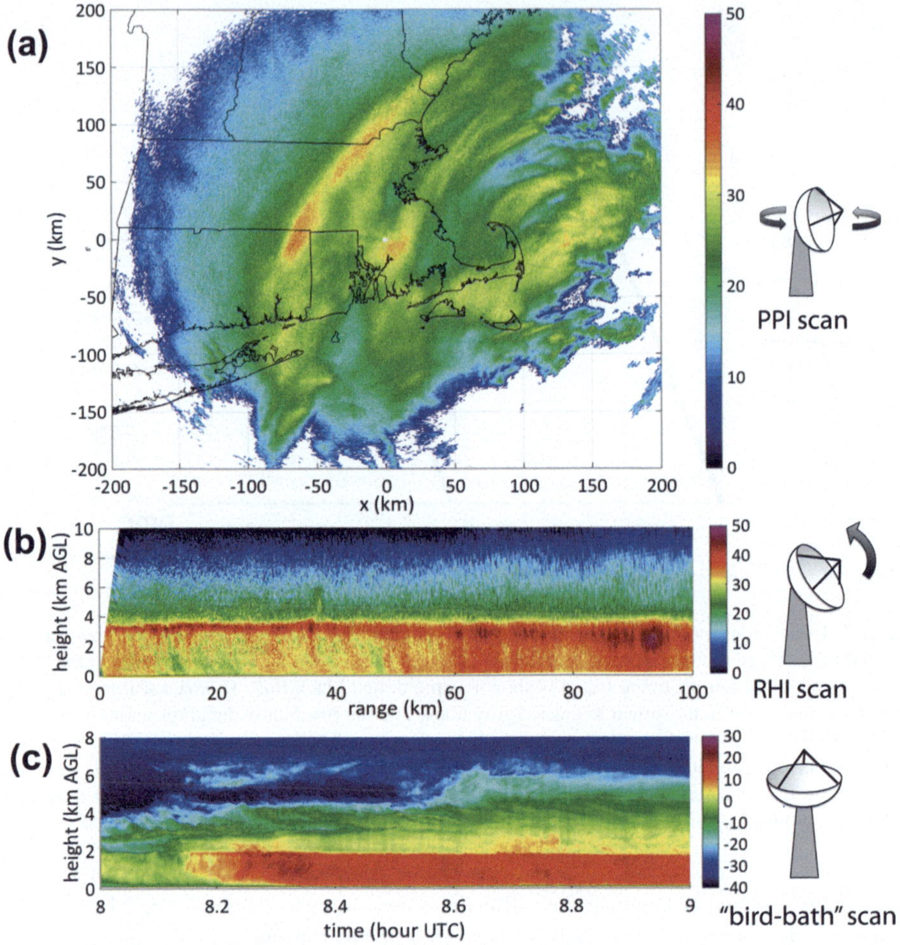

Fig. 8 Example displays from various types of radar scans with a cartoon showing the antenna scanning strategy to the right of each panel. (**a**) A PPI or surveillance scan of equivalent reflectivity factor at horizontal polarization (Z_H, in dBz, shaded according to scale) in which the radar antenna is at fixed elevation angle and varies in azimuth; the radar is located at the origin of the coordinate system. (**b**) A RHI scan of Z_H (in dBz, shaded according to scale) in which the radar antenna is at fixed azimuth and varies in elevation. (**c**) A zenith-pointing or "bird-bath" scan of Z_e (in dBz, shaded according to scale) in which the radar antenna is pointing vertically. Sometimes, the antenna is rotated in azimuth while vertically pointing

such surveillance scans at different elevation angles in collectively what is known as a *volume coverage pattern* or VCP. For example, a typical VCP employed by the U.S. National Weather Service for precipitation surveillance includes 10–15 elevation angles between 0.5° and 19.5°. Data from a VCP can be interpolated onto a single altitude for displays in what is known as a *constant-altitude plan-position indicator* (CAPPI).

If azimuth angle is held constant while the radar antenna scans in elevation, the radar can collect data from a vertical cross section through clouds and precipitation known as a *range-height indicator* (RHI) scan (Fig. 8b) in which radar pixel locations are specified by range and height. Other radars point vertically and sample clouds and precipitation as they move over the radar location. The resulting data are displayed in time-height plots (Fig. 8c). Such zenith-pointing radars may have an immobile antenna, or it may rotate in azimuth while pointing vertically. The latter is often done for calibration purposes. Because the zenith-pointing parabolic reflector antenna resembles a bowl or bird bath, such scans are colloquially referred to as "bird-bath" scans. Radar meteorologists may obtain different views of clouds and precipitation by combining one or more of these types of scans into a scanning strategy.

6 Doppler Effect

Anyone who has witnessed a passing train blowing its whistle or a passing ambulance blaring its siren has experienced the Doppler effect: the tone of the whistle or siren changes as its source moves towards or away from you. The changing tone indicates the sound waves changing frequency: an increase in frequency for sources moving towards the observer, and a decrease in frequency for sources moving away from the observer.

To understand the Doppler effect in terms of electromagnetic wave scattering, a conceptual model following [34] is presented here. Consider a stationary insect that is about to be irradiated by a transmitted pulse at a time t_0 (Fig. 9a, left). At some brief instant later ($t_0 + \delta t$; Fig. 9, right), the bug has been illuminated by some number of electromagnetic wave cycles (in the cartoon, it is 2). This time-varying electromagnetic field forces polar molecules in the bug to vibrate at the same frequency, scattering radiation at the same frequency back towards the radar. In this case, there is no Doppler shift.

Now consider a bug flying towards the radar (Fig. 9b, left). Relative to a stationary bug, the bug flying towards the radar "experiences" more rapid oscillations of the electromagnetic field: the cartoon shows the bug having experienced three complete cycles (Fig. 9b, right) in the same time period δt. More cycles over a given amount of time means a higher frequency. Because the bug has been irradiated by more cycles per second, its molecules are forced to vibrate at this higher frequency and thus scatter radiation at the higher frequency back towards the radar. This frequency shift is determined by how far the target moves in a given amount of time, or its radial speed v_r, relative to the distance over which the radar wave completes one cycle, or its wavelength λ. The Doppler frequency shift is therefore $f_D = 2v_r/\lambda$, with the factor of two arising from (1) the bug approaching the radar experiences more cycles per second of the transmitted wave, and (2) in the reference frame of the moving bug, the radar is approaching the bug at v_r and thus "experiences" more cycles per second of the scattered wave. Thus, the scattered signal has its frequency

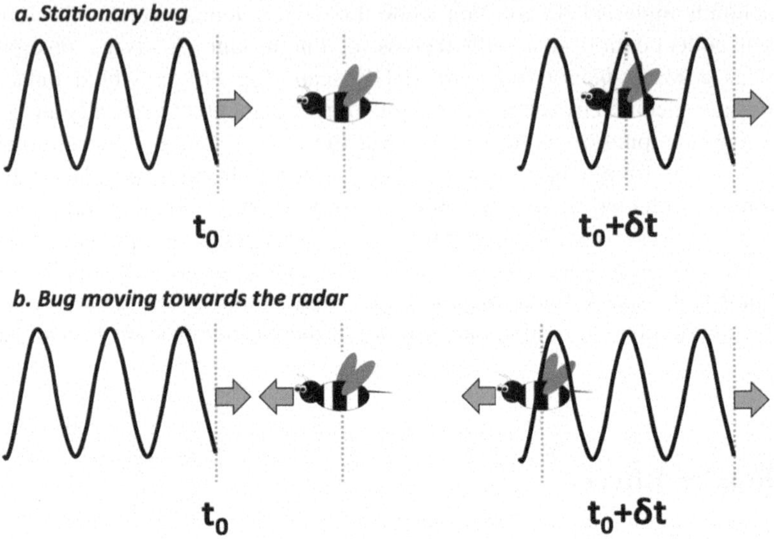

Fig. 9 Schematic explaining the physical mechanism behind the Doppler shift in the context of electromagnetic scattering. Two times are shown: t_0 (left column) and an instant later $t_0 + \delta t$ (right column). In (**a**), the bug is stationary as the wave propagates towards and irradiates it. In (**b**), the bug is traveling towards the propagating wave

shifted relative to the transmitted wave. *This is the Doppler shift.* Targets moving towards the radar produce a frequency increase, whereas targets moving away from the radar produce a frequency decrease.

Because of this frequency shift, the Doppler-shifted scattered wave will be out of sync with the transmitted wave at a given point. Thus, the Doppler-shifted wave will accumulate a phase shift (relative to the transmitted wave) at a rate of $2\pi f_D$ radians per second. By tracking phase shift over time, the radar can in principle estimate the Doppler frequency shift. In practice, however, the Doppler shift for weather applications is too small to be measured over the duration of a single pulse. For example, consider an atmospheric target moving with $v_r = 10\,\mathrm{m\,s^{-1}}$. With a 10-cm-wavelength (S-band) radar, the Doppler frequency shift $f_D = 200\,\mathrm{Hz}$. Over the duration of a typical 1-μs pulse, the resulting phase shift is only 0.072°. Thus, the phase shift is instead measured from pulse to pulse, or every PRT. For many modern weather radars, the PRT is on the order of 10^{-3} s, a much longer amount of time that allows the target to move a distance that causes a measurable phase shift.

But, just as there is a limit to the maximum range at which a radar can detect clouds and precipitation, there is a maximum radial velocity that can be detected between pulses, known as the *Nyquist velocity*. If the target moves towards the radar a distance spanning half a wavelength between pulses, then the radar cannot distinguish this from if a target moves away from the radar a distance spanning half a wavelength. In other words, the maximum velocity a radar can detect within consecutive pulses is equivalent to that which creates a phase shift of ±180°. This

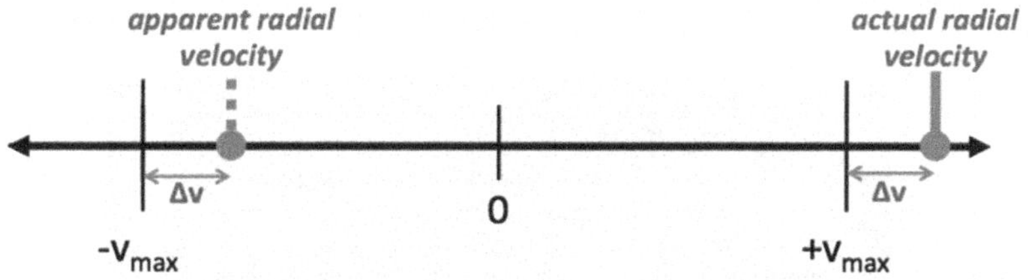

Fig. 10 Schematic illustrating Doppler velocity aliasing. If the true radial velocity (gray line) exceeds the magnitude of the Nyquist velocity ($\pm v_{max}$), then the radar interprets the apparent radial velocity (dashed gray line). Note that in this case the aliased Doppler velocity is of opposite sign to the actual radial velocity

so-called *maximum unambiguous velocity* is given as $v_{\max} = \pm\lambda/(4 \times \mathrm{PRT})$. Either sign is possible given that Doppler velocity can be positive or negative; for weather radars, the typical convention is negative velocities for targets approaching the radar and positive velocities for targets moving away from the radar. If the true radial velocity of a target exceeds $v_{\max}$ in magnitude, its apparent velocity is aliased (Fig. 10).

Figure 11 shows an example of aliased Doppler velocities in a tornado sampled by an X-band Doppler on Wheels radar. The image shows a clear low-Z_H eye at the center of the tornado arising from centrifuging of debris and precipitation particles. Because the particles in the tornado have speeds towards or away from the radar in excess of the Nyquist velocity (which was about $\pm 55\,\mathrm{m\,s^{-1}}$ in this case), the velocities are aliased. This is visible as the sudden change from large positive (outbound) to large negative (inbound) velocities on the tornado's northeast side, and to a lesser extent the change of large negative to large positive velocities on the southwest side.

Notice that the PRT is found in both $r_{\max}$ and $v_{\max}$ expressions: a larger PRT (indicating a longer time between pulses) leads to a larger $r_{\max}$ but a smaller $v_{\max}$. This is known as the Doppler dilemma, which can be expressed mathematically as $|v_{\max}| \times r_{\max} = c\lambda/8 =$ constant. Modern radar systems use various techniques to mitigate these ambiguities, including sampling clouds and precipitation with multiple or staggered PRTs (e.g., [10, 159, 160]).

Recall that the received signal is the result of the sum of contributions from all the scatterers within the sampling volume. Given that sample volumes are typically much larger than the radar wavelength (tens to hundreds of meters versus centimeters for precipitation radars), the total signal's phase measurement at a given location or time is meaningless for weather targets. The pulse-to-pulse change in this received phase is related to the mean radial velocity of particles within the sampling volume. Note that longer-timescale changes in phase from stationary ground clutter targets can be related to changing atmospheric conditions, which affect the atmospheric index of refraction or refractivity (e.g., [14, 38]). In general, hydrometeors in the radar sampling volume may be moving at different radial velocities owing to differences in their fall speeds, wind shear across the

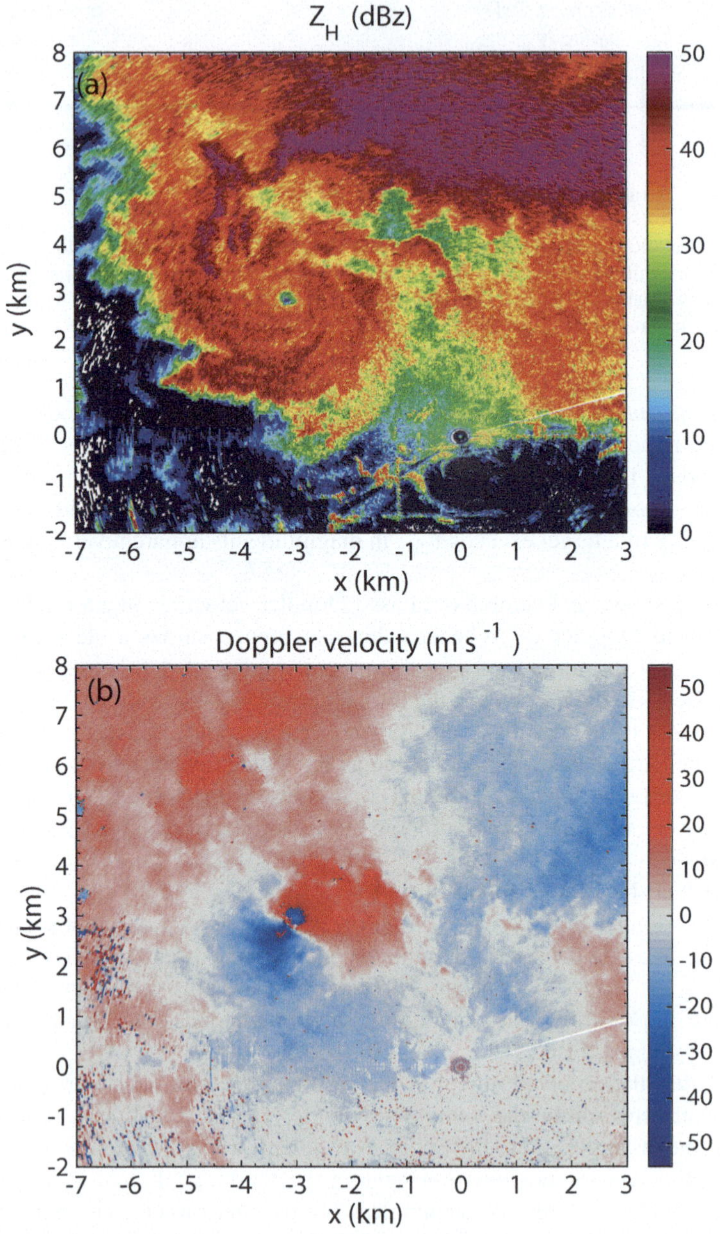

Fig. 11 Example of Doppler velocity aliasing in a tornado. Shown are surveillance scans of (**a**) Z_H (in dBz, shaded according to scale) and (**b**) Doppler velocity (in m s^{-1}, shaded according to scale). Data collected on 19 May 2013 at 0045 UTC by a Doppler on Wheels X-band mobile radar, courtesy of Dr. Karen Kosiba (Center for Severe Weather Research). Positive velocities (reds) are outbound, negative (blues) are inbound. The radar is located at $x = 0$ km, $y = 0$ km, and the tornado at $x = -3$ km, $y = 3$ km

sampling volume, turbulence, etc. Because the total backscattered signal is the sum of each particle's scattered wavelet, it therefore may contain a spectrum of Doppler frequency shifts. Fourier analysis of the time series of the received signal allows for the recovery of this spectrum of frequency shifts, known as the *Doppler spectrum*.

The Doppler spectrum shows how much power scatterers contribute to the total signal as a function of their Doppler velocity. Integrating over the entire spectrum provides the total power, which is related to Z_e. The power-weighted mean value of this spectrum is the mean Doppler velocity: this is what is assigned to a given radar pixel in graphical displays. The variance of the Doppler velocities within the sampling volume is characterized by the Doppler *spectrum width*. For example, broad spectra can be associated with highly turbulent regions of storms and/or regions of enhanced wind shear, and can be useful for identifying convergent boundaries and tornadoes (e.g., [150]). Others have used Doppler spectral information to estimate turbulence and entrainment in clouds (e.g., [1, 74, 109]). Recent studies using vertically pointing cloud radars have also explored other spectral quantities such as skewness and kurtosis and their relationship to microphysical processes (e.g., [43, 75, 103, 161]).

Figure 12 shows example data collected with a vertically pointing radar during stratiform precipitation. The time-height depiction reveals a clear increase of Z_e below about 2000 m (Fig. 12a), along with an increase in magnitude of the mean Doppler velocity (Fig. 12b). This marks the transition from weaker-scattering, slower-falling snow that melts into stronger-scattering, faster-falling rain. This transition is also evident as an increase in the Doppler spectrum width (Fig. 12c), which demonstrates that raindrops have greater variance in their fall speeds than the snow and ice particles above the melting layer. The Doppler spectrum at the point given by the magenta marker is presented in Fig. 12d. It shows that most of the power is contributed by particles with Doppler velocities between about -2 and $0\,\mathrm{m\,s^{-1}}$. The dashed gray line indicates the power-weighted mean Doppler velocity ($-1.2\,\mathrm{m\,s^{-1}}$), and the gray arrows indicate two times the Doppler spectrum width ($0.5\,\mathrm{m\,s^{-1}}$). Doppler spectra can provide highly detailed information about clouds and precipitation.

7 The Radar System

Though radar technology continues to improve, many components of modern weather radars have remained fundamentally the same for several decades. Herein we describe a simplified block diagram for a pulsed Doppler radar system (Fig. 13); more technical treatments are available in standard radar engineering texts (e.g., [20, 34, 147]). The creation of the radar signal begins with the *stabilized local oscillator* (STALO), which generates a pure sinusoidal signal of frequency f_S, and the *coherent oscillator* (COHO), which generates a pure sinusoidal signal of frequency f_C that is locked in phase with the STALO signal. The COHO frequency f_C is typically much lower than the radar's operating frequency. For example, for many S-band radars

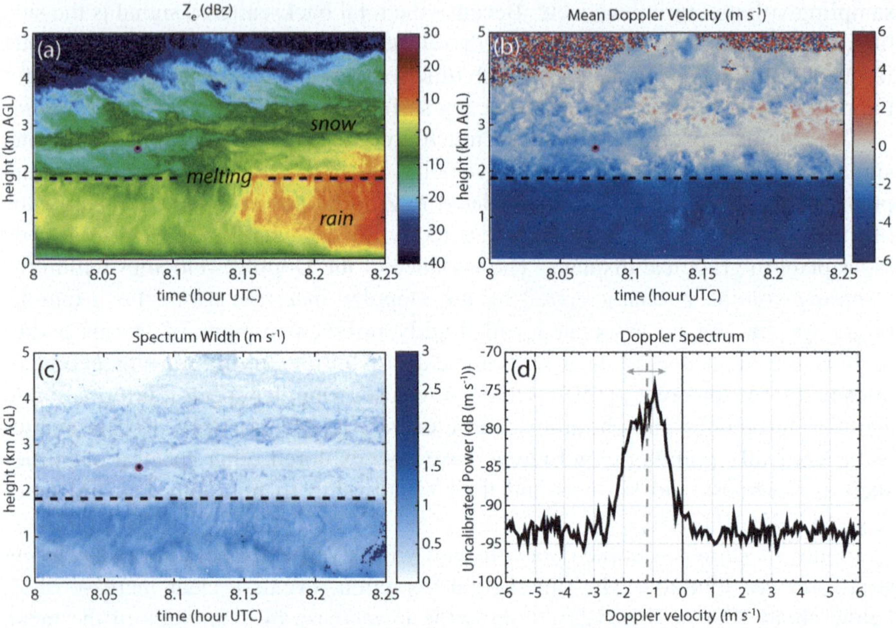

Fig. 12 Example data from the Atmospheric Radiation Measurement (ARM) Ka-band zenith-pointing radar in north-central Oklahoma, taken during a stratiform precipitation case on 27 April 2011. Shaded fields shown are time-height depictions of (**a**) equivalent reflectivity factor Z_e (dBz); (**b**) mean Doppler velocity v_r (m s^{-1}); (**c**) spectrum width (m s^{-1}). Panel (**d**) shows the Doppler spectrum taken at the point indicated in panels (**a**)–(**c**) as a magenta and black circle. The vertical dashed gray line shows the mean Doppler velocity (-1.2 m s^{-1}), and the double-sided arrow indicates twice the Doppler spectral width (0.5 m s^{-1}). The negative Doppler velocities indicate descending hydrometeors.

(2–4 GHz), f_C is about 30–60 MHz. This lower frequency is called the *intermediate frequency* (IF). The radar's operating frequency f_0 (sometimes called the *carrier frequency*) is usually taken to be the sum of the COHO and STALO frequencies ($f_0 = f_S + f_C$), and is obtained from mixing the COHO and STALO signals. Figure 13 indicates mixers as circled × symbols.

This continuous wave signal is modified by the *pulse modulator*, which controls the timing and width of the signal to generate a series of pulses. The pulses are then increased in intensity in the *power amplifier*. In many modern weather radars, the pulses are amplified by the *klystron*, which is essentially an electron gun that "bunches" electrons together by applying an oscillating voltage across the beam. Another power amplifier sometimes used in weather radars is the *travelling wave tube* (TWT). Historically, many radars used power oscillators known as *magnetrons*. Magnetrons are smaller and cheaper for a given output power, but generate signals with random phase. This means that the signal phase must be tracked from pulse to pulse in order to have a reference to utilize in Doppler velocity estimation. More

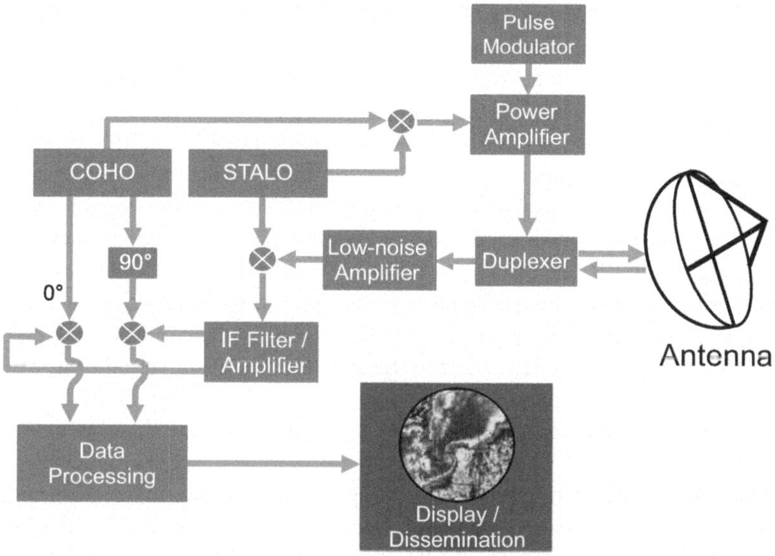

Fig. 13 Simplified block diagram showing the main components of a pulsed Doppler radar. See text for details

recently, lower-cost *solid-state transmitters* are becoming increasingly popular. Selection of transmitter type involves considering these trade-offs between power, cost, and performance.

After the pulses are amplified, they are sent to the *duplexer*. The duplexer has the important job of sending the high-power signal to the antenna during transmit times, but allowing the low-power received signal from the antenna to enter the receiving chain at other times. Sensitive electronics in the receiving chain could be severely damaged or destroyed if the high-power transmitted signal leaks into them. When transmitting, the pulsed signal is sent from the duplexer to the radar's *antenna*: the component that directs, focuses, transmits, and receives radiation. Many modern weather radars use a parabolic reflector that focuses the radiation into a narrow beam (typically, $\sim 1°$ or less is desired). Radiation is sent into the parabolic reflector by the *feed horn*, which is located at the focal point of the reflector. This feed horn is typically held in place by one or more struts. Other geometries are used for cloud and precipitation radars, including the dual-offset Gregorian antenna of the CSU-CHILL radar (e.g., [69]). Recent work has focused on developing phased-array radar technology, in which an array of small microwave transmit/receive elements in a planar or cylindrical configuration is used (e.g., [176]). Such phased-array radars have the advantage of being able to direct or steer the beam electronically through phase differences applied to each of the transmitter elements, rather than needing to move the reflector to direct the radiation (e.g., [178]). However, such radars are quite expensive compared to radars employing traditional reflector antennas.

For *monostatic* radars, the same antenna that transmits radiation to the atmosphere often also serves to receive the signals backscattered from clouds, precipitation, or other targets.[1] The received signal is sent back through the duplexer, this time into the receiving chain. A principle role of the receiving chain is to amplify this signal of interest and filter out unwanted noise. Note that the received signal amplitude is often many orders of magnitude weaker than the transmitted signal, thus necessitating a *low-noise amplifier*. The received signal frequency f_R is the sum of the transmitted frequency plus any Doppler frequency shift: $f_R = f_0 \pm f_D = f_S + f_C \pm f_D$. The electronics used in weather radars work more effectively with signals of lower frequency than the one transmitted (i.e., lower than f_0). Therefore, the amplified received signal is mixed with a reference signal from the STALO in a process known as *heterodyning*. The resulting signal now is said to be *downconverted* to a frequency $f_C \pm f_D$, which is the IF plus any Doppler shift. An IF filter (sometimes called a *matched filter*) is applied to isolate this downconverted signal, which is also further amplified.

To isolate the Doppler frequency, the IF signal (with frequency $f_C \pm f_D$) is again mixed in what is often called the *phase detector*, this time with two reference signals of frequency f_C from the COHO: one at 0° phase shift with respect to the transmitted signal, and one that has been phase-shifted by 90°. Recall that identifying differences in the signal phase from pulse to pulse allows for retrieval of the Doppler velocity. Both the in-phase (0°) and quadrature (90°) signals are needed to determine whether the target is moving towards or away from the radar. These signals are then sent to the data processor, where traditional radar quantities of interest for operational and research meteorologists and hydrologists are determined. Finally, these quantities are sent for dissemination and/or display.

8 Dual-Polarization Radar Variables

Operational weather radar networks around the world have or will undergo upgrades to dual-polarization capabilities, meaning that they transmit and/or receive electromagnetic waves at two orthogonal polarizations. This is in part due to the rich information available from radars with such capabilities. More detailed reviews of the principles and applications of dual-polarization radars are available in the literature (see reviews by Zrnić and Ryzhkov [177], Kumjian [77–79] and Chandrasekar et al. [26]). Dual-polarization or "polarimetric" radars (note: "dual-polarimetric" is redundant and therefore incorrect terminology) transmit horizontally (H) and vertically (V) polarized waves, either simultaneously or alternately. In the mode of simultaneous transmission and reception (STAR) of H- and V-polarized waves, the electric field vector in general traces an ellipse in the plane perpendicular to the direction of wave propagation. Assume the amplitudes of the H- and V-polarization

[1]For bistatic radars, there are separate transmitting and receiving antennas; e.g., [170, 173].

signals are equal. If the radar's differential phase between H and V polarizations on transmit is $\pm 90^\circ$, the electric field vector traces out a circle, and the wave is said to be *circularly polarized*. If the phase difference is 0° or $\pm 180^\circ$, the electric field vector traces a line slanted between the H and V polarization directions (i.e., at 45° or 135°). This is sometimes called *slant-*45° *linear polarization*. In the alternate transmission and reception mode, one polarization is transmitted at a time, while *both* polarizations are received. Various configurations exist for this mode: transmitting only H polarization, only V polarization, or both H and V polarizations alternately (though still one at a time). Most existing operational weather radar networks utilize STAR mode. The benefits and drawbacks of these two modes of operation are discussed in [35, 60–62, 179].

By comparing the backscattered radiation at both polarizations, one can obtain information about the shapes and types of hydrometeors present. This information is contained in the *polarimetric radar variables*, which are discussed individually in the subsections below. More comprehensive treatments are found in [20, 34, 77].

8.1 Differential Reflectivity Z_{DR}

Section 3 introduced how nonspherical particles will scatter radiation differently depending on the polarization of the illuminating radiation. Recall that Z_e is proportional to the amount of power backscattered from hydrometeors within the radar sampling volume. By comparing equivalent radar reflectivity factors at different polarizations, one can obtain information about the shape of a particle. The differential reflectivity Z_{DR} was first introduced by Seliga and Bringi [142] and is defined as the ratio of equivalent radar reflectivity factor at H polarization (Z_H) to that at V polarization (Z_V) when they are expressed in $\mathrm{mm}^6\,\mathrm{m}^{-3}$, or the difference $Z_H - Z_V$ when they are expressed in dBz. Thus, particles that scatter radiation equally at both polarizations have $Z_{\mathrm{DR}} = 0\,\mathrm{dB}$. For electromagnetically small particles, those that have more mass aligned in the horizontal have $Z_{\mathrm{DR}} > 0\,\mathrm{dB}$. Electromagnetically small particles with more mass aligned in the vertical have $Z_{\mathrm{DR}} < 0\,\mathrm{dB}$. This is because the near-field interactions between tiny finite scattering elements in the particle constructively interfere to enhance the total internal electric field at one polarization and destructively interfere to reduce the total internal electric field at the other polarization, as described in Sect. 3. For example, highly nonspherical hydrometeors like dendrites and planar crystals that tend to fall with their maximum dimensions in the horizontal can produce enormous Z_{DR} values ($>6\,\mathrm{dB}$ for extreme aspect ratios). In contrast, conical graupel particles that fall with their maximum dimension in the vertical produce negative Z_{DR} values. Raindrops become increasingly deformed by aerodynamic drag with increasing size, leading to positive Z_{DR}. For spherical particles or those that tumble randomly, radiation is equally scattered at H and V polarizations, leading to $Z_{\mathrm{DR}} = 0\,\mathrm{dB}$. For typical ranges of values at S band in different types of precipitation, see [77].

For a given particle shape, increasing the relative permittivity enhances the magnitude of the excited electric field within the particle, and thus enhances the constructive and destructive near-field interactions of tiny scattering elements, which leads to greater Z_{DR} magnitudes. For example, a nonspherical raindrop will have a larger Z_{DR} than an ice particle of identical size and shape. Particles with highly dispersed mass like fluffy snow aggregates (which in the scientific literature are sometimes characterized by a sphere or spheroidal particle with very low relative permittivity) have very weak constructive/destructive near-field interactions, and thus Z_{DR} near 0 dB. This is true even though snow aggregates tend to have highly nonspherical and irregular shapes (e.g., [67]). Aggregates also tend to have large variability in their orientation angles when falling, also leading to near-0 Z_{DR}.

If particles are electromagnetically large, resonance scattering effects obfuscate Z_{DR} interpretation, as it is no longer directly related to the particle shape. For precipitation radars, this is particularly relevant in the case of large hail. For example, calculations performed at S band (e.g., [86]) reveal negative Z_{DR} values for oriented wet, oblate hail with maximum dimension > 5 cm. Given that large hailstones tend to tumble, however, shape effects tend to be averaged out in the sampling volume, leading to observed Z_{DR} values typically near 0 dB. Z_{DR} is Z_e-weighted for a radar sampling volume filled with hydrometeors. Thus, the observed Z_{DR} is biased towards those particles that contribute the most to the overall Z_e. For populations of different particle types within a sampling volume, the observed Z_{DR} may fall somewhere between the "intrinsic" Z_{DR} of each population, depending on each population's contribution to the total Z_e.

8.2 *Differential Phase Shift* Φ_{DP}

As an electromagnetic wave propagates through a dielectric medium such as liquid or ice, it acquires a phase shift relative to a wave propagating the same distance in a vacuum. This is because the phase speed decreases for a wave propagating through liquid or ice (which are characterized by a refractive index greater than 1.0); equivalently, the wavelength of the radiation decreases within such a medium. Thus, the wave passing through the medium undergoes more of its 360° oscillation than a wave outside the medium at that range, leading to a phase shift relative to the freely propagating wave. This problem is often thought of conceptually as a slab of the dielectric material that extends infinitely in the directions orthogonal to the wave propagation direction but is of finite thickness in the direction of wave propagation. For this "infinite slab" model, the wave passing through the material emerges with a phase shift relative to a freely propagating wave traveling the same distance.

For particles small compared to the wavelength, one can replace the slab of the dielectric material with a slab of particles and achieve the same result (Fig. 14; see also [16]). However, the physics are different. Consider a radar pulse sampling this slab of particles. Assume the radar sampling volume extends much further radially than the radar wavelength (note that this is not shown in Fig. 14 for graphical

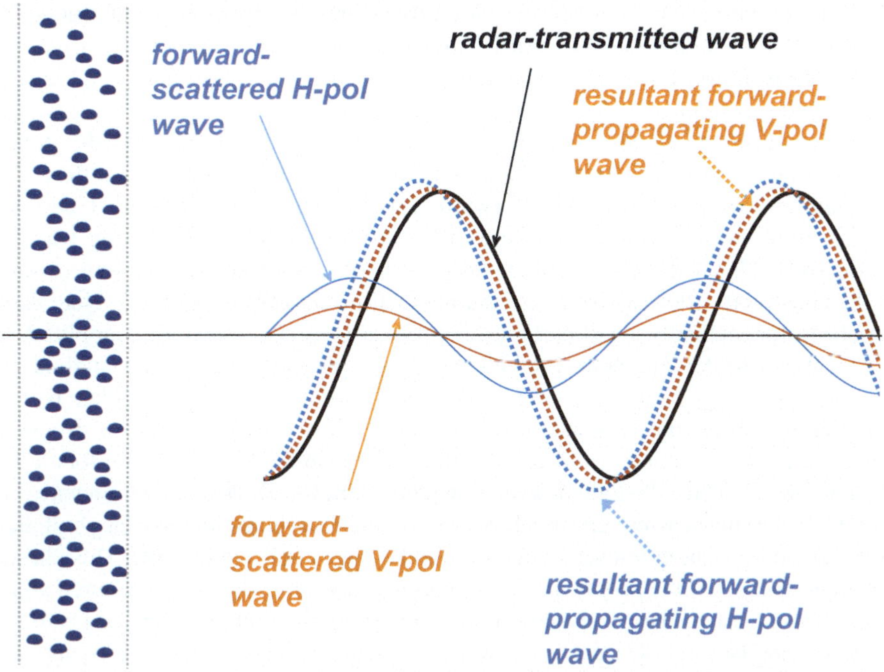

Fig. 14 Schematic showing waves propagating through an infinite slab of identical raindrops (each considered to be electromagnetically small compared to the wavelength). The solid black line is the original wave transmitted by the radar. The thin blue and orange lines represent the forward-scattered waves at H and V polarization, respectively, from the slab of raindrops, shown with a 90° phase shift. Note that the amplitude is greater for the H-polarization wave because the raindrops are oblate with their maximum dimension horizontally aligned. The total forward-propagating wave is the sum of these two, shown as the dotted lines. Note that the resultant H-polarization wave (blue dotted line) has a larger phase lag than the resultant V-polarization wave (orange dotted line); this difference is one half of Φ_{DP}. Adapted from Bohren and Huffman [16], with changes to show the polarization dependence

purposes). Thus, depending on the particles' positions within the sampling volume, dipole oscillations of particles' tiny scattering elements will be excited by different portions of the incident wave cycle. Bohren and Huffman [16] show that, for a sampling volume uniformly filled with hydrometeors, the expected mean phase shift for the forward-scattered wave at some distance beyond the slab is $90° + \Theta$ relative to the freely propagating wave, where Θ is the phase shift between the incident wave and the oscillations of the tiny scattering elements themselves. For weather radars, Θ is small and will be neglected here. In Fig. 14, the forward-scattered waves are depicted by the thin colored lines and the freely propagating incident wave is the thick black line. The total propagating wave is the sum of the phase-shifted forward-scattered waves from each of the particles illuminated in a sampling volume and the original transmitted wave (dotted colored lines in Fig. 14). The total propagation phase shift relative to the transmitted wave propagating in free space is thus between 0° and 90°, depending on the forward-scattered wavelets' contributions. In turn, the contributions of these wavelets depend on the size and concentration of particles in the sampling volume.

For electromagnetically small particles, we expect no difference in this total propagation phase shift between H and V polarizations unless there is a difference in amplitude of these particles' forward-scattered radiation between H and V polarizations, which would contribute differently to the total propagating wave's phase shift. This will only occur if the particles are nonspherical, arising owing to the constructive/destructive near-field interactions described in Sect. 3. Note that for electromagnetically small particles, the forward- and backscattering amplitudes are identical, so the arguments applied above for backscattering are also applied to forward scattering. Therefore, for a population of horizontally aligned oblate particles like raindrops, the H-polarization forward-scattered wave provides a stronger contribution to the total forward-propagating wave than the V-polarization forward-scattered wave. Thus, the total propagation phase shift for the H-polarization wave is weighed more heavily towards this forward-scattered wave, leading to a larger phase shift than at V polarization (cf. dotted blue and orange curves in Fig. 14). This phase difference is half of the *propagation differential phase shift* Φ_{DP}; the measured Φ_{DP} comes from the two-way propagation out to the sampling volume range and back.

Seliga and Bringi [143], Jameson [65], and Sachidananda and Zrnić [136] argued for the use of Φ_{DP} for rainfall estimation. Φ_{DP} tends to increase monotonically with range for electromagnetically small horizontally aligned particles, such as raindrops at longer wavelengths. As described above, the magnitude of Φ_{DP} increases with the concentration, size, and relative permittivity of *nonspherical* particles. Unlike Z_H or Z_V, however, Φ_{DP} is not affected by the presence of spherical particles or those that scatter equally at H and V polarizations. This makes Φ_{DP} quite useful for identifying mixtures of precipitation types, such as rain mixed with hail or pristine ice crystals mixed with snow aggregates/graupel. For the latter case, Z_e may be dominated by the much larger aggregates or graupel, leading to Z_{DR} near 0 dB. However, these larger, isotropically scattering particles do not contribute much to Φ_{DP}, whereas the smaller nonspherical crystals do, leading to increasing Φ_{DP} in the presence of Z_{DR} near 0 dB (e.g., [81, 140]).

Meteorologists often use one half[2] the range derivative of Φ_{DP}, known as the *specific differential phase* K_{DP}, which provides the phase shift per unit radial distance. This allows meteorologists to more easily identify regions of heavy precipitation containing nonspherical particles, like rain. One can think of K_{DP} as a measure of the amount and/or size of nonspherical particles in the sampling volume. Because they are based on phase measurements and not power, Φ_{DP} and K_{DP} are unaffected by attenuation or differential attenuation, except in the case of total extinction of the radar signal. For this reason, these variables are often used in estimating and correcting for attenuation (e.g., [18, 22, 49, 133, 134, 148, 154]).

---

[2] One-half the derivative is taken because Φ_{DP} is the two-way propagation differential phase shift: out to the sampling volume and back.

In practice, the radar measures the *total differential phase shift* Ψ_{DP}, which is a combination of the propagation differential phase shift Φ_{DP}, any differential phase of the transmitted wave caused by the radar system (sometimes called the "system differential phase"), and any differential phase shift imparted on backscatter, known as the *backscatter differential phase* δ. Such backscatter differential phase generally arises from electromagnetically large, nonspherical particles. Recent studies have investigated the microphysical information content of δ in rain and melting snow (e.g., [162, 163]).

8.3 Co-polar Correlation Coefficient ρ_{hv}

The *co-polar correlation coefficient* ρ_{hv} is the correlation between the received co-polar signals at H and V polarizations (i.e., transmitted H and received H; transmitted V and received V). It can be thought of as a measure of particle variability within the radar sampling volume, where values of 1.0 indicate perfectly homogeneous precipitation and values below 1.0 indicate particle diversity. Specifically, ρ_{hv} decreases with increasing diversity of particle shapes, orientation angles, and/or relative permittivities within the sampling volume. The simplest way to think about ρ_{hv} is as a measure of the variability of particles' intrinsic Z_{DR} within the radar sampling volume: if particles in the sampling volume have a narrow distribution of intrinsic Z_{DR} values, then ρ_{hv} is close to 1.0. When particle Z_{DR} varies dramatically within the sampling volume, ρ_{hv} is reduced. Additionally, ρ_{hv} is reduced when significant variability of differential phase shift Ψ_{DP} exists within the sampling volume. This may arise from the existence of particles producing δ in the sampling volume, or gradients of Φ_{DP} within the sampling volume (e.g., when the beam is not filled uniformly; [79, 128]). Such nonuniform beam filling is common in convective storms, resulting in radially oriented streaks of reduced ρ_{hv} extending rearward from heavy precipitation cores.

At S band, ρ_{hv} is very high (>0.98) for uniform precipitation such as pure rain or pure aggregated snow. For pristine snow crystals in the presence of aggregates, the enhanced shape diversity leads to slightly reduced ρ_{hv} (>0.95; [77]). Because of resonance scattering effects for 5–6 mm raindrops at C band, which contribute to δ, ρ_{hv} in rain at C band can fall as low as 0.93. In contrast, ρ_{hv} can be much lower (<0.85) in melting snow, melting hail, and is extremely low (<0.7) in cases of nonmeteorological scatterers like insects, birds, smoke and ash, military chaff, and tornado debris. For this reason, ρ_{hv} is often used to discriminate between precipitation and nonprecipitation echoes in operational weather radar data, and for detecting the melting layer [41, 42]. Further, it is useful in identifying ongoing tornadoes [15, 82, 129, 141, 164], as discussed in the next section.

8.4 Linear Depolarization Ratio LDR

Dual-polarization weather radars that transmit and receive H- and V-polarization signals alternately (e.g., transmit H, receive H and V; transmit V, receive H and V, etc.) can measure the amount of depolarization that occurs in precipitation. If a radar transmits radiation at H polarization, the power of the received signal at V polarization is known as the *cross-polar* power. In terms of equivalent radar reflectivity factor, this is typically written as Z_{VH} (the standard convention is that the second subscript indicates the transmitted signal polarization and the first subscript indicates the received signal polarization). The ratio of Z_{VH} to Z_H (or their difference $Z_{\mathrm{VH}} - Z_H$ if expressed in logarithmic units) is known as the *linear depolarization ratio*, or LDR. Because the cross-polar component of the backscattered signal is typically very weak, values (in dB) in most hydrometeors are negative. Radar system noise limits the lowest detectable LDR to about −40 dB. LDR is only available from radars that operate in the mode of alternate transmission and reception of H and V waves: if only H polarization is transmitted, then received V-polarization signals come from depolarization. In STAR mode, it is impossible to determine what portion of the received signal of one polarization comes from co-polar or cross-polar contributions. Thus, LDR is not available from the U.S. WSR-88D radar network, which operates in STAR mode.

LDR is enhanced when nonspherical particles have a mean orientation angle with respect to the plane of polarization. This is referred to as the *canting angle*. In most situations, hydrometeors have approximately zero mean canting angle, making LDR rather low in rain, aggregated snow, etc. However, LDR is enhanced for electromagnetically large, wobbling nonspherical particles, especially those of irregular shape like wet hail (e.g., [72]). There is also evidence that wobbling, freezing raindrops in convective storm updrafts can produced noticeably larger LDR values (e.g., [59, 87]). The melting layer "bright band" typically features locally enhanced LDR values as well (Fig. 15). Recent work has shown LDR is also useful for detecting columnar ice crystals in data collected with zenith-pointing radar [116].

9 Weather Radar Applications

The information available with dual-polarization and Doppler weather radar has led to a plethora of insights and novel applications of remote sensing data in clouds and precipitation. This section provides a limited outline of some of these applications; others will be discussed elsewhere in this volume.

Much attention has focused on detection of hazards in severe convective storms. Doppler velocity information allows operational meteorologists to identify severe winds and downbursts in convective storms (e.g., [91, 104, 125]), as well as regions of strong azimuthal shear associated with the circulations of mesocyclones and

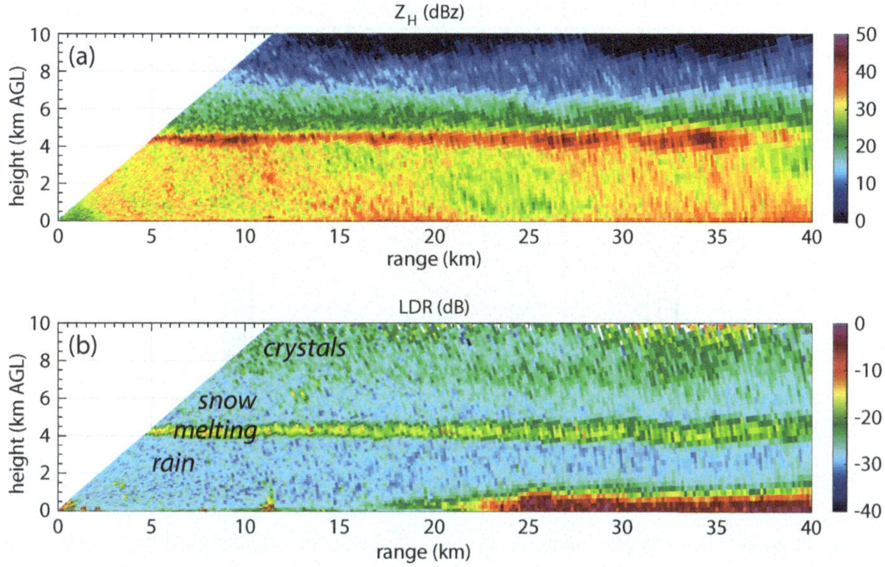

Fig. 15 Vertical cross section of (**a**) Z_H and (**b**) LDR through stratiform precipitation, collected using the SPolKa radar during a field campaign in the Indian Ocean. The melting layer bright band is clear in both Z_H and LDR. Hydrometeor types are annotated in the LDR image. LDR is enhanced in the melting layer, and slightly enhanced in a region of pristine ice crystals aloft; in snow aggregates and rain, LDR values are rather low

tornadoes (e.g., [23, 24]). Further, dynamical retrievals from weather radars have substantially improved our understanding of hazardous storms, including severe convection and tropical cyclones. When two or more Doppler radars probe the same storm from different viewing angles at approximately the same time, they measure different components of the particles' motion. In particular, they measure the radial projection of the particles' three-dimensional motion vector. The particles' horizontal motion vector usually is assumed to be equal to the horizontal wind vector, although this is a poor assumption in cases of strong vertical wind shear, or in strong circulations where centrifuging becomes important (e.g., [36, 114, 169]). With the mass continuity equation to couple horizontal and vertical motion and an estimate of particle fall speeds (typically parameterized rather crudely as a function of Z_e for scanning radars), one may estimate the three-dimensional winds. Numerous so-called *dual-Doppler* or *multiple-Doppler* synthesis techniques exist in the scientific literature (e.g., [3, 29, 40, 76, 144], among many others). These three-dimensional wind syntheses are more accurate when the radar beams sampling a particular point are separated by an angle $>30°$ and $<150°$, when both beams are at lower elevation angles, and when the radars are sampling common regions of the storm at the same time. Kinematic quantities of interest can be calculated utilizing estimates of the three-dimensional wind field obtained from multiple-Doppler syntheses to provide insights into hazardous storms. In Fig. 16, vertical

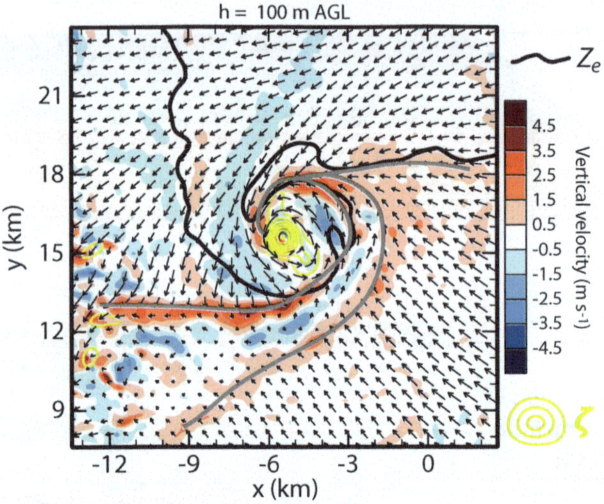

Fig. 16 Example dual-Doppler analysis showing retrieved two-dimensional horizontal wind vectors, vertical velocity (shaded in m s^{-1}, according to scale), vertical vorticity (goldenrod contours), and the outline of the Z_e echo (black) at 100 m AGL. Gray lines indicate convergence lines associated with the primary and secondary rear-flank gust fronts. Image courtesy of Dr. Karen Kosiba (Center for Severe Weather Research)

velocity and vertical vorticity at 100 m AGL are shown for a tornadic supercell. Regions of updrafts and downdrafts are seen spiraling around the tornado, which is indicated by the vertical vorticity maximum.

The advent of dual-polarization radars has led to burgeoning operational algorithms aiding forecasters in hydrometeor classification (e.g., [97, 98, 117, 151, 157, 168], among many others). In these algorithms, the information content from the different radar measurements is distilled into a best-guess classification of each radar pixel into one of a set of hydrometeor or non-meteorological echo types (Fig. 17). In addition, the added information on drop sizes combined with echo classification has significantly improved quantitative precipitation estimation (QPE), although continued efforts to further refine algorithms are ongoing (e.g., [28, 41, 48, 134]). Such improvements to QPE have lead to better detection of floods (e.g., [17, 27]).

The combination of Doppler velocity information and polarimetry has allowed for improved tornado detection: strong rotation evident in Doppler velocities combined with dramatically reduced ρ_{hv} is indicative of an ongoing tornado lofting debris, as first discovered by Ryzhkov et al. [129]. The combination of irregular shapes and orientations of debris elements leads to ρ_{hv} values much lower than expected for precipitation (Fig. 18). This *tornado debris signature* has been the focus of numerous recent studies (e.g., [15, 82, 141, 164]). In some significant tornado cases, debris may be lofted to great heights and fall out on the left flank of the storm, as evident in Fig. 18. Ongoing research is exploring the utility of the tornadic debris signature for real-time information on tornado severity (e.g., [15]).

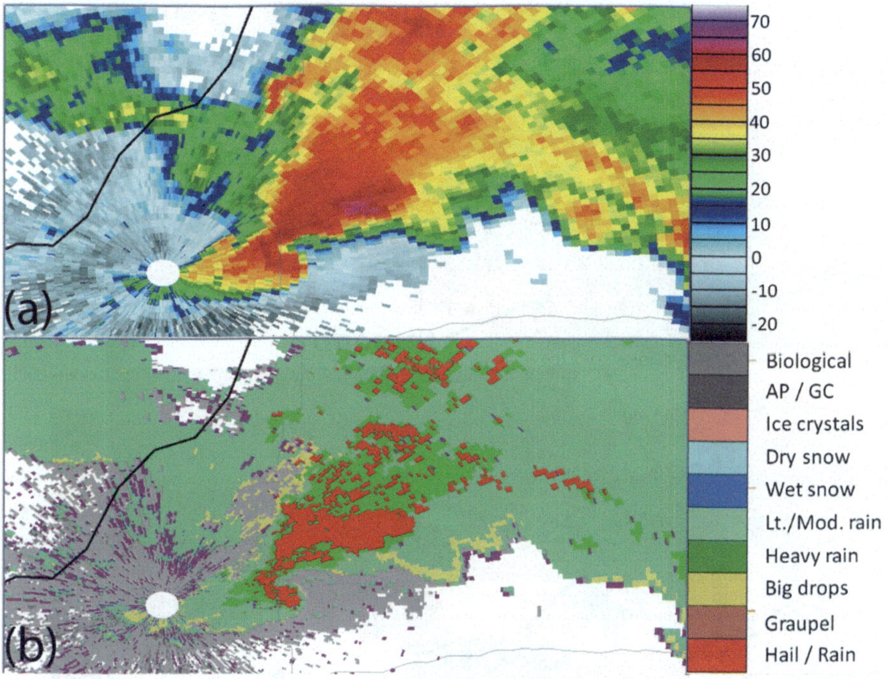

Fig. 17 Example low-level surveillance scan of (**a**) Z_H (shaded according to scale in dBz) and (**b**) output from the operational hydrometeor classification algorithm (color-coded according to scale) for a tornadic storm on 1 March 2017 in southwestern Indiana, USA, collected with the dual-polarization WSR-88D radar near Evansville (KVWX) at 0432 UTC. Magenta colors in the hydrometeor classification indicate "unknown"

The detection of damaging hail is of great interest to operational meteorologists. Dual-polarization radar can aid in hail detection owing to the differences in scattering behaviors of raindrops versus hailstones (e.g., [5, 21]). In particular, large hail tends to fall with a more chaotic orientation than raindrops, leading to Z_{DR} values near 0 dB regardless of shape. This is paired with generally large Z_H (>55 dBz), and reduced ρ_{hv} values. More recently, research efforts have focused on improving the determination of hail size [115, 132, 133]. Discrimination of small, typically non-damaging hail (<2.5 cm in diameter) from large hail is relatively straightforward with S-band dual-polarization radars (e.g., [52, 133]). This is because small hail tends to acquire a significant liquid water coating during melting [124], leading to scattering resembling large raindrops with high Z_{DR} and K_{DP} values. However, distinguishing between severe (>2.5 cm) and significantly severe (>5 cm) hail has proven challenging (e.g., [11, 115]). Some studies have suggested identifying significantly severe hail aloft, identifiable as large reductions in ρ_{hv} in the storm updraft above the environmental 0 °C level (e.g., [86, 119]), thought to arise from the presence of irregular shapes and lobes resulting from wet growth, and/or significant

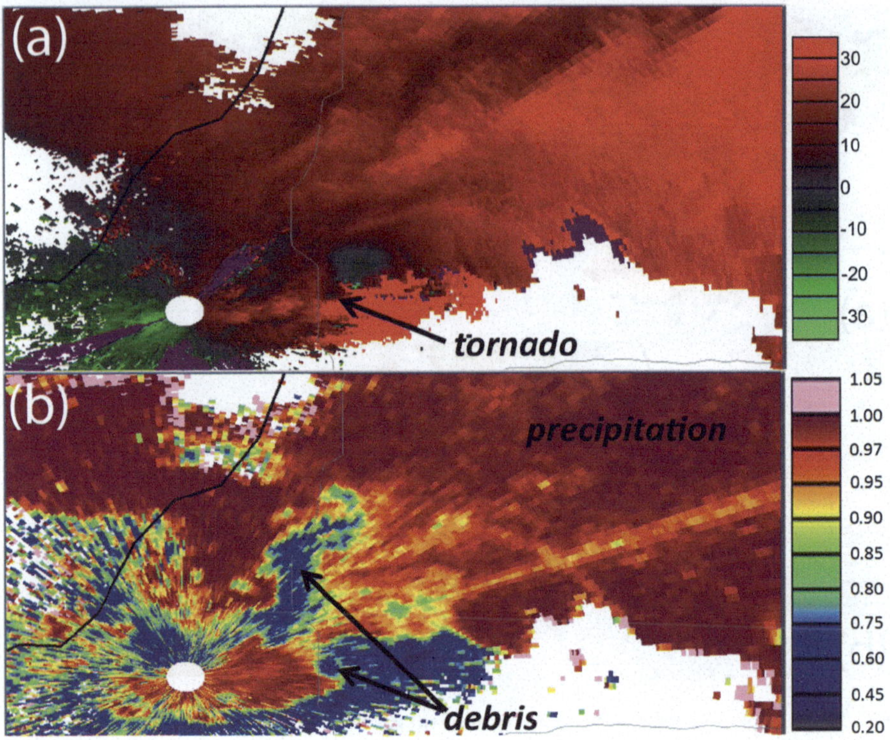

Fig. 18 As in Fig. 17, but (**a**) Doppler velocity (m s^{-1}, shaded according to scale) and (**b**) ρ_{hv} (shaded according to scale) are shown. Arrows indicate the regions of Doppler-velocity-indicated rotation in (**a**) and tornadic debris as extremely low ρ_{hv} values in (**b**). The higher ρ_{hv} values indicate precipitation

resonance effects and backscatter differential phase δ. Recent work has also found anomalously high (>10 deg km^{-1}) S-band K_{DP} values in cases of extremely high concentrations of small hail that can lead to accumulations several cm deep [80]. These topics are all areas of ongoing research.

Monitoring convective storm behavior is also of interest, as weather radars can offer clues into the near-term evolution of a storm and its associated hazards. Since first discovered in the 1980s, dual-polarization radar observations of convective storms have repeatedly revealed the presence of a columnar region of positive Z_{DR} values that extends above the environmental 0 °C level (e.g., [50]), known as the Z_{DR} *column*. Figure 19 is an RHI scan through a convective storm exhibiting two Z_{DR} columns. Whereas the surrounding regions of primarily graupel and snow exhibit near-zero Z_{DR}, the positive Z_{DR} values in Z_{DR} columns indicate the updraft lofting liquid particles (raindrops or wet ice with large liquid water content) above the 0 °C level. For surveillance scans like those used by the WSR-88D radar network, Z_{DR} columns can be identified as localized enhanced Z_{DR} regions above

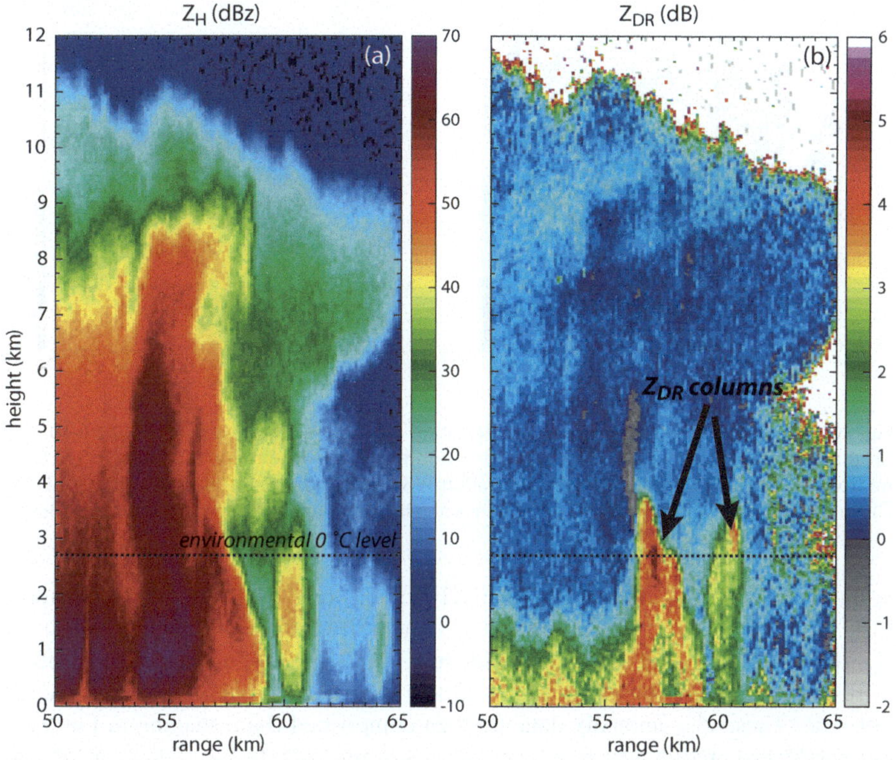

Fig. 19 Vertical cross sections of (**a**) Z_H and (**b**) Z_{DR} (shading according to scales) through a convective storm in Oklahoma on 10 February 2009. Two Z_{DR} columns are evident as positive Z_{DR} values above the environmental 0 °C level, which is annotated on both panels as the dashed line. Data from the KOUN radar in Norman, Oklahoma, courtesy of Dr. Valery Melnikov (National Severe Storms Laboratory)

the environmental 0 °C level (Fig. 20). The Z_{DR} column serves as a marker of the convective storm's updraft. In addition, the behavior of the Z_{DR} column is correlated to updraft intensity and evolution [89]. In particular, increases in the Z_{DR} column height above the 0 °C level are correlated with increases in updraft intensity, as well as subsequent increases in precipitation intensity near the surface. An operational algorithm is being developed to monitor these changes in Z_{DR} column height [149]. The breadth of the Z_{DR} column may be an important determinant of the storm's capability to produce large hail [33, 78]. Similar to Z_{DR} columns, enhanced K_{DP} columns (e.g., [59, 82, 166]) also provide diagnostic information on storm behavior and short-term evolution.

Another recent algorithm is being developed to detect regions of ongoing hydrometeor size sorting in convective storms. Such size sorting can be associated with developing convective updrafts as small drops are lofted, whereas larger, faster-falling drops and small melting ice particles are able to fall through the updraft,

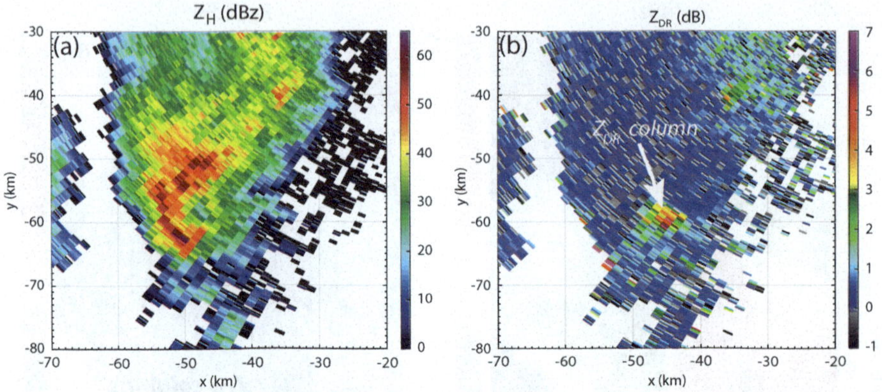

Fig. 20 Example surveillance scan at 3.2° elevation angle showing fields of (**a**) Z_H and (**b**) Z_{DR} (shading according to scales) from a severe hailstorm in southwestern South Dakota, USA. Data are from 21 June 2013, from the dual-polarization WSR-88D radar near Rapid City, SD (KUDX). The radar sampling volume at the Z_{DR} column is at a height of about 4.5 km above the radar level

resulting in large Z_{DR} values with relatively low Z_H (e.g., [84]). The Thunderstorm Risk Estimation and Nowcasting Development from Size Sorting (TRENDSS) algorithm [121] is being developed to identify such regions (Fig. 21). The algorithm works by first identifying positive Z_{DR} outliers for a given range of Z_H values in each PPI scan. These Z_{DR} anomaly data are then composited (blue shading in Fig. 21a) as a product to predict near-term convective storm trends. In Fig. 21a, areas of size sorting are detected on both the north and south flanks of the left-most cell; 20 min later, this storm has split (Fig. 21b).

Anticipating a storm's potential to produce a tornado is a key concern for operational meteorologists, as it improves the warning lead time and can save lives. A promising area of active research is in exploring the relationship between low-level dual-polarization radar signatures and the near-storm environment. In supercells and some nonsupercell tornadic storms, researchers have discovered low-level enhanced Z_{DR} and K_{DP} regions of special diagnostic value called the Z_{DR} *arc* and K_{DP} *foot*, respectively [82, 127]. The Z_{DR} enhancement region tends to be centered on a Z_H gradient, and offset from the K_{DP} enhancement region. The K_{DP} enhancement region tends to be near the storm's Z_H core in an area of heavy precipitation. This arrangement of low-level polarimetric features arises from the size sorting of hydrometeors in the presence of nonzero storm-relative flow over the lowest several km of the troposphere, which is typically associated with vertical wind shear [31, 32, 82, 83]. When the Z_{DR} enhancement region's major axis is aligned parallel to storm motion, this often implies substantial low-level storm-relative helicity (Fig. 22), an important parameter used by forecasters to assess a storm's potential to produce tornadoes (e.g., [30, 106, 123, 155, 156]). Research efforts to quantify the separation and orientation of the Z_{DR} and K_{DP} enhancement regions with respect to storm motion are ongoing (e.g., [99, 100]).

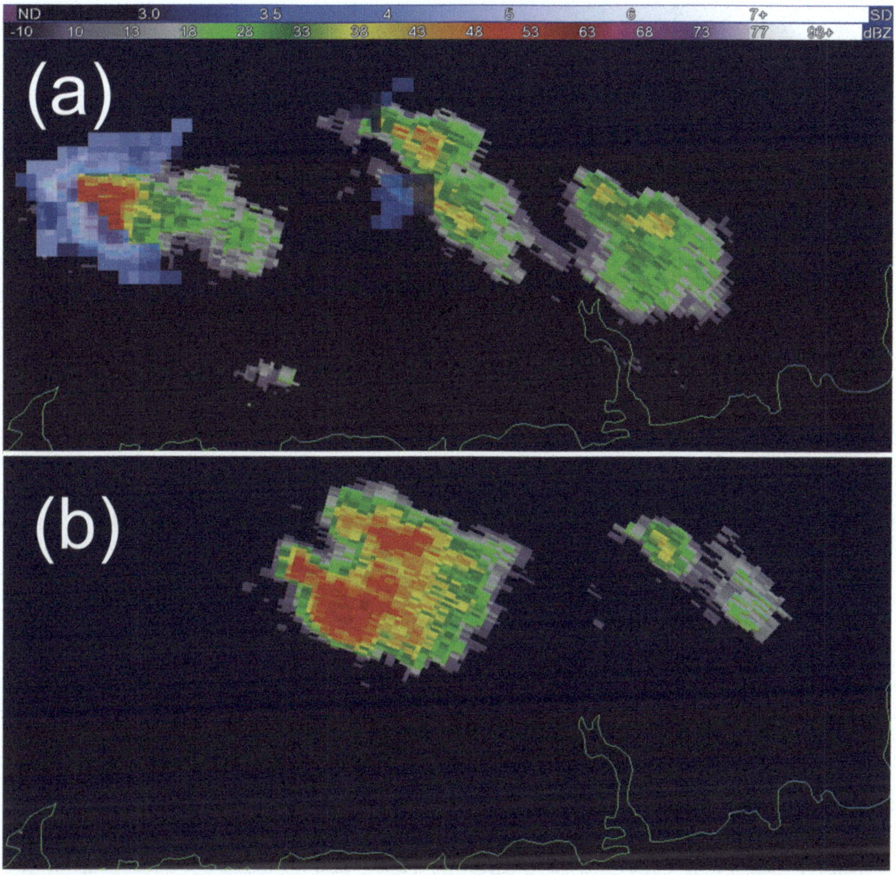

Fig. 21 Example surveillance scans of Z_H (in dBz, shaded according to scale) at two different separated by 20 min. The top panel (**a**) also includes output from the TRENDSS algorithm (blue shading), indicating areas of size sorting and potential growth on both flanks of the left-most cell. Indeed, the storm splits after 20 min (**b**). Data from multiple WSR-88D radars shown over southern Connecticut on 23 June 2015. Image courtesy of Joey Picca (Storm Prediction Center)

Polarimetric radar also provides important insights into cold-season precipitation hazards, including precipitation-type transitions and heavy snowfall. Figure 23 shows a series of surveillance scans of the ρ_{hv} field during a precipitation-type transition over Long Island, New York. Recall that a diversity of hydrometeor shapes and compositions leads to reductions in ρ_{hv}. Therefore, in mixed-phase precipitation regions when hydrometeors are melting or freezing, ρ_{hv} can be reduced. In the case shown in Fig. 23, the precipitation type reported at the Islip airport near the radar transitioned from snow to a snow/rain/ice pellets mix as the band of reduced ρ_{hv} values passed overhead (Fig. 23a, b). Once the band of reduced ρ_{hv} values moved to the north of the radar, precipitation was entirely rain (Fig. 23c, d). Operational

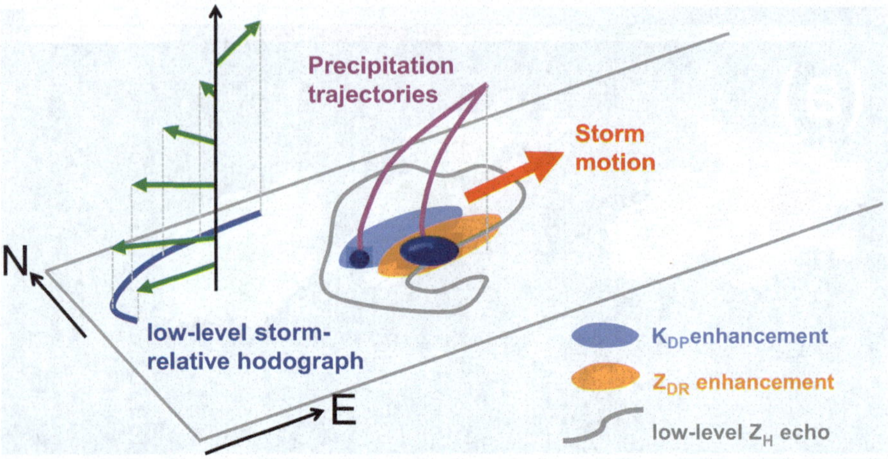

Fig. 22 Schematic showing the relationship between low-level storm-relative helicity and the offset of Z_{DR} and K_{DP} enhancement signatures at low levels. The low-level Z_H echo is shown in gray, whereas the Z_{DR} and K_{DP} enhancements are shown in orange and blue, respectively. The vertical wind profile is shown by green vectors at different altitudes, and the low-level storm-relative hodograph is shown as a thick blue curve to the left. Storm motion is indicated by the red vector. Precipitation trajectories are shown in fuchsia, indicating the fallout locations of large and small drops

forecasters make use of these important signatures from dual-polarization radar to assess precipitation transitions in real time (e.g., [120]). In addition, recent research has uncovered signatures in Z_{DR} associated with ice pellet formation and transitions from ice pellets to freezing rain [85, 88, 165]. Though this so-called *refreezing signature* has been observed in numerous cases and is a robust indicator of ice pellets, the exact underlying microphysical processes responsible for its appearance remain unknown and are a topic of current research.

Identifying regions of heavy snowfall helps forecasters monitor for locations receiving large snow accumulations. Recently, a series of studies have found that enhancements in K_{DP} at altitudes near $-15\,°\mathrm{C}$ are associated with increased precipitation rates near the surface [2, 8, 71, 111, 138–140]. Sometimes referred to as dendritic growth zone signatures, the K_{DP} (and sometimes Z_{DR}) enhancement signifies vigorous planar crystal growth and subsequent highly efficient aggregation that together lead to heavier snowfall beneath the signatures. Depictions of such signatures in time-height format known as quasi-vertical profiles (QVPs; e.g., [135]) show a clear association between enhanced K_{DP} aloft and heavy snowfall (large Z_H) below (Fig. 24). The QVP technique was recently extended to include Doppler velocity and estimates of mesoscale divergence/convergence over the radar by Kumjian and Lombardo [81]; such depictions also show a clear connection between enhanced mesoscale ascent and vigorous planar crystal growth signatures near $-15\,°\mathrm{C}$ (Fig. 24). In the case shown in Fig. 24, snowfall rates beneath the enhanced K_{DP} signature exceeded $15\,\mathrm{cm\,h^{-1}}$ at times.

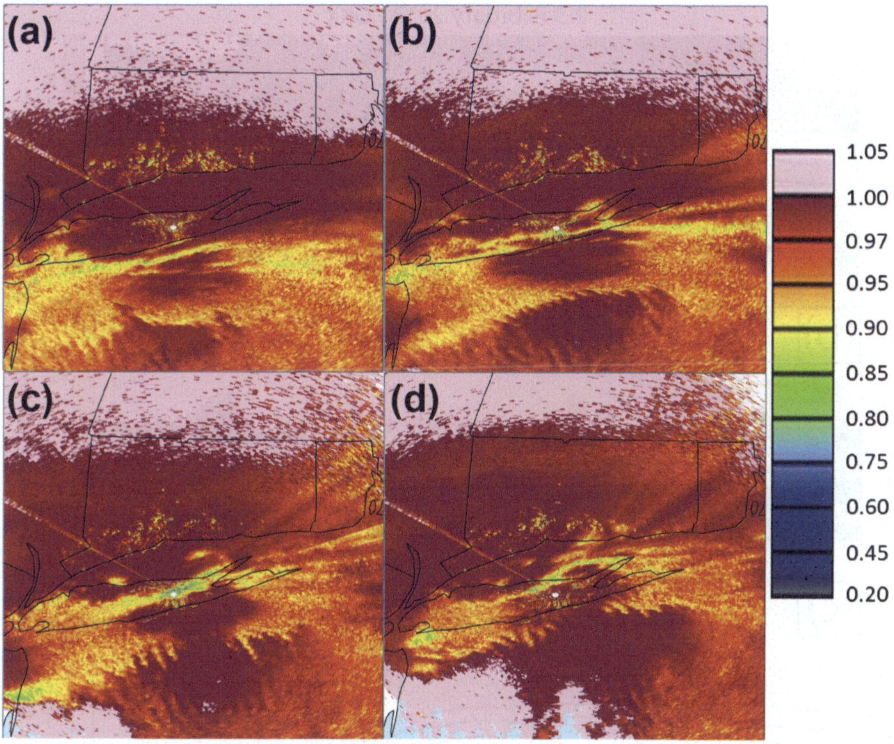

Fig. 23 Time series of 0.5° elevation angle surveillance scans of ρ_{hv} (shaded according to scale) from a winter storm precipitation transition over Long Island, New York, USA, on 13 February 2014. Data are from the dual-polarization WSR-88D radar near Upton, NY (KOKX) from (**a**) 1426 UTC, (**b**) 1456 UTC, (**c**) 1526 UTC, and (**d**) 1556 UTC

Supercooled liquid water in cold clouds can pose a serious icing hazard to aircraft. However, owing to their small sizes, cloud droplets cannot be detected in the presence of other much larger precipitation particles directly with low-frequency precipitation radars. Thus, detection of supercooled liquid water remains a significant challenge for weather radars. Instead, researchers are looking for indirect indicators of supercooled liquid water, such as the presence of ongoing riming and secondary ice production. A promising signature in dual-polarization weather radar observations may be enhanced K_{DP} collocated with near-zero Z_{DR} at temperatures between about −3 and −8 °C, which are thought to indicate the presence of quasi-spherical rimed particles (graupel) and/or aggregates with large concentrations of needle-like crystals (e.g., [43, 81, 90, 146]). Coordinated in situ measurements and polarimetric radar data are needed to verify the relationship between these signatures and the presence of supercooled liquid water.

One of the greatest challenges involving weather radar is combining advanced weather radar observations and numerical models. In particular, assimilation of

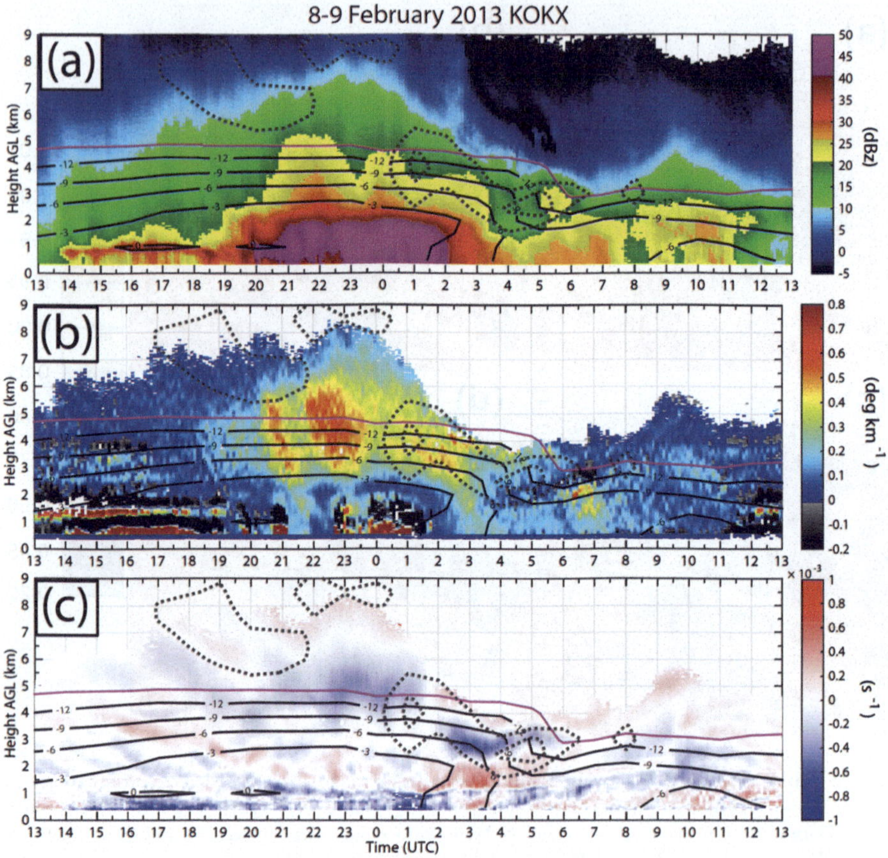

Fig. 24 Time series of quasi-vertical profiles (QVPs) of (**a**) Z_H, (**b**) K_{DP}, and (**c**) large-scale divergence during the 8–9 February 2013 blizzard, values shaded according to scales. Data from the dual-polarization WSR-88D radar near New York, NY (KOKX). Overlaid are temperature contours from the Rapid Refresh model (black curves, -12 to $0\,^{\circ}$C in $3\,^{\circ}$C increments, and magenta line indicating $-15\,^{\circ}$C) and mesoscale ascent (black dotted lines, starting at $-1\,\mathrm{Pa\,s^{-1}}$ in 1-$\mathrm{Pa\,s^{-1}}$ increments). Adapted from Kumjian and Lombardo [81], with changes

dual-polarization radar data is an area of ongoing research (e.g., [68, 122, 175]). This is difficult given that many operational numerical weather prediction (NWP) models employ simplified microphysics schemes that provide no information about particle shapes, fall behaviors, or the water content in mixed-phase hydrometeors. Crucially, these particle properties exert the strongest influences on the observed radar quantities. In turn, the radar observables are strongly biased by the largest particles in the sampling volume, and exhibit a highly uncertain relationship with typical model prognostic variables like total number concentration and total mass content, which are not as sensitive to the large-size tail of the particle size distribution. Currently, many *forward operators* (the bridge between model-predicted variables and observed radar variables) have built-in assumptions about many of these important particle properties (e.g., [68, 131]). Many current forward operators for use in radar data assimilation are overly simplistic (e.g., [95, 96, 158]), leading to large uncertainties [25]. However, recent advances in numerical modeling reveal a push towards predicting particle properties (e.g., [51, 66, 112, 113]), which may facilitate the blending of NWP models and weather radar data (e.g., [152, 153]). Future work in advanced electromagnetic scattering calculations for complex particle shapes such as snow crystals will also lead to improvements in forward operators (e.g., [102]).

10 Summary and Conclusion

This chapter highlights the fundamentals of radar remote sensing of clouds and precipitation and recent advances in hydrometeorological applications of weather radar. Radar remains the most important tool for examining clouds and precipitation. In particular, dual-polarization Doppler weather radar observations have facilitated novel insights into cloud and precipitation physics, as well as operationally important advances in the detection and monitoring hazardous weather. This includes severe convective storms (hail, tornadoes, damaging winds, floods) and winter storms (heavy snow, precipitation-type transitions). Algorithms for classification of radar echoes and quantitative precipitation estimation have also made substantial improvements in recent decades. Future breakthroughs undoubtedly will involve improvements to our understanding of fundamental physical processes in clouds and precipitation and the blending of numerical weather prediction models and polarimetric Doppler weather radar observations via data assimilation. The remainder of this volume covers many of these important applications of radar remote sensing.

Acknowledgements The author is grateful for the opportunity to contribute to this volume and thanks Dr. Constantin Andronache for facilitating the process. In addition, he would like to acknowledge the fruitful conversations and material provided by Joseph Picca (National Weather Service-Storm Prediction Center), Israel Silber (Penn State University), Marcus van Lier-Walqui (Columbia University and National Aeronautics and Space Administration Goddard Institute for Space Studies), Hughbert Morrison (National Center for Atmospheric Research), and Olivier Prat (North Carolina Institute for Climate Studies, North Carolina State University). In particular, he thanks his Penn State colleagues for numerous discussions over the last several years: Kültegin Aydin, Craig Bohren, Anthony Didlake, Eugene Clothiaux, Steven Greybush, Jerry Harrington, Paul Markowski, James Marquis, Yvette Richardson, and Johannes Verlinde. Finally, he would like to thank his awesome students in the Penn State RADAR, INC. group: Robert Schrom, Dana Tobin, Kyle Elliott, Scott Loeffler, Steven Naegele, John Banghoff, Charlotte Martinkus, and Zhiyuan Jiang. Radar data examples used in this chapter were obtained courtesy of the National Oceanic and Atmospheric Administration and the U.S. Department of Energy Atmospheric System Research program. The author would like to thank Scott Collis (Argonne National Laboratory), Karen Kosiba (Center for Severe Weather Research), and Angela Rowe (University of Washington) for their helpful critiques of this work.

References

1. Albrecht, B., M. Fang, and V. Ghate. 2016. Exploring stratocumulus cloud-top entrainment processes and parameterizations by using Doppler cloud radar observations. *Journal of the Atmospheric Sciences* 73: 729–742.
2. Andrić, J., M.R. Kumjian, D.S. Zrnić, J.M. Straka, and V.M. Melnikov. 2013. Polarimetric signatures above the melting layer in winter storms: an observational and modeling study. *Journal of Applied Meteorology and Climatology* 52: 682–700.
3. Armijo, L. 1969. A theory for the determination of wind and precipitation velocities with Doppler radars. *Journal of the Atmospheric Sciences* 26: 570–573.
4. Atlas, D. 2002. Radar calibration: Some simple approaches. *Bulletin of the American Meteorological Society* 83: 1313–1316.
5. Aydin, K., T.A. Seliga, and V. Balaji. 1986. Remote sensing of hail with a dual linear polarized radar. *Journal of Climate and Applied Meteorology*, 25: 1475–1484.
6. Battaglia, A., C.D. Westbrook, S. Kneifel, P. Kollias, N. Humpage, U. Löhnert, J. Tyynelä, and G.W. Petty. 2014. G band atmospheric radars: a new frontier in cloud physics. *Atmospheric Measurement Techniques* 7: 1527–1546.
7. Battan, L.J. 1973. *Radar Observation of the Atmosphere*, 324. Chicago: University of Chicago Press,
8. Bechini, R., L. Baldini, and V. Chandrasekar. 2013. Polarimetric radar observations of the ice region of precipitation clouds at C-band and X-band radar frequencies. *Journal of Applied Meteorology and Climatology* 52: 1147–1169.
9. Beck, J., and O. Bousquet. 2013. Using gap-filling radars in mountainous regions to complement a national radar network improvements in multiple-Doppler wind syntheses. *Journal of Applied Meteorology and Climatology* 52: 1836–1850.
10. Bharadwaj, N., and V. Chandrasekar. 2007. Phase coding for range ambiguity mitigation in dual-polarized Doppler weather radars. *Journal of Atmospheric and Oceanic Technology* 24: 1351–1363.
11. Blair, S.F., D.R. Deroche, J.M. Boustead, J.W. Leighton, B.L. Barjenbruch, and W.P. Gargan. 2011. A radar-based assessment of the detectability of giant hail. *Electronic Journal of Severe Storms Meteorology* 6 (7): 1–30.
12. Bluestein, H.B., W.P. Unruh, D.C. Dowell, T.A. Hutchinson, T.M. Crawford, A.C. Wood, and H. Stein. 1997. Doppler radar analysis of the Northfield, Texas, tornado of 25 May 1994. *Monthly Weather Review* 125: 212–230.
13. Bluestein, H.B., M.M. French, I. PopStefanija, R.T. Bluth, and J.B. Knorr. 2010. A mobile, phased-array Doppler radar for the study of severe convective storms The MWR-05XP. *Bulletin of the American Meteorological Society* 91: 579–600.
14. Bodine, D.J., D. Michaud, R.D. Palmer, P.L. Heinselman, J. Brotzge, N. Gasperoni, B.L. Cheong, M. Xue, and J. Gao. 2011. Understanding radar refractivity sources of uncertainty. *Journal of Applied Meteorology and Climatology* 50: 2543–2560.
15. Bodine, D.J., M.R. Kumjian, R.D. Palmer, P.L. Heinselman, and A.V. Ryzhkov. 2012. Tornado damage estimation using polarimetric radar. *Weather and Forecasting* 28: 139–158.
16. Bohren, C., and D. Huffman. 1983. *Absorption and Scattering of Light by Small Particles*, 530. Wiley, New York.
17. Boodoo, S., D. Hudak, A. Ryzhkov, P. Zhang, N. Donaldson, D. Sills, and J. Reid. 2015. Quantitative precipitation estimation from a C-band dual-polarized radar for the 8 July 2013 flood in Toronto, Canada. *Journal of Hydrometeorology* 16: 2027–2044.
18. Borowska, L., A.V. Ryzhkov, D.S. Zrnić, C. Simmer, and R.D. Palmer. 2011. Attenuation and differential attenuation of 5-cm-wavelength radiation in melting hail. *Journal of Applied Meteorology and Climatology* 50: 59–76.
19. Bremer, J., and U. Berger. 2002. Mesospheric temperature trends derived from ground-based LF phase-height observations at mid-latitudes comparison with model simulations. *Journal of Atmospheric and Solar - Terrestrial Physics* 64: 805–816.

20. Bringi, V., and V. Chandrasekar. 2001. *Polarimetric Doppler Weather Radar Principles and Applications*, 636. Cambridge: Cambridge University Press.
21. Bringi, V.N., J. Vivekanandan, and J.D. Tuttle. 1986. Multiparameter radar measurements in Colorado convective storms. Part II: Hail detection studies. *Journal of the Atmospheric Sciences* 43: 2564–2577.
22. Bringi, V.N., V. Chandrasekar, N. Balakrishnan, and D.S. Zrnić. 1990. An examination of propagation effects on radar measurements at microwave frequencies. *Journal of Atmospheric and Oceanic Technology* 7: 829–840.
23. Brown, R.A., L.R. Lemon, and D.W. Burgess. 1978. Tornado detection by pulsed Doppler radar. *Monthly Weather Review* 106: 29–38.
24. Burgess, D.W., L.D. Hennington, R.J. Doviak, and P.S. Ray. 1976. Multimoment Doppler display for severe storm identification. *Journal of Applied Meteorology* 15: 1302–1306.
25. Carlin, J.T., A.V. Ryzhkov, J.C. Snyder, and A.P. Khain. 2016. Hydrometeor mixing ratio retrievals for storm- scale radar data assimilation Utility of current relations and potential benefits of polarimetry. *Monthly Weather Review* 144: 2981–3001.
26. Chandrasekar, V., R. Keranen, S. Lim, and D. Moisseev. 2013. Recent advances in classification of observations from dual polarization weather radars. *Atmospheric Research* 119: 97–111.
27. Chang, W.-Y., J. Vivekanandan, K. Ikeda, and P.-L. Lin. 2016. Quantitative precipitation estimation of the epic 2013 Colorado flood event: Polarization radar-based variational scheme. *Journal of Applied Meteorology and Climatology* 55: 1477–1495.
28. Cifelli, R., V. Chandrasekar, S. Lim, P.C. Kennedy, Y. Wang, and S.A. Rutledge. 2011. A new dual-polarization radar rainfall algorithm application in Colorado precipitation events. *Journal of Atmospheric and Oceanic Technology* 28: 352–364.
29. Collis, S., A. Protat, P.T. May, and C. Williams. 2013. Statistics of storm updraft velocities from TWP-ICE including verification with profiling measurements. *Journal of Applied Meteorology and Climatology* 52: 1909–1922.
30. Davies-Jones, R., D.W. Burgess, and M. Foster. 1990. Test of helicity as a forecast parameter. In *16th Conference on Severe Local Storms*, Kananaskis Park, AB, Canada. 588–592. Boston: American Meteor Society (Preprints).
31. Dawson, D.T., E.R. Mansell, Y. Jung, L.J. Wicker, M.R. Kumjian, and M. Xue. 2014. Low-level ZDR signatures in supercell forward flanks: The role of size sorting and melting of hail. *Journal of the Atmospheric Sciences* 71: 276–299.
32. Dawson, D.T., E.R. Mansell, and M.R. Kumjian. 2015. Does wind shear cause hydrometeor size sorting? *Journal of the Atmospheric Sciences* 72: 340–348.
33. Dennis, E.J., and M.R. Kumjian. 2017. The impact of vertical wind shear on hail growth in simulated supercells. *Journal of the Atmospheric Sciences* 74: 641–663.
34. Doviak, R.J., and D.S. Zrnić. 1993. *Doppler Radar and Weather Observations*, 562. San Diego: Academic Press.
35. Doviak, R.J., V. Bringi, A. Ryzhkov, A. Zahrai, and D.S. Zrnić. 2000. Considerations for polarimetric upgrades to operational WSR-88D radars. *Journal of Atmospheric and Oceanic Technology* 17: 257–278.
36. Dowell, D.C., C.R. Alexander, J. Wurman, and L.J. Wicker. 2005. Centrifuging of hydrometeors and debris in tornadoes: radar-reflectivity patterns and wind-measurement errors. *Monthly Weather Review* 133: 1501–1524.
37. Fabry, F. 2015. *Radar Meteorology Principles and Practice*, 256. Cambridge: Cambridge University Press.
38. Fabry, F., C. Frush, I. Zawadzki, and A. Kilambi .1997. On the extraction of near-surface index of refraction using radar phase measurements from ground targets. *Journal of Atmospheric and Oceanic Technology* 14: 978–987.
39. French, M.M., D.W. Burgess, E.R. Mansell, and L.J. Wicker. 2015. Bulk hook echo raindrop sizes retrieved using mobile, polarimetric Doppler radar observations. *Journal of Applied Meteorology and Climatology* 54: 423–450.
40. Gal-Chen, T. 1978. A method for the initialization of the anelastic equations: implications for matching models with observations. *Monthly Weather Review* 106: 587–606.

41. Giangrande, S.E., and A.V. Ryzhkov. 2008. Estimation of rainfall based on the results of polarimetric echo classification. *Journal of Applied Meteorology and Climatology* 47: 2445–2462.
42. Giangrande, S.E., J.M. Krause, and A.V. Ryzhkov. 2008. Automatic designation of the melting layer with a polarimetric prototype of the WSR-88D radar. *Journal of Applied Meteorology and Climatology* 47: 1354–1364.
43. Giangrande, S.E., T. Toto, A. Bansemer, M.R. Kumjian, S. Mishra, and A.V. Ryzhkov. 2016. Insights into riming and aggregation processes as revealed by aircraft, radar, and disdrometer observations for a 27 April 2011 widespread precipitation event. *Journal of Geophysical Research-Atmospheres* 121: 5846–5863.
44. Girardin-Gondeau, J., F. Baudin, and J. Testud. 1991. Comparison of coded waveforms for an airborne meteorological Doppler radar. *Journal of Atmospheric and Oceanic Technology* 8: 234–246.
45. Goddard, J.W.F., J. Tan, and M. Thurai. 1994. Technique for calibration of meteorological radar using differential phase. *Electronics Letters* 30: 166–167.
46. Gorgucci, E., G. Scarchilli, and V. Chandrasekar. 1992. Calibration of radars using polarimetric techniques. *IEEE Transactions on Geoscience and Remote Sensing* 30: 853–858.
47. Gourley, J.J., A.J. Illingworth, and P. Tabary. 2009. Absolute calibration of radar reflectivity using redundancy of the polarization observations and implied constraints on drop shapes. *Journal of Atmospheric and Oceanic Technology* 26: 689–703.
48. Gourley, J.J., Y. Hong, Z.L. Flamig, J. Wang, H. Vergara, and E.N. Anagnostou. 2011. Hydrologic evaluation of rainfall estimates from radar, satellite, gauge, and combinations on Ft. Cobb Basin, Oklahoma. *Journal of Hydrometeorology* 12: 973–988.
49. Gu, J.-Y., A.V. Ryzhkov, P. Zhang, P. Neilley, M. Knight, B. Wolf, and D.-I. Lee. 2011. Polarimetric attenuation correction in heavy rain at C band. *Journal of Applied Meteorology and Climatology* 50: 39–58.
50. Hall, M.P.M., S.M. Cherry, J.W.F. Goddard, and G.R. Kennedy. 1980. Rain drop sizes and rainfall rate measured by dual-polarization radar. *Nature* 285: 195–198.
51. Harrington, J.Y., K. Sulia, and H. Morrison. 2013. A method for adaptive habit prediction in bulk microphysical models. Part I: Theoretical development. *Journal of the Atmospheric Sciences* 70: 349–364.
52. Heinselman, P.L., and A.V. Ryzhkov. 2006. Validation of polarimetric hail detection. *Weather and Forecasting* 21: 839–850.
53. Heymsfield, G.M., L. Tian, L. Li, M. McLinden, and J. Cervantes. 2013. Airborne radar observations of severe hailstorms Implications for future spaceborne radar. *Journal of Applied Meteorology and Climatology* 52: 1851–1867.
54. Hibbins, R.E., J.D. Shanklin, P.J. Epsy, M.J. Jarvis, D.M. Riggin, D.C. Fritts, and F.-J. Lübken. 2005. Seasonal variations in the horizontal wind structure from 0–100 km above Rothera station, Antarctica (67S, 68W). *Atmospheric Chemistry and Physics* 5: 2973–2980.
55. Hildebrand, P.H., and R.K. Moore. 1990. Meteorological radar observations from mobile platforms. In *Radar in Meteorology*, ed. D. Atlas, 287–314. Boston: American Meteorological Society.
56. Hildebrand, P.H., W.-C. Lee, C.A. Walther, C. Frush, M. Randall, E. Loew, R. Neitzel, R. Parsons, J. Testud, F. Baudin, and A. LeCornec. 1996. The ELDORA/ASTRAIA Airborne Doppler weather radar: high-resolution observations from TOGA COARE. *Bulletin of the American Meteorological Society* 77: 213–232.
57. Hocking, W.K. 1999. Temperatures using radar-meteor decay times. *Geophysical Research Letters* 26: 3297–3300. https://doi.org/10.1029/1999GL003618.
58. Hocking, W.K., B. Fuller, and B. Vandepeer. 1999. Real-time determination of meteor-related parameters utilizing digital technology. *Journal of Atmospheric and Solar—Terrestrial Physics* 63: 155–169.
59. Hubbert, J.C., V.N. Bringi, L.D. Carey, and S. Bolen. 1998. CSU-CHILL polarimetric radar measurements from a severe hail storm in eastern Colorado. *Journal of Applied Meteorology* 37: 749–775.

60. Hubbert, J.C., S.M. Ellis, M. Dixon, and G. Meymaris. 2010. Modeling, error analysis, and evaluation of dual- polarization variables obtained from simultaneous horizontal and vertical polarization transmit radar. Part I: Modeling and antenna errors. *Journal of Atmospheric and Oceanic Technology* 27: 1583–1598.
61. Hubbert, J.C., S.M. Ellis, M. Dixon, and G. Meymaris. 2010. Modeling, error analysis, and evaluation of dual-polarization variables obtained from simultaneous horizontal and vertical polarization transmit radar. Part II: Experimental data. *Journal of Atmospheric and Oceanic Technology* 27: 1599–1607.
62. Hubbert, J.C., S.M. Ellis, W.-Y. Chang, S. Rutledge, and M. Dixon. 2014. Modeling and interpretation of S-band ice crystal depolarization signatures from data obtained by simultaneously transmitting horizontally and vertically polarized fields. *Journal of Applied Meteorology and Climatology* 53: 1659–1677.
63. Illingworth, A., and T. Blackman. 2002. The need to represent raindrop size spectra as normalized gamma distributions for the interpretation of polarization radar observations. *Journal of Applied Meteorology* 41: 286–297.
64. Isom, B., R.D. Palmer, R. Kelley, J. Meier, D. Bodine, M. Yeary, B.L. Cheong, Y. Zhang, T.-Y. Yu, and M.I. Biggerstaff. 2013. The atmospheric imaging radar: Simultaneous volumetric observations using a phased array weather radar. *Journal of Atmospheric and Oceanic Technology* 30: 655–675.
65. Jameson, A.R. 1985. Microphysical interpretation of multiparameter radar measurements in rain. Part III: Interpretation and measurement of propagation differential phase shift between orthogonal linear polarizations. *Journal of the Atmospheric Sciences* 42: 607–614.
66. Jensen, A.A., and J.Y. Harrington. 2015. Modeling ice crystal aspect ratio evolution during riming: a single-particle growth model. *Journal of the Atmospheric Sciences* 72: 2569–2590.
67. Jiang, Z., M. Oue, J. Verlinde, E.E. Clothiaux, K. Aydin, G. Botta, and Y. Lu. 2017. What can we conclude about the real aspect ratios of ice particle aggregates from two-dimensional images? *Journal of Applied Meteorology and Climatology* 56: 725–734.
68. Jung, Y., G. Zhang, and M. Xue. 2008. Assimilation of simulated polarimetric radar data for a convective storm using the ensemble Kalman filter. Part I: Observation operators for reflectivity and polarimetric variables. *Monthly Weather Review* 136: 2228–2245.
69. Junyent, F., V. Chandrasekar, V.N. Bringi, S.A. Rutledge, P.C. Kennedy, D. Brunkow, J. George, and R. Bowie. 2015. Transformation of the CSU-CHILL radar facility to a dual-frequency, dual-polarization Doppler system. *Bulletin of the American Meteorological Society* 96: 975–996.
70. Kalesse, H., W. Szyrmer, S. Kneifel, P. Kollias, and E. Luke. 2016. Fingerprints of a riming event on cloud radar: Doppler spectra observations and modeling. *Atmospheric Chemistry and Physics* 16: 2997–3012. https://doi.org/10.5194/acp-16-2997-2016.
71. Kennedy, P.C., and S.A. Rutledge. 2011. S-band dual-polarization radar observations of winter storms. *Journal of Applied Meteorology and Climatology* 50: 844–858.
72. Kennedy, P.C., S.A. Rutledge, W.A. Petersen, and V.N. Bringi. 2001. Polarimetric radar observations of hail formation. *Journal of Applied Meteorology* 40: 1347–1366.
73. Kneifel, S., A. von Lerber, J. Tiira, D. Moisseev, P. Kollias, and J. Leinonen. 2015. Observed relations between snowfall microphysics and triple-frequency radar measurements. *Journal of Geophysical Research-Atmospheres* 120: 6034–6055. https://doi.org/10.1002/2015JD023156.
74. Kollias, P., B.A. Albrecht, R. Lhermitte, and A. Savtchenko. 2001. Radar observations of updrafts, downdrafts, and turbulence in fair-weather cumuli. *Journal of the Atmospheric Sciences* 58: 1750–1766.
75. Kollias, P., J. Remillard, E. Luke, and W. Szyrmer. 2011. Cloud radar Doppler spectra in drizzling stratiform clouds 1. Forward modeling and remote sensing applications. *Journal of Geophysical Research* 116: D13201.
76. Kosiba, K.A., J. Wurman, P. Markowski, Y. Richardson, P. Robinson, and J. Marquis. 2013. Genesis of the Goshen County, WY tornado on 05 June 2009 during VORTEX2. *Monthly Weather Review* 141: 1157–1181.

77. Kumjian, M.R. 2013. Principles and applications of dual-polarization weather radar. Part 1: Description of the polarimetric radar variables. *Journal of Operational Meteorology* 1 (19): 226–242.
78. Kumjian, M.R. 2013. Principles and applications of dual-polarization weather radar. Part 2: Warm and cold season applications. *Journal of Operational Meteorology* 1 (20): 243–264.
79. Kumjian, M.R. 2013. Principles and applications of dual-polarization weather radar. Part 3: Artifacts. *Journal of Operational Meteorology* 1 (21): 265–274.
80. Kumjian, M.R., and Z.J. Lebo. 2016. Large accumulations of small hail. In *28th Conference on Severe Local Storms*. Portland, OR: American Meteorological Society. 8A.4.
81. Kumjian, M.R., and K.A. Lombardo. 2017. Insights into the evolving microphysical and kinematic structure of northeastern U.S. winter storms from dual-polarization Doppler radar. *Monthly Weather Review* 145: 1033–1061.
82. Kumjian, M.R., and A.V. Ryzhkov. 2008. Polarimetric signatures in supercell thunderstorms. *Journal of Applied Meteorology and Climatology* 47: 1940–1961.
83. Kumjian, M.R., and A.V. Ryzhkov. 2009. Storm-relative helicity revealed from polarimetric radar measurements. *Journal of the Atmospheric Sciences* 66: 667–685.
84. Kumjian, M.R., and A.V. Ryzhkov. 2012. The impact of size sorting on the polarimetric radar variables. *Journal of the Atmospheric Sciences* 69: 2042–2060.
85. Kumjian, M.R., and A.D. Schenkman. 2014. The curious case of ice pellets in Middle Tennessee on 1 March 2014. *Journal of Operational Meteorology—Image of Note* 2 (17): 209–213.
86. Kumjian, M.R., J.C. Picca, S.M. Ganson, A.V. Ryzhkov, J. Krause, D.S. Zrnić, and A.P. Khain. 2010. Polarimetric radar characteristics of large hail. In *25th Conference on Severe Local Storms*, Denver, CO. 11.2. Boston: American Meteorological Society (Preprints).
87. Kumjian, M.R., S. Ganson, and A.V. Ryzhkov. 2012. Raindrop freezing in deep convective updrafts: a microphysical and polarimetric model. *Journal of the Atmospheric Sciences* 69: 3471–3490.
88. Kumjian, M.R., A.V. Ryzhkov, H.D. Reeves, and T.J. Schuur. 2013. A dual-polarization radar signature of hydrometeor refreezing in winter storms. *Journal of Applied Meteorology and Climatology* 52: 2549–2566.
89. Kumjian, M.R., A.P. Khain, N. BenMoshe, E. Ilotoviz, A.V. Ryzhkov, and V.T.J. Phillips. 2014. The anatomy and physics of Z_{DR} columns Investigating a polarimetric radar signature with a spectral bin microphysical model. *Journal of Applied Meteorology and Climatology* 53: 1820–1843.
90. Kumjian, M.R., S. Mishra, S.E. Giangrande, T. Toto, A.V. Ryzhkov, and A.R. Bansemer. 2016. Polarimetric radar and aircraft observations of saggy bright bands during MC3E. *Journal of Geophysical Research-Atmospheres* 121: 3584–3607.
91. Kuster, C.M., P.L. Heinselman, and T.J. Schuur. 2016. Rapid-update radar observations of downbursts occurring within an intense multicell thunderstorm on 14 June 2011. *Weather and Forecasting* 31: 827–851.
92. Lhermitte, R.M. 1987. A 94 GHz Doppler radar for cloud observations. *Journal of Atmospheric and Oceanic Technology* 4: 36–48.
93. Lhermitte, R.M. 1988. Observation of rain at vertical incidence with a 94 GHz Doppler radar: an insight on Mie scattering. *Geophysical Research Letters* 15: 1125–1128.
94. Lhermitte, R.M. 1990. Attenuation and scattering of millimeter wavelength radiation by clouds and precipitation. *Journal of Atmospheric and Oceanic Technology* 7: 464–479.
95. Li, X., and J.R. Mecikalski. 2012. Impact of the dual-polarization Doppler radar data on two convective storms with a warm-rain radar forward operator. *Monthly Weather Review* 140: 2147–2167.
96. Li, X., J.R. Mecikalski, and D. Posselt. 2017. An ice-phase microphysics forward model and preliminary results of polarimetric radar data assimilation. *Monthly Weather Review* 145: 683–708.

97. Lim, S., V. Chandrasekar, and V.N. Bringi. 2005. Hydrometeor classification system using dual-polarization radar measurements: model improvements and in situ verification. *IEEE Transactions on Geoscience and Remote Sensing* 43: 792–801.
98. Liu, H., and V. Chandrasekar. 2000. Classification of hydrometeors based on polarimetric radar measurements: development of fuzzy logic and neuro-fuzzy systems, and in situ verification. *Journal of Atmospheric and Oceanic Technology* 17: 140–164.
99. Loeffler, S.M., and M.R. Kumjian. 2016. Analysis of polarimetric radar signatures in tornadic non-supercellular storms. In *28th Conference on Severe Local Storms*, 12A.4. Boston: American Meteorological Society.
100. Loeffler, S.M., and M.R. Kumjian. 2016. Quantifying $Z_{DR} - K_{DP}$ separation in severe convective storms to assess tornadic potential. In *28th Conference on Severe Local Storms*, Poster 169. Boston: American Meteorological Society.
101. Lu, Y., E.E. Clothiaux, K. Aydin, G. Botta, and J. Verlinde. 2013. Modeling variability in dendritic ice crystal backscattering cross sections at millimeter wavelengths using a modified Rayleigh-Gans theory. *Journal of Quantitative Spectroscopy and Radiation Transfer* 131: 95–104.
102. Lu, Y., Z. Jiang, K. Aydin, J. Verlinde, E.E. Clothiaux, and G. Botta. 2016. A polarimetric scattering database for non-spherical ice particles at microwave wavelengths. *Atmospheric Measurement Techniques* 9: 5119–5134.
103. Luke, E., and P. Kollias. 2013. Separating cloud and drizzle radar moments during precipitation onset using Doppler spectra. *Journal of Atmospheric and Oceanic Technology* 30: 1656–1671.
104. Mahale, V.N., G. Zhang, and M. Xue. 2016. Characterization of the 14 June 2011 Norman, Oklahoma downburst through dual-polarization radar observations and hydrometeor classification. *Journal of Applied Meteorology and Climatology* 55: 2635–2655.
105. Maki, M., et al. 2012. Tokyo metropolitan area convection study for extreme weather resilient cities (TOMACS). Extended Abstracts. In *7th European Conference on Radar in Meteorology and Hydrology*, Toulouse, France.
106. Markowski, P.M., and Y.P. Richardson. 2010. *Mesoscale Meteorology in Midlatitudes*. 1st ed., 407 pp. Oxford: Wiley-Blackwell.
107. Matrosov, S.Y. 2017. Characteristic raindrop size retrievals from measurements of differences in vertical Doppler velocities at Ka- and W-band radar frequencies. *Journal of Atmospheric and Oceanic Technology* 34: 65–71.
108. McLaughlin, D., et al. 2009. Short-wavelength technology and the potential for distributed networks of small radar systems. *Bulletin of the American Meteorological Society* 90: 1797–1817.
109. Melnikov, V.M., and R.J. Doviak. 2009. Turbulence and wind shear in layers of large Doppler spectrum width in stratiform precipitation. *Journal of Atmospheric and Oceanic Technology* 26: 430–443.
110. Mie, G. 1908. Beiträge zur Optik trüber Medien, speziell kolloidaler Metallösungen. *Annals of Physics* 330: 377–445.
111. Moisseev, D.N., S. Lautaportti, J. Tyynela, and S. Lim. 2015. Dual-polarization radar signatures in snowstorms: role of snowflake aggregation. *Journal of Geophysical Research-Atmospheres* 120: 12644–12655.
112. Morrison, H., and J.A. Milbrandt. 2015. Parameterization of cloud microphysics based on the prediction of bulk ice particle properties. Part I: Scheme description and idealized tests. *Journal of the Atmospheric Sciences* 72: 287–311.
113. Morrison, H., J.A. Milbrandt, G.H. Bryan, K. Ikeda, S.A. Tessendorf, and G. Thompson. 2015. Parameterization of cloud microphysics based on the prediction of bulk ice particle properties. Part II: Case study and comparisons with observations and other schemes. *Journal of the Atmospheric Sciences* 72: 312–339.
114. Nolan, D.S. 2013. On the use of Doppler radar-derived wind fields to diagnose the secondary circulations of tornadoes. *Journal of the Atmospheric Sciences* 70: 1160–1171.

115. Ortega, K.L., J.M. Krause, and A.V. Ryzhkov. 2016. Polarimetric radar characteristics of melting hail. Part III: Validation of the algorithm for hail size discrimination. *Journal of Applied Meteorology and Climatology* 55: 829–848.
116. Oue, M., M.R. Kumjian, Y. Lu, J. Verlinde, K. Aydin, and E. Clothiaux. 2015. Linear depolarization ratios of columnar ice crystals in a deep precipitation system over the Arctic observed by zenith-pointing Ka-band Doppler radar. *Journal of Applied Meteorology and Climatology* 54: 1060–1068.
117. Park, H.S., A.V. Ryzhkov, D.S. Zrnić, and K.-E. Kim. 2009. The hydrometeor classification algorithm for the polarimetric WSR-88D: description and application to an MCS. *Weather and Forecasting* 24: 730–748.
118. Pazmany, A.L., J.B. Mead, H.B. Bluestein, J.C. Snyder, and J.B. Houser. 2013. A mobile, rapid-scanning X-band polarimetric (RaXPol) Doppler radar system. *Journal of Atmospheric and Oceanic Technology* 30: 1398–1413.
119. Picca, J., and A.V. Ryzhkov. 2012. A dual-wavelength polarimetric analysis of the 16 May 2010 Oklahoma City extreme hailstorm. *Monthly Weather Review* 140: 1385–1403.
120. Picca, J.C., D.M. Schultz, B.A. Colle, S. Ganetis, D.R. Novak, and M.J. Sienkiewicz. 2014. The value of dual-polarization radar in diagnosing the complex microphysical evolution of an intense snowband. *Bulletin of the American Meteorological Society* 95: 1825–1834.
121. Picca, J.C., D.M. Kingfield, and A.V. Ryzhkov. 2017. Utilizing a polarimetric size sorting signature to develop a convective nowcasting algorithm. In *18th Conference on Aviation, Range, and Aerospace Meteorology*, Seattle, WA. Boston: American Meteorological Society, 14.5.
122. Posselt, D.J., X. Li, S.A. Tushaus, and J.R. Mecikalski. 2015. Assimilation of dual-polarization radar observations in mixed- and ice-phase regions of convective storms: information content and forward model errors. *Monthly Weather Review* 143: 2611–2636.
123. Rasmussen, E.N., and D.O. Blanchard. 1998. A baseline climatology of sounding-derived supercell and tornado forecast parameters. *Weather and Forecasting* 13: 1146–1164.
124. Rasmussen, R.M., and A.J. Heymsfield. 1987. Melting and shedding of graupel and hail. Part I: Model physics. *Journal of the Atmospheric Sciences* 44: 2754–2763.
125. Richter, H., J. Peter, and S. Collis. 2014. Analysis of a destructive wind storm on 16 November 2008 in Brisbane, Australia. *Monthly Weather Review* 142: 3038–3060.
126. Rinehart, R.E. 2004. *Radar for Meteorologists*, 482 pp. Columbia, MO: Rinehart Publications.
127. Romine, G.S., D.W. Burgess, and R.B. Wilhelmson. 2008. A dual-polarization-radar-based assessment of the 8 May 2003 Oklahoma City area tornadic supercell. *Monthly Weather Review* 136: 2849–2870.
128. Ryzhkov, A.V. 2007. The impact of beam broadening on the quality of radar polarimetric data. *Journal of Atmospheric and Oceanic Technology* 24: 729–744.
129. Ryzhkov, A.V., T.J. Schuur, D.W. Burgess, and D.S. Zrnić. 2005. Polarimetric tornado detection. *Journal of Applied Meteorology* 44: 557–570.
130. Ryzhkov, A.V., S.E. Giangrande, V.M. Melnikov, and T.J. Schuur. 2005. Calibration issues of dual-polarization radar measurements. *Journal of Atmospheric and Oceanic Technology* 22: 1138–1155.
131. Ryzhkov, A.V., M. Pinsky, A. Pokrovsky, and A.P. Khain. 2011. Polarimetric radar observation operator for a cloud model with spectral microphysics. *Journal of Applied Meteorology and Climatology* 50: 873–894.
132. Ryzhkov, A.V., M.R. Kumjian, S.M. Ganson, and A.P. Khain. 2013. Polarimetric radar characteristics of melting hail. Part I: Theoretical simulations using spectral microphysical modeling. *Journal of Applied Meteorology and Climatology* 52: 2849–2870.
133. Ryzhkov, A.V., M.R. Kumjian, S.M. Ganson, and P. Zhang. 2013. Polarimetric radar characteristics of melting hail. Part II: Practical implications. *Journal of Applied Meteorology and Climatology* 52: 2871–2886.

134. Ryzhkov, A.V., M. Diederich, P. Zhang, and C. Simmer. 2014. Potential utilization of specific attenuation for rainfall estimation, mitigation of partial beam blockage, and radar networking. *Journal of Atmospheric and Oceanic Technology* 31: 599–619.
135. Ryzhkov, A.V., P. Zhang, H. Reeves, M. Kumjian, T. Tschallener, S. Trömel, and C. Simmer. 2016. Quasi-vertical profiles – a new way to look at polarimetric radar data. *Journal of Atmospheric and Oceanic Technology* 33: 551–562.
136. Sachidananda, M., and D.S. Zrnić. 1986. Differential propagation phase shift and rainfall rate estimation. *Radio Science* 21: 235–247.
137. Schenkman, A.D., M. Xue, A. Shapiro, K. Brewster, and J. Gao. 2011. Impact of CASA radar and Oklahoma Mesonet data assimilation on the analysis and prediction of tornadic mesovortices in an MCS. *Monthly Weather Review* 139: 3422–3445.
138. Schneebeli, M., N. Dawes, M. Lehning, and A. Berne. 2013. High-resolution vertical profiles of X-band polarimetric radar observables during snowfall in the Swiss Alps. *Journal of Applied Meteorology and Climatology* 52: 378–394,
139. Schrom, R.S., and M.R. Kumjian. 2016. Connecting microphysical processes in Colorado winter storms with vertical profiles of radar observations. *Journal of Applied Meteorology and Climatology* 55: 1771–1787.
140. Schrom, R.S., M.R. Kumjian, and Y. Lu. 2015. Polarimetric radar signatures of dendritic growth zones within Colorado winter storms. *Journal of Applied Meteorology and Climatology* 54: 2365–2388.
141. Schultz, C.J., et al. 2012. Dual-polarization tornadic debris signatures. Part I: Examples and utility in an operational setting. *Electronic Journal of Operational Meteorology* 13 (9): 120–137.
142. Seliga, T.A., and V.N. Bringi. 1976. Potential use of radar differential reflectivity measurements at orthogonal polarizations for measuring precipitation. *Journal of Applied Meteorology* 15: 69–76.
143. Seliga, T.A., and V.N. Bringi. 1978. Differential reflectivity and differential phase shift: applications in radar meteorology. *Radio Science* 13: 271–275.
144. Shapiro, A.M., C.K. Potvin, and J. Gao. 2009. Use of a vertical vorticity equation in variational dual-Doppler wind analysis. *Journal of Atmospheric and Oceanic Technology* 26: 2089–2106.
145. Silber, I., and C. Price. 2016. On the use of VLF narrowband measurements to study the lower ionosphere and the mesosphere-lower thermosphere. *Surveys in Geophysics*. https://doi.org/10.1007/s10712-016-9396-9.
146. Sinclair, V.A., D. Moisseev, and A. von Lerber. 2016. How dual-polarization radar observations can be used to verify model representation of secondary ice. *Journal of Geophysical Research-Atmospheres* 121: 10954–10970.
147. Skolnik, M.I. 2001. *Introduction to Radar Systems*, 772 pp. New York: McGraw Hill.
148. Snyder, J.C., H.B. Bluestein, G. Zhang, and S.J. Frasier. 2010. Attenuation correction and hydrometeor classification of high-resolution, X-band, dual-polarized mobile radar measurements in severe convective storms. *Journal of Atmospheric and Oceanic Technology* 27: 1979–2001.
149. Snyder, J.C., A.V. Ryzhkov, M.R. Kumjian, A.P. Khain, and J.C. Picca. 2015. A Z_{DR} column detection algorithm to examine convective storm updrafts. *Weather and Forecasting* 30: 1819–1844.
150. Spoden, P.J., R.A. Wolf, and L.R. Lemon. 2012. Operational uses of spectrum width. *Electronic Journal of Severe Storms Meteorology* 7 (2): 1–28.
151. Straka, J.M., and D.S. Zrnić. 1993. An algorithm to deduce hydrometeor types and contents from multiparameter radar data. In *26th Conference on Radar Meteorology*, Norman, OK, 513–516. Boston: American Meteorological Society (Preprints).

152. Sulia, K.J., and M.R. Kumjian. 2017. Simulated polarimetric fields of ice vapor growth using the adaptive habit model. Part I: Large-eddy simulations. *Monthly Weather Review* 145: 2281–2302.
153. Sulia, K.J., and M.R. Kumjian. 2017. Simulated polarimetric fields of ice vapor growth using the adaptive habit model. Part II: A case study from the FROST experiment. *Monthly Weather Review* 145: 2303–2323.
154. Testud, J., E. Le Bouar, E. Obligis, and M. Ali-Mehenni .2000. The rain profiling algorithm applied to polarimetric weather radar. *Journal of Atmospheric and Oceanic Technology* 17: 332–356.
155. Thompson, R.L., R. Edwards, J.A. Hart, K.L. Elmore, and P.M. Markowski. 2003. Close proximity soundings within supercell environments obtained from the Rapid Update Cycle. *Weather and Forecasting* 18: 1243–1261.
156. Thompson, R.L., C.M. Mead, and R. Edwards. 2007. Effective storm-relative helicity and bulk shear in supercell thunderstorm environments. *Weather and Forecasting* 22: 102–115.
157. Thompson, E.J., S.A. Rutledge, B. Dolan, V. Chandrasekar, and B.L. Cheong. 2014. A dual-polarization radar hydrometeor classification algorithm for winter precipitation. *Journal of Atmospheric and Oceanic Technology* 31: 1457–1481.
158. Tong, M., and M. Xue. 2005. Ensemble Kalman filter assimilation of Doppler radar data with a compressible nonhydrostatic model OSS experiments. *Monthly Weather Review* 133: 1789–1807.
159. Torres, S., Y.F. Dubel, and D.S. Zrnić. 2004. Design, implementation, and demonstration of a staggered PRT algorithm for the WSR-88D. *Journal of Atmospheric and Oceanic Technology* 21: 1389–1399.
160. Torres, S., R. Passarelli Jr., A. Siggia, and P. Karhunen. 2010. Alternating dual-pulse, dual-frequency techniques for range and velocity ambiguity mitigation on weather radars. *Journal of Atmospheric and Oceanic Technology* 27: 1461–1475.
161. Tridon, F., and A. Battaglia. 2015. Dual-frequency radar Doppler spectral retrieval of rain drop size distributions and entangled dynamics variables. *Journal of Geophysical Research-Atmospheres* 120: 5585–5601.
162. Trömel, S., M.R. Kumjian, A.V. Ryzhkov, C. Simmer, and M. Diederich. 2013. Backscatter differential phase—estimation and variability. *Journal of Applied Meteorology and Climatology* 52: 2529–2548.
163. Trömel, S., A.V. Ryzhkov, P. Zhang, and C. Simmer. 2014. Investigations of backscatter differential phase in the melting layer. *Journal of Applied Meteorology and Climatology* 53: 2344–2359.
164. Van Den Broeke, M.S., and S.T. Jauernic. 2014. Spatial and temporal characteristics of tornadic debris signatures. *Journal of Applied Meteorology and Climatology* 53: 2217–2231.
165. Van Den Broeke, M.S., D.M. Tobin, and M.R. Kumjian. 2016. Polarimetric radar observations of precipitation type and rate from the 2–3 March 2014 winter storm in Oklahoma and Arkansas. *Weather and Forecasting* 31: 1179–1196.
166. Van Lier-Walqui, M., A.M. Fridlind, A.S. Ackerman, S. Collis, J. Helmus, D.R. MacGorman, K. North, P. Kollias, and D.J. Posselt. 2016. On polarimetric radar signatures of deep convection for model evaluation: columns of specific differential phase observed during MC3E. *Monthly Weather Review* 144: 737–758.
167. Vincent, R.A., and I.M. Reid. 1983. HF Doppler measurements of mesospheric gravity waves momentum fluxes. *Journal of the Atmospheric Sciences* 40: 1321–1333.
168. Vivekanandan, J., D.S. Zrnić, S. Ellis, D. Oye, A.V. Ryzhkov, and J.M. Straka. 1999. Cloud microphysics retrieval using S-band dual-polarization radar measurements. *Bulletin of the American Meteorological Society* 80: 381–388.
169. Wakimoto, R.M., P. Stauffer, W.-C. Lee, N.T. Atkins, and J. Wurman. 2012. Finescale structure of the LaGrange, Wyoming, tornado during VORTEX2: GBVTD and photogrammetric analyses. *Monthly Weather Review* 140: 3397–3418.
170. Wurman, J. 1994. Vector winds from a single-transmitter bistatic dual-Doppler radar network. *Bulletin of the American Meteorological Society* 75: 983–994.

171. Wurman, J.M., and K.A. Kosiba. 2013. Finescale radar observations of tornado and mesocyclone structures. *Weather and Forecasting* 28: 1157–1174.
172. Wurman, J., and M. Randall. 2001. An inexpensive, mobile rapid-scan radar. In *30th Conference on Radar Meteorology*, Munich, Germany, P3.4. Boston: American Meteorological Society (Preprints).
173. Wurman, J., S. Heckman, and D. Boccippio. 1993. A bistatic multiple-Doppler radar network. *Journal of Applied Meteorology* 32: 1802–1814.
174. Wurman, J.M., J. Straka, E. Rasmussen, M. Randall, and A. Zahrai. 1997. Design and deployment of a portable, pencil-beam, pulsed, 3-cm Doppler radar. *Journal of Atmospheric and Oceanic Technology* 14: 1502–1512.
175. Yokota, S., H. Sako, M. Kunii, H. Yamauchi, and H. Niino. 2016. The tornadic supercell on the Kanto Plain on 6 May 2012: polarimetric radar and surface data assimilation with EnKF and ensemble-based sensitivity analysis. *Monthly Weather Review* 144: 3133–3157.
176. Zhang, G., R.J. Doviak, D.S. Zrnić, R.D, Palmer, L. Lei, and Y. Al-Rashid. 2011. Polarimetric phased-array radar for weather measurement: a planar or cylindrical configuration? *Journal of Atmospheric and Oceanic Technology* 28: 63–73.
177. Zrnić, D.S., and A.V. Ryzhkov. 1999. Polarimetry for weather surveillance radars. *Bulletin of the American Meteorological Society* 80: 389–406.
178. Zrnić, D.S., J.F. Kimpel, D.E. Forsyth, A. Shapiro, G. Crain, R. Ferek, J. Heimmer, W. Benner, T.J. McNellis, and R.J. Vogt. 2007. Agile-beam phased array radar for weather observations. *Bulletin of the American Meteorological Society* 88: 1753–1766.
179. Zrnić, D.S., R.J. Doviak, G. Zhang, and A.V. Ryzhkov. 2010. Bias in differential reflectivity due to cross coupling through the radiation patterns of polarimetric weather radars. *Journal of Atmospheric and Oceanic Technology* 27: 1624–1637.

Ground-Based Radar Technologies for Tornado Observations

David J. Bodine and James M. Kurdzo

1 The Role of Radar for Tornado Studies

Tornadoes produce the most intense winds observed on the Earth, producing winds that can exceed 135 m s^{-1} [174, 203]. Tornadoes cause an average of 70 fatalities per year in the United States, and can cause hundreds of fatalities during major tornado outbreaks (e.g., 27 April 2011). Individual tornado outbreaks can also cause billions of dollars in damage. In a given year, an average of 1250 tornadoes are reported in the United States. Other regions of the world also have substantial numbers of tornadoes, including Europe, Argentina, and Bangladesh [72, 75].

The observation of severe local storms and tornadoes is limited to relatively few techniques, including both in situ and remote measurements as well as damage surveys. Due to the relatively low density of in situ data [84], radar has contributed a significant amount of detail to our current understanding of these phenomena [118]. While in situ measurements provide direct measurements of meteorological quantities of interest, in situ measurements are primarily taken at the surface and only in a few locations within a tornadic supercell. In situ measurements in tornadoes are especially difficult and are relatively rare because instruments must be placed in the tornado's path. Radar measurements, on the other hand, provide high spatial and temporal resolution mapping of three-dimensional wind and precipitation patterns throughout the parent storm and tornado, and can be collected at a relatively safe distance from the tornado. Damage surveys provide

D. J. Bodine (✉)
University of Oklahoma, Norman, OK, USA
e-mail: bodine@ou.edu

J. M. Kurdzo
MIT Lincoln Laboratory, Lexington, MA, USA
e-mail: James.Kurdzo@ll.mit.edu

C. Andronache (ed.), *Remote Sensing of Clouds and Precipitation*,
Springer Remote Sensing/Photogrammetry,
https://doi.org/10.1007/978-3-319-72583-3_3

important information on tornado wind speeds and diameter and can elucidate smaller scale structure of the tornado that might be unresolved by radar, such as suction vortices. However, damage survey information provides an indirect measurement of tornado kinematic structure whereas radar measurements measure wind speeds more directly. Thus, radar observations have provided wind speed measurements that form the basis of our knowledge about the typical range of tornado wind speeds and diameters.

Tornado researchers have worked toward several important goals to improve tornado forecasting and mitigate tornado risks and hazards. To better forecast tornadoes, scientists must first understand how tornadoes form and then apply this understanding to techniques that identify tornado formation from meteorological measurements (including radar) or numerical weather prediction forecasts (some of which assimilate radar data). Doppler radar observations have provided valuable information about how winds and precipitation patterns within a thunderstorm change during tornadogenesis, and Doppler radar measurements can therefore be used to detect tornadogenesis processes in many cases. Another important focus of tornado research is examining the three-dimensional distribution of winds and debris in tornadoes. Knowledge about the three-dimensional characteristics of winds and debris allows scientists and engineers to evaluate how tornadoes will impact different structures and then mitigate the risks to life and property by improving structural codes.

New developments in ground-based radar technology over several decades have greatly increased scientific understanding of tornadoes and tornadic thunderstorms. These research efforts have led to major improvements in tornado warnings, contributing to a modern-day lead time of approximately 14 min compared to 5 min or less before Doppler radars were deployed in the NEXRAD network [6, 169]. Since the NEXRAD network was deployed, tornado fatalities and injuries have decreased by nearly 50% [169]. Although these improvements are impressive, many significant challenges remain to mitigating societal vulnerability to tornadoes, and continued scientific advancements in ground-based radar systems are critical to addressing these challenges.

New radar technologies have the potential to address many of these challenges for improving tornado forecasts and understanding the risks posed by tornadoes to society. As weather radar technology has matured, the time required for a Doppler radar to complete a volumetric measurement of a storm (called update time) has improved from about 5 min to just 5–10 s. New radar technologies enabling these fast scan rates will be discussed in this chapter, including their feasibility for future operational radar networks. New radar systems have also incorporated dual-polarization radar technology, in which the radar transmits and receives two electromagnetic waves at orthogonal polarizations. Dual-polarization technology has enabled scientists and forecasters to better diagnose precipitation distributions in tornadic storms as well as adding the capability to detect tornado debris remotely [22, 97, 158].

In this chapter, we will discuss ground-based radar systems used to study tornadoes, and explore how scientists have used these radar systems to advance scientific knowledge about tornadoes and improve tornado forecasting. A historical perspective of ground-based radar systems will be presented, including an overview of the radar technology, a discussion of how these radars have been incorporated into scientific field experiments, and a detailed exploration of the scientific discoveries about tornadogenesis and tornado dynamics that have been obtained using different radar systems. The chapter concludes by examining the future of ground-based radar systems, including how future radar systems will help scientists address important questions remaining about tornadoes.

2 Theory and Simulations of Tornadoes

Before discussing ground-based radar observations of tornadoes, theory and simulations of tornadoes are briefly described here to provide appropriate context. A more in-depth treatment of these topics can be found in review articles on tornado dynamics [53, 151]. Based on theory, vortex chamber experiments, and numerical simulations, conceptual models for the flow structure of the tornado were developed. Examples of the flow regions within a simulated tornado from a Large-Eddy Simulation (LES) model [24, 126] are shown in Fig. 1. The outer flow region is characterized by constant angular momentum (Γ=vr), and thus tangential velocities decrease proportionally to 1/r. Within the core region, angular momentum increases radially outward toward the outer flow region. The flow exhibits cyclostrophic balance, in which the radial pressure gradient directed inward is balanced by the centrifugal force acting outward.

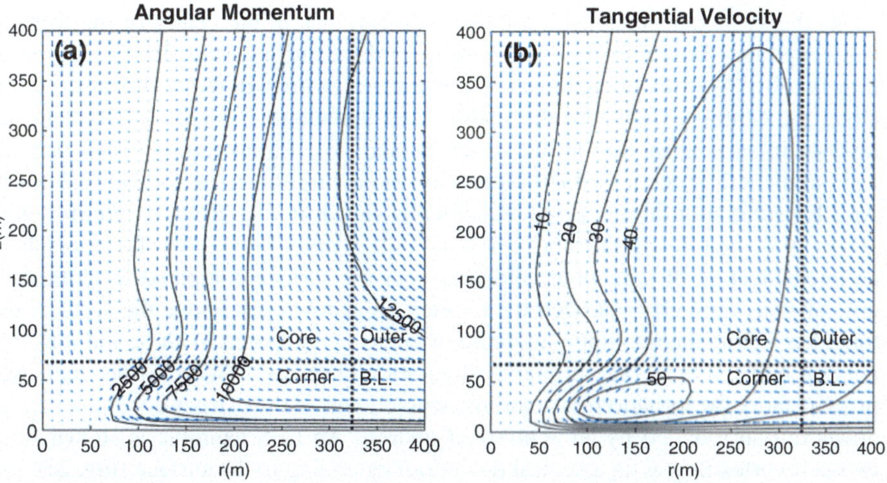

Fig. 1 Example tornado simulation from the Large-Eddy Simulation model with wind vectors in the r–z plane and contours of (**a**) angular momentum and (**b**) tangential velocity. The corner, boundary layer (B. L.), core, and outer flow regions are labelled on each panel

For the corner flow and boundary layer regions, friction causes a breakdown of cyclostrophic balance. Friction causes a reduction in tangential velocities and a decrease in centrifugal forces, while the strong pressure deficit in the center of the tornado remains largely unaffected. Thus, the near-surface flow becomes directed inward toward the tornado, driven by the large radial pressure gradient. Although frictional effects are largest near the surface, the largest tangential velocities are often observed within the corner flow region because the strong radially inward jet transports angular momentum into a smaller radius compared to aloft, producing a more intense tornado [61, 111, 113]. The corner flow and boundary layer regions of the tornadoes are quite shallow, with a maximum height of 10–100 m above ground level. Very close to the surface (e.g., <10 m AGL), a decrease in angular momentum is observed due to frictional dissipation, and this region is called the friction layer (Fig. 1b). Doppler radar observations of these regions are challenging since the radar beam height increases with range, and is above these regions except at very close ranges (e.g., less than a few km).

Early laboratory studies identified that the tornado flow type changes dramatically as the swirl ratio changes [46, 51, 150]. In particular, changes in swirl ratio tend to affect the secondary circulation which is comprised of the radial and vertical winds. The swirl ratio is a non-dimensional parameter that is computed by taking the ratio between the tangential velocity (or circulation) and the vertical velocity (or mass flux). At low swirl ratios, the radial flow converges to the central axis and forms an upward jet of rotating air in the center of the tornado, producing a one-cell vortex. The evolution of tornado structure from low to high swirl ratios is shown in Fig. 2. The LES one-cell vortex simulation (Fig. 2a) shows a narrow core diameter tornado with an elevated region of high tangential velocities and updraft throughout the tornado.

As the swirl ratio increases, a downdraft develops in the center of the tornado and penetrates progressively downward toward the surface (Fig. 2b). This type of flow is called a vortex breakdown [5], and the vortex breakdown occurs at the point along the central axis where the updraft and downdraft meet. In addition to the reversal in vertical motion at the vortex breakdown, a transition between laminar and turbulent flow occurs with turbulent flow occurring above the vortex breakdown. An LES simulation of a vortex breakdown is shown in Fig. 2b with updraft below 100 m AGL and downdraft above 100 m AGL. Vortex breakdown flows result in an intensification of tangential velocities near the surface compared to aloft. At a critical swirl ratio characterized by a vortex breakdown just above the surface, the tornado exhibits the maximum increase in intensity compared to aloft, and thus this is often referred to as the optimal configuration [111, 151]. With further increases in swirl ratio, the tornado flow exhibits two-cell vortex flow structure with a central downdraft surrounded by an annulus of updraft. An LES example is shown for a two-cell vortex flow with a central downdraft extending to the surface (Fig. 2c).

For very high swirl ratios, the tornado develops multiple intense vortices that rotate around the parent vortex, called suction vortices or subvortices. The number of subvortices can range from 2 to 6 at a given time, and subvortices can exist on multiple scales or have flow structures similar to the aforementioned flow types.

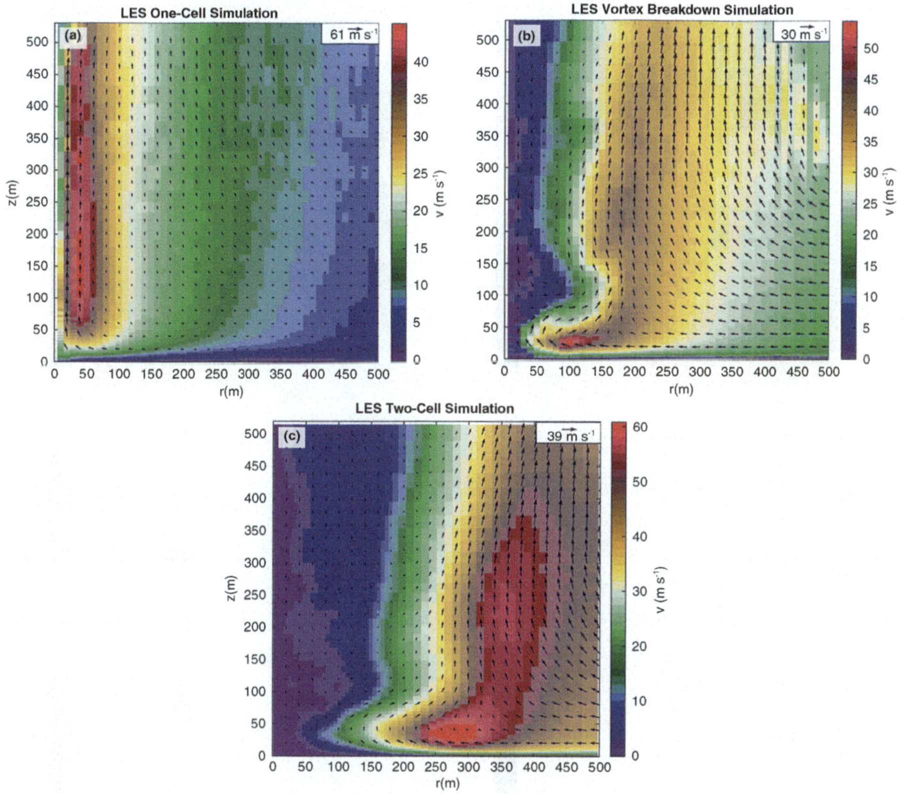

Fig. 2 Large-Eddy Simulations of a (**a**) one-cell, (**b**) vortex breakdown, and (**c**) two-cell vortex flows with axisymmetric means of wind data computed every 20 m. Wind vectors are shown for the flow components in the *r*–*z* plane, and tangential velocity is shaded

For example, subvortices can contain smaller subvortices or possess one- or two-cell vortex flows. An example of a multiple vortex tornado simulated with the LES model is shown in Fig. 3, and the simulation was initialized using Doppler radar observations from three-dimensional axisymmetric wind analyses of the 10 May 2010 EF-4 Moore, Oklahoma tornado. The simulation produces four large subvortices that produce the most intense horizontal and vertical velocities and the highest turbulent kinetic energy (TKE) in the simulation domain. On the larger scale, the tornado is characterized by a narrow band of intense updraft that surrounds the broad central downdraft. The tornado subvortices form along the intense horizontal gradient in vertical velocity at the updraft-downdraft interface. Subvortices can lead to horizontal wind speeds that are two to three times more intense than the mean flow within the tornado [60, 113, 136], and are often responsible for increased damage [67, 68].

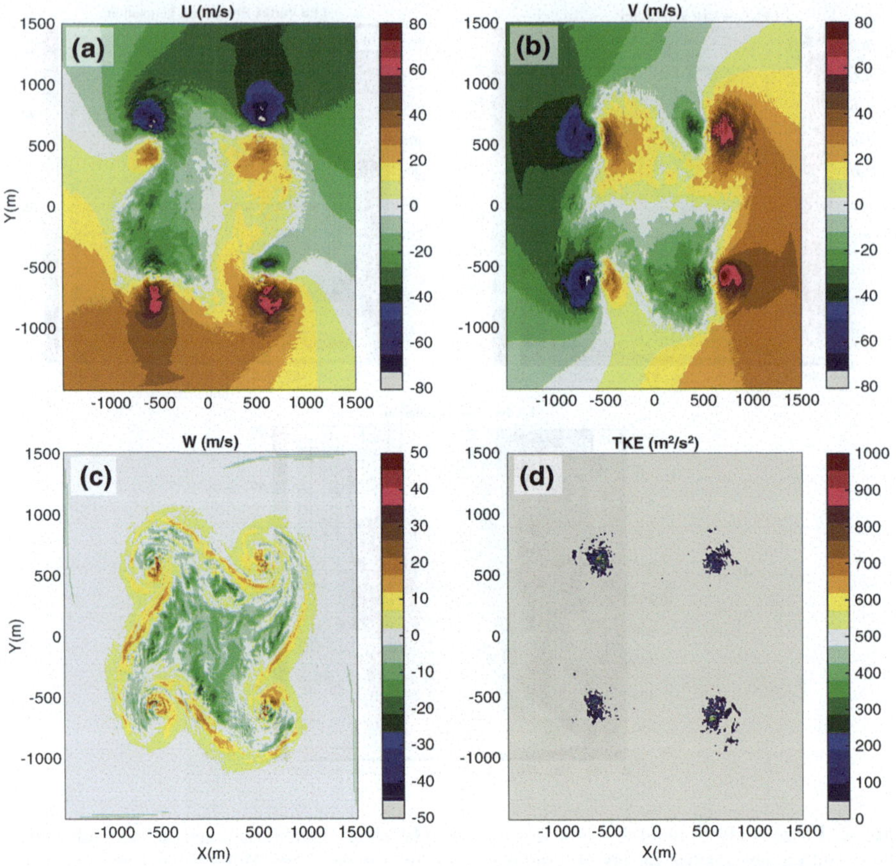

Fig. 3 Large-Eddy Simulation of the 10 May 2010 Moore, Oklahoma tornado using Doppler radar data to initialize the lateral and upper boundary conditions. Instantaneous (**a**) zonal (U), (**b**) meridonal (V), (**c**) vertical velocity (W), and (**d**) 1.2-s average turbulent kinetic energy are shown at 64 m AGL. The simulation produces large subvortices consistent with single- and dual-Doppler analyses in [74]

3 History of Ground-Based Tornado-Observing Radar Technology

Tornadoes have been observed by radar since the 1940s, resulting in a continuously growing database of tornadic datasets and knowledge gleaned from their collection. Radar architectures and capabilities have evolved significantly over this time, leading to a wide and varied series of case examples with different spatial and temporal resolutions, sensitivities, Doppler capabilities, polarimetry types, antenna designs, transmitter varieties, and wavelengths. Early tornado observations were mainly collected by fixed-site radars, including military systems, the original National Severe Storms Laboratory (NSSL) radars, and several Weather Surveillance Radars (WSR-57, 74, and 88D systems). Fixed-site radars, however, have the disadvantage of limited spatial resolution and lack of low-level coverage if a storm of interest is not serendipitously close to the radar site. Mobile radars have historically provided many of the best opportunities to observe and analyze convective storms and tornadoes due to their ability to be positioned relatively close to a storm of interest [10, 207, 209].

3.1 Early Radar Observations

The need for advance warning of tornadoes drove many of the early observations and studies of tornadoes using radar. Before the existence of weather radar, military radars were often used to prepare bases for incoming storms. This process was described by Bradford [28] while detailing the first tornado forecast by Major Ernest J. Fawbush and Captain Robert C. Miller at Tinker Air Force Base in Oklahoma City, Oklahoma on 25 March 1948. Fawbush and Miller issued their forecast based on atmospheric similarities to a tornado that hit the Base 5 days prior, combined with the knowledge that thunderstorms had formed sixty miles northwest to one hundred miles southwest of Oklahoma City using a military radar [28]. This warning, albeit widely considered exceptionally lucky given the unlikely chance that a tornado would hit near Tinker AFB again 5 days later, gave hope to the idea that tornado forecasting and warning was indeed possible.

This idea was furthered significantly 5 years later on 9 April 1953 when researchers at the Illinois State Water Survey observed the first-known hook echo on a re-built weather radar near Champaign, Illinois [11]. After discovering that a tornado had been confirmed near the curious shape on radar, hook echoes were, for the first time, associated with tornadoes. Just a month later, on 11 May 1953, Texas A&M researchers utilizing a modified SO-7N S-band radar observed similar characteristics in a later-to-be-determined tornadic thunderstorm [28]. Waco, Texas suffered considerable damage from the storm, and 114 people were killed. The researchers' connection between the hook echo and the resulting tornado led to the eventual goal of a tornado detection and warning network across the southern and central plains using radar, including the WSR-1/A, WSR-3, and WSR-4, all variants of the AN/APS-2F military radar [198]. One of these radars was used for the first radar-based tornado warnings via Texas A&M meteorologists on 5 April 1956 near Bryan and College Station, Texas [8, 198]. Combined with efforts by the Miami Hurricane Forecast Center, these types of observations and their societal benefits were used to develop the Nation's first large-scale weather radar network: the WSR-57 C-band radar system, which consisted of 66 AN/FPS-41 radars across the United States [149, 198].

Multiple tornadoes were observed over the following 1–2 decades by the WSR-57 network, but there was a critical piece missing for tornado warnings: velocity data. It was discovered that not every hook echo produced a tornado, and not every tornado was associated with a hook echo, leading researchers to believe that other information was necessary for tornado warnings to be effective. Multiple pulsed-Doppler radars for weather purposes were being developed during the 1960s and 1970s, but it wasn't until 24 May 1973 that an NSSL S-band Doppler radar discovered the next link in the radar-tornado chain [39]. During that storm, the NSSL radar detected a tight couplet of gate-to-gate wind shear co-located with a tornado near Union City, Oklahoma. This observation concluded that, at least in some cases, the existence of strong rotational shear at low levels on a Doppler radar could indicate the existence of a tornado. Of course, the radar beam is generally situated above the ground, so the level at which the rotation exists may be well above ground level and may be indicative of a strong mesocyclone rather than a tornado [33]. Nevertheless, this observation led to the Tornado Vortex Signature (TVS) as described in [33], and gave additional fuel to the desire for a network of ground-based Doppler radars in the United States.

During the 1970s and early 1980s, a lot was happening in the world of weather radars. The United States first upgraded their national weather radar network in the mid-1970s with the non-Doppler combination of WSR-74 C- and S-band radars. The WSR-74s were added to the existing WSR-57s to create a network of 128 radars across the United States. The 71 WSR-74 systems were combined with 12 more added from 1981–1985, bringing the total to 83 National Weather Service (NWS) "local-warning" radars [198]. Additionally, many of the WSR-57 radars were kept online through the mid-1990s. Around the same time as the development of the WSR-74 network, weather radar networks began appearing in other countries around the world, as the immense benefits to public safety were realized [47, 82].

Due in part to the observations near Union City in 1973, however, Doppler radar was on the horizon, with numerous research centers developing new radar techniques. Doppler radar designed specifically for meteorological uses was being developed at NSSL, the National Center for Atmospheric Research (NCAR), the Air Force Cambridge Research Laboratory, and the Cornell Aeronautical Laboratory, among others [11]. In the mid-1970s, NSSL added a second Doppler radar near Oklahoma City, resulting in the ability to provide dual-Doppler capabilities. Since a single radar can only provide Doppler velocities, which are the component of velocity along the radar beam, two Doppler radars are necessary to estimate two-dimensional wind fields. A series of dual-Doppler cases were collected in central Oklahoma [29, 32, 173], and some of these cases included timeseries data that were used to examine Doppler spectra in tornadoes [11, 220–222].

3.2 First Portable Doppler Observations

While ground-based radar for meteorological purposes was well under development by the late 1980s, it was becoming obvious that most tornadoes don't occur close to a fixed-site radar. Even though fixed-site radars were important for sustainable warning networks, they wouldn't provide the high-quality data en masse necessary to learn more about tornado dynamics, tornadogenesis, and the tornadic lifecycle. The best way to get high spatial resolution data low to the ground would be to go to the tornadoes with a mobile platform. A team at the University of Oklahoma (OU), led by Howie Bluestein, worked with the Los Alamos National Laboratory (LANL) to convert a continuous-wave, X-band Doppler radar into a portable meteorological radar system for tornado observations [11]. The portable nature of the LANL system allowed researchers to get exceptionally close to tornadoes, enabling collection of Doppler spectra and wind speeds over $120\,\mathrm{m\,s^{-1}}$ at low elevations [16]. The LANL radar also allowed for frequency modulation, resulting in the ability to determine spectra across multiple range gates [13]. These were the first steps toward gaining an advanced knowledge of tornado dynamics via portable-Doppler radar wind measurements.

The follow-on to the LANL radar was a W-band fully scannable Doppler radar on the bed of a truck, built by the University of Massachusetts (UMASS) at Amherst and again utilized by the Bluestein team at OU. The use of W band allowed for a relatively small dish to be mountable on a truck, while still providing excellent beamwidths better than 0.6° (and better than 0.2° after an upgrade in 1999) [11, 12]. However, at W band, attenuation is exceptional, leading to the need to get very close to strong tornadoes. As will be seen in the upcoming subsections, the choice between frequencies for mobile radar systems is of great importance to the mission at hand, and the choices vary quite substantially. During a prolific 1999 tornado season, and with such high spatial resolutions, the UMASS W-band radar was able to discern Weak-Echo Holes (WEHs) or “eyes”, spiral bands, and multiple vortices and trochoidal motions in tornadoes [12, 17, 19, 179]. This radar was used from the mid-1990s through 2010 in VORTEX2, bringing many ground-breaking observations and findings to the radar and severe storms communities.

In the same time frame, during the mid-1990s, Josh Wurman and his team at OU completed work on the first X-band Doppler on Wheels (DOW) [209, 210]. The DOW was instrumental during the first Verification of the Origins of Rotation in Tornadoes Experiment (VORTEX) [147], collecting the first three-dimensional estimates of reflectivity and wind fields in strong tornadoes [206]. The first DOW collected data led to a better understanding of centrifuging debris in tornadoes [58] and the structure of wind fields and their relation to damage in tornadoes [1, 205]. Shortly after the success of the first DOW was realized, and in similar fashion to the original NSSL dual-Doppler network, a second DOW was constructed for dual-Doppler studies. DOW2 provided the ability to place mobile radars in multiple locations, resulting in better chances of near-storm data collection [41] and the capability for dual-Doppler studies at unprecedented spatial and temporal resolutions [3, 123, 211, 212]. The DOWs were updated through many iterations in the early- to mid-2000s, cementing X-band radars as a staple in mobile Doppler weather radar due to its relatively acceptable tradeoffs between spatial resolution and dish size for mounting on a portable platform.

3.3 *Current Mobile Doppler Radar Systems*

Other frequencies were employed in the 2000s in order to observe storm-scale phenomena and mitigate severe attenuation issues. The Shared Mobile Atmospheric Research and Teaching Radar, or SMART-R program, was a cooperative effort between OU, Texas A&M, Texas Tech, and NSSL [7]. The SMART-Rs, run since the late 2000s by Mike Biggerstaff and his team at OU, consisted of a pair of radars that utilized C-band frequencies to collect dual-Doppler wind fields in severe storms [188], hurricanes [88], and other phenomena [140]. Due to their longer wavelength, relatively large dishes were necessary, making the SMART-R trucks rather large. Although they weren't necessarily nimble enough to get as close to tornadoes as some other mobile platforms, the SMART-Rs played a large role in the collection of storm-scale data near supercells and tornadoes during VORTEX2 [170, 171, 214]. One of the SMART-R trucks was converted for dual-polarization capabilities in the early 2010s, making it the only polarimetric C-band mobile weather radar currently in use in the United States.

Another system that brought polarimetric observations to the table was the UMASS polarimetric X-band radar [20] (Fig. 4). Used by the Bluestein team at OU, this radar was a driving force behind the proliferation of polarimetric X-band weather radars. A number of key polarimetric findings were supported by the UMASS X-band, including polarimetric signatures in supercells [176] (at X band; such observations were reported at S band by Ryzhkov, Kumjian, and others previously). Additionally, the first EF5 tornado (based on the Enhanced Fujita Scale) was observed by this radar, resulting in a long-duration dataset that provided insight into the transition from cyclic tornado production to long-track tornado production [180].

On the opposite side of the frequency spectrum, the Texas Tech Ka-band (TTUKa) radars were designed to collect extremely high spatial resolution data in supercells and tornadoes [78, 197]. The 3-dB beamwidth of the Ka-band radars is 0.33°, which provides three times better azimuthal resolution than most X-band mobile radars and exceptional clear-air sensitivity. Run by Chris Weiss and his team at Texas Tech, this pair of radars, similar to the SMART-Rs and DOWs, work in tandem to collect dual-Doppler observations whenever possible. One significant disadvantage of Ka-band radars is significant attenuation that can lead to complete extinction in heavy rainfall.

A plethora of mobile, polarimetric, X-band radars took hold by the early 2010s. The NSSL NOXP, a 1°-beamwidth X-band polarimetric system, was active for VORTEX2, Plains Elevated Convection at Night (PECAN) [70], and numerous smaller field projects. Led primarily by Don Burgess and Ted Mansell, NOXP has observed tornadic dynamics and tornadogenesis, mainly during VORTEX2 [42, 164], and was the source of determining raindrop sizes in hook echoes during several tornadic and nontornadic supercells [65]. Numerous iterations of the DOWs led to DOWS3-4 [203], and eventually to DOW6 and DOW7 [213], polarimetric X-band systems that were also used in VORTEX2, PECAN, and Radar Observations

Fig. 4 The UMASS X-Pol mobile radar scanning a tornado. Photograph provided by Jeff Snyder

of Tornadoes and Thunderstorms Experiment (ROTATE) campaigns. Led by Josh Wurman at the Center for Severe Weather Research (CSWR), DOWs 6 and 7 were primary contributors to the VORTEX2 catalog of supercell and tornado cases [2, 92, 125, 190, 191, 215]. The CSWR DOWs have been prolific contributors to the severe convective storm, tornado, and radar communities, with many cases in the literature [193, 207, 216].

Additional X-band polarimetric systems were developed in different forms and for different purposes around the same time period. The Advanced Radar Research Center (ARRC) at OU developed the PX-1000, a transportable, X-band, polarimetric weather radar with a 1.8° beamwidth [44]. Led by Boon Leng Cheong at OU, PX-1000 took part in PECAN and various other smaller field projects, while also staying situated in Norman, Oklahoma as a "stationary" platform during severe storm seasons. Since PX-1000 is a trailer-based platform, it is not ideal for tornado chasing; however, it was a primary observer of the EF5 Moore, Oklahoma tornado on 20 May 2013, leading to new discoveries regarding tornadic debris ejections and "failed" occlusions [102]. PX-1000 also served as a proving ground for pulse compression in weather radar systems, a technique now used on multiple mobile

Fig. 5 The Center for Severe Weather Research Rapid DOW observing the Goshen County, Wyoming tornado on 5 June 2009. Photograph provided by Paul C. Robinson

and stationary tornado-scale and other weather radars around the world [101]. The University of Alabama at Huntsville (UAH) Mobile Alabama X-band (MAX) radar is another dual-polarized, X-band radar developed for hurricane and tornado studies [142]. MAX has primarily been used for lightning research, nocturnal storms (during PECAN), and tornado studies (during VORTEX-SE).

Of key importance to tornado researchers, specifically, is extremely rapid volumetric update rates. The current WSR-88D network, for example, completes its fastest volume of 4.2 min in Volume Coverage Pattern (VCP) 12 [37], while the advective time scale of tornadoes is on the order of seconds. In many instances, tornadoes do not even last for 4.2 min, and therefore the entire evolution of the tornado is missed. For this reason, a small subset of mobile radar systems have been developed with scanning speed as the key design criteria, with scanning capabilities on the order of seconds. There are currently four radars that fall under this category, all operating at X band. The first to be developed was the CSWR Rapid DOW [208], originally called DOW5 (Fig. 5). The Rapid DOW achieves 7-s volumes with six pencil beams (0.8–0.9° in width) in the vertical by scanning across frequencies ranging from 9.3 to 9.75 GHz. It can increase the vertical resolution by adding a second set of six pencil beams on a second scan, bringing volume update times to 14 s with 12 beams. Combined with its range resolution of 11 m, the Rapid DOW spatiotemporal resolution has led to multiple fine-scale studies of tornadoes in three dimensions [91, 193, 215]. The second rapid-scan mobile radar used for tornado studies is the MWR-05XP, a modified Naval Postgraduate School radar used in the

Fig. 6 The University of Oklahoma Atmospheric Imaging Radar (AIR) scanning a tornado

field by Howie Bluestein in the late 2000s to mid 2010s [21]. The MWR-05XP obtains 7-s 90° × 20° volumes with a 2.0° effective beamwidth in each dimension, and 24-s 360° × 20° volumes with the same spatial resolution. Although its spatial resolution (150 m in range) is a bit coarser than other rapid-scan radars, it has the highest sensitivity of the rapid-scan radar group, with −15 dBZ sensitivity at 10 km range. MWR-05XP has been used to study the vertical development of TVSs in supercells [63, 64].

The third rapid-scan mobile radar to be developed was the Rapid X-band Polarimetric radar, or RaXPol [141]. As the only polarimetric radar in the group of rapid-scan systems, RaXPol has been used by Howie Bluestein and his team since 2011 to observe extremely rapid changes in polarimetric characteristics of tornadoes at exceptionally close range. RaXPol is a dish-based platform that can rotate at 180° s^{-1}, resulting in a 360° × 20° scan time of 40 s (although the spacing and number of elevations can be changed to create faster scans). In order to deal with the widening of the effective beamwidth due to the rapid rotation of the dish, RaXPol utilizes frequency hopping to produce independent samples more rapidly. To date, RaXPol has been instrumental in analysis of the El Reno tornadoes of 2011 [79] and 2013 [192, 193], including the observation of debris tails and ejections similar to PX-1000.

The final rapid-scanning mobile radar is the Atmospheric Imaging Radar (AIR; Fig. 6), built by Brad Isom and several ARRC collaborators and used by Jim Kurdzo and David Bodine and their teams at OU since 2012 [81]. The AIR utilizes a technique known as digital beamforming, or radar imaging [129]. Although imaging had been used previously in the atmospheric wind profiling community [43, 129, 137], the AIR was the first attempt at using imaging for severe convective storms. The

AIR transmits a wide beam in the vertical dimension (20° tall by 1° wide) and uses a vertical, 36-element phased array to form 1° by 1° beams on receive. In this fashion, individual range-height indicators (RHIs) are formed with each pulse, and the antenna only has to be rotated in the azimuthal dimension. Without the need to scan vertically with a pencil beam, the AIR is capable of forming a 90° × 20° sector volume in under 6 s, with the added benefit of collecting RHI-style data at every azimuth. The AIR has been used for analysis of severe convective storms and tornadoes [103] as well as gust fronts and density currents [115].

3.4 NEXRAD and Other Fixed-Site Radars

The WSR-88D, also commonly referred to as NEXRAD ("Next-Generation Radar"), brought widespread, ground-based, fixed-site, Doppler radar coverage to the vast majority of the continental United States during the late 1980s and early 1990s [48]. Across the plains, midwest, and southeastern portions of the nation especially, the relatively densely populated network of Doppler weather radars has been responsible for saving an inestimable number of lives due to advanced warning and forecasting capabilities for severe convective storms and tornadoes [143]. Both the number of tornadoes warned and the average tornado warning lead time nearly doubled with the installation of the WSR-88D network, while the expected number of fatalities and injured fell by 40–45% [169]. The WSR-88Ds operate at S band, with a short-pulse peak power of 750 kW, a 0.9° beamwidth, and a volumetric update rate that varies between 4.2 and 10 min. This combination of power, Doppler capabilities, and spatiotemporal resolution led to numerous algorithmic enhancements over the years, including storm cell tracking [83], mesocyclone detection [105, 178], and tornado detection [131].

Despite the fact the WSR-88D network is approaching it's 30th year of life [219], a series of significant updates to scanning strategies, algorithms, and capabilities have been implemented, including a major service life extension program [49]. The WSR-88D employs a series of VCPs for different weather scenarios. From the severe storms perspective, scanning at the proper rate for acceptable data quality mixed with ideal update times and elevation cuts is critical for the performance of severe storm algorithms [34, 35]. The implementation of VCP 12 [37] was particularly important to the WSR-88D network, given its faster scanning and a denser vertical sampling at lower elevation levels. Another important update to the WSR-88D was the super-resolution oversampling technique employed in the mid-to-late 2000s. This technique was exploited for better mesocyclone and tornado detection [36, 202], as well as enhancements to various areas of hydrology [165]. In terms of rapid update rates, advancements such as Automated Volume Scan Evaluation and Termination (AVSET) and Supplemental Adaptive Intra-Volume Low-Level Scans (SAILS) have allowed the revisitation of lower elevation angles more often for updates in severe storms [49, 50]. The most critical update to the WSR-88D, in similar fashion to the mobile systems described previously, was the capability for dual-polarized observations in the early 2010s [55]. The ability to transmit a slant-45 pulse and receive on both horizontal and vertical channels opened new possibilities for rainfall estimation [157], hydrometeor type determination [139, 159], and tornado detection [158].

In addition to the WSR-88Ds, a variety of S-, C-, and X-band fixed-site radars were introduced around the United States in the 1990s and 2000s at research institutions, universities, and as part of large-scale research initiatives funded by the National Science Foundation and other agencies. The University of Alabama Huntsville procured a polarimetric C-band radar known as ARMOR that has observed numerous large hail and tornado cases [89] and is involved in VORTEX-SE. The University of Oklahoma debuted the dual-polarization C-band OU-PRIME in 2009, a year before it collected 0.5° native-resolution observations of EF4 tornadoes at close range [138]. OU-PRIME has been involved in polarimetric properties in supercell hook echoes [93] as well as studies regarding tornadic debris signatures [23]. Additionally, NSSL and the Radar Operations Center (ROC) have run multiple fixed-site radars over the past several decades, including KOUN (NSSL) and KCRI (ROC), both polarimetric versions of a WSR-88D and located at University of Oklahoma Westheimer Airport in Norman, Oklahoma. Both radars have observed numerous strong and violent tornadoes over their history, and have been used in a variety of relevant studies [22, 98, 158, 181, 195, 217, 218]. Finally, the Terminal Doppler Weather Radar (TDWR) is a network of 45 C-band weather radars situated at major airports across the United States that was deployed in the 1990s in response to microburst and windshear aviation disasters [59, 130]. Use of the TDWRs as an augmentation to the WSR-88Ds for tornado detection and warning was discussed by [187].

As the WSR-88Ds age, a number of concepts have exhibited their merits as potential future designs for a national weather radar network in the United States. Among the competing interests, two primary areas of concern have been identified for tornado forecasters and researchers: distance to tornadoes (i.e., low-level scanning capabilities) and temporal resolution [31]. The first of these concepts has been taken on as part of the Collaborative Adaptive Sensing of the Atmosphere (CASA) project [128]. CASA aimed to use a large number (likely thousands) of X-band, dual-polarization radars across the United States to fill the weather radar need in the nation. The advantage of having so many radars would be low-level scanning near tornadoes [100], with the possibility of multiple-Doppler data assimilation for the warn-on-forecast initiative [177]. The first CASA test network was deployed in southwest Oklahoma, and the second network is currently deployed in the Dallas area. A series of tornado cases led to studies regarding severe weather operations and data assimilation [114, 145, 160].

The second concept for a next-generation weather radar network in the United States is the Multi-Function Phased Array Radar (MPAR) [196, 224]. The MPAR concept argues that a phased array antenna can scan storms rapidly due to the lack of need to steer a large dish in azimuth and elevation. As part of MPAR studies, a surplus SPY-1A United States Navy radar was procured by NSSL; it was named the National Weather Radar Testbed (NWRT; Fig. 7). The NWRT used electronically

Fig. 7 The National Weather Radar Testbed (NWRT) in Norman, Oklahoma. The testbed will be replaced by the MPAR Advanced Technology Demonstrator in 2018

steered pencil beams to scan volumes in 30–60 s [76, 77] and observed multiple severe storm and tornado cases [104, 134, 181]. During the Phased Array Radar Innovative Sensing Experiments (PARISE), forecasters were invited to the Hazardous Weather Testbed to evaluate NWRT PAR data compared to NEXRAD data and assess the impacts of higher temporal resolution radar data on their forecasts. Forecasters noted that the rapid-scan data helped them identify severe weather precursors earlier and improved the accuracy of their tornado warnings [26, 27].

3.5 Major Tornado Field Projects Using Radar

Before the mid-1990s, most field work involving tornado observations was focused on visual [71, 132] and in-situ measurements [4] made by storm chasers and field researchers attempting to learn more about severe convective storm structure. Some of the visual observations made during the Tornado Intercept Project at NSSL were combined with radar data [133] from the same Union City storm mentioned in the previous sections. Additional combinations of visual observations and radar data were reported by Lemon and Doswell [109], as summarized by Bluestein [10].

During the Tornado Intercept Project years, the Joint Doppler Operational Project (JDOP) also worked to combine visual observations with NSSL radar signatures [10]. The in-situ measurements consisted mostly of data from the Totable Tornado Observatory (TOTO), a device that was placed in the path of a tornado in order to collect observations of wind speed and direction, pressure, temperature, and corona discharge [9, 10]. Additional in-situ measurements were made via radiosonde launches into supercells and tornadoes [14, 15, 153–155, 189, 200].

As mentioned in previous sections, the first ground-based portable radar observations of tornadoes were from the LANL system by Howie Bluestein and his students in the late 1980s and early 1990s [13]. By the mid-1990s, the UMASS W-band radar was also in use by the Bluestein teams, but aside from sustained funding from the National Science Foundation (NSF) for use by teams at OU, there had never been an organized effort to use multiple radars to observe tornadoes in a coordinated way. The first Verification of the Origins of Rotation in Tornadoes Experiment (VORTEX), which took place in 1994 and 1995, was the first attempt at real-time coordination of multiple radars (and other instruments) in the field to study tornadoes [147]. During the first year of VORTEX, in 1994, only the LANL and UMASS W-band were available to researchers [10]. Additional observing platforms included radiosondes, airborne radar, in-situ tornado observation probes, and mobile mesonets. VORTEX was the first time that all of these platforms were used in concert in order to discern links between the kinematic and thermodynamic aspects of severe convective storms and tornadoes. In the second year of VORTEX (1995), the first DOW was added to the fleet of instruments. Some of the first relatively high-resolution radar-based datasets in tornadoes were collected during VORTEX in 1995 by the DOW, including 16 May [209], 2 June, and 8 June. These first radar observations collected in tandem as part of a large field project are summarized in [10].

After VORTEX, a series of smaller field projects to study tornadoes took place; many led by OU and eventually CSWR. In the late 1990s, organized coordination between the first DOW and a second DOW resulted in the first dual-Doppler studies of tornadoes. Two related but smaller follow-on projects, SUBVORTEX (1998) and VORTEX-99 (1999) utilized similar instrumentation packages, and included data collected during the 3 May 1999 Moore, Oklahoma tornado. Howie Bluestein's group conducted experiments with the UMass X-Pol and the UMass W-band in the 1990s through the early 2010s, and the DOWs participated in annual severe weather field experiments over the same timeframe. Although the large-scale, organized field campaigns provide a more comprehensive view on tornadogenesis and tornado dynamics with the array of instruments available, these smaller-scale mobile radar studies have yielded valuable understanding of tornadoes and their parent storms. In particular, these studies have significantly increased the sample size of tornado and supercell cases observed by mobile radars because major tornado-related field projects are only conducted once every 10–20 years. These experiments have also incorporated several emerging mobile radar technologies, including polarimetric and rapid-scan radars, providing new insights into microphysical characteristics of supercells and significant advancements in understanding of tornadogenesis and tornado dynamics through much improved temporal sampling.

Fig. 8 Some of the mobile radars that took part in VORTEX2

After VORTEX and its follow-on projects, the second iteration of the full project took shape as VORTEX2 in 2009 and 2010 [214]. This time around, an array of 11 mobile radars (SMART-R1, SMART-R2, DOW 6, DOW 7, RSDOW, NOXP, UMASS X-Pol, TTU Ka 1 and 2, UMASS-W, and MWR-05XP; Fig. 8) and numerous fixed-site radars (WSR-88D radars, NWRT, KOUN, and CASA) took part in the project. Many other instrument teams were involved, ranging from tornado pods to sticknets, unmanned aerial systems, mobile radiosondes, mobile mesonets, disdrometers, and photogrammetry teams. VORTEX2 was billed as the largest tornado field project in history, and it collected multiple groundbreaking datasets. The most notable of these was the 5 June 2009 Goshen/LaGrange tornado in Southeast Wyoming [2, 92, 121, 125, 190, 191], where virtually all of the instrument teams collected data on a large, long-track tornado. Storm-scale and tornado-scale radar observations captured tornadogenesis, the tornado maintenance, and the tornadic dissipation phases with multiple Doppler lobes and varying resolutions, distances, and capabilities. The Goshen case is widely considered to be the most-studied and well-observed supercell/tornado in history. With the unprecedented suite of instruments and radars, VORTEX2 helped define much of the modern strategy for scientific deployments near tornadoes, especially instrument locations.

After VORTEX2, the primary structured radar-based tornado research project in the United States was the Radar Observations of Tornadoes and Thunderstorms Experiment (ROTATE), led by Josh Wurman and Karen Kosiba at CSWR. ROTATE was primarily focused on the lowest levels of tornadoes, necessitating exceptionally close observations. ROTATE also focused on airborne debris and the differences between strong tornadoes and weaker tornadoes. DOWs 6, 7, and 8 were used, along with a number of mobile mesonets and tornado pods for in-situ measurements. Multiple findings came out of ROTATE, including the combination of in-situ and remote sensing datasets [215], new finescale observations of tornadoes [207], and observations of multiple-vortex tornadoes [216]. ROTATE also collected numerous

Fig. 9 The RaXPol is deployed on the Sand Mountain plateau in northeast Alabama during VORTEX-SE, and collected data on a supercell thunderstorm that passed over the plateau

datasets before VORTEX2. In addition to ROTATE, teams at OU and TTU ran annual small-scale field projects in the 2010s. Chris Weiss continues to lead annual work with the TTU Ka-band radars, while Howie Bluestein works primarily with RaXPol each season. The ARRC at OU manages the Atmospheric Imaging Radar, led by Jim Kurdzo in the early 2010s and David Bodine and Casey Griffin since 2016. Mike Biggerstaff runs occasional field programs with the polarimetric SMART-R, while Don Burgess leads the NOXP radar during annual storm operations. Robin Tanamachi at Purdue still utilizes the UMASS X-band radar during storm season. The NWRT, KOUN, KCRI, PX-1000, and other radars are routinely run in rapid-scanning modes during severe weather in the Spring on the plains. Finally, since the spring of 2016, CSWR has transitioned from ROTATE to TWIRL; Tornadic Winds: In situ and Radar observations at Low levels. TWIRL aims to better understand how tornadoes form and how the lowest-level winds do damage. The three current DOW trucks and a stable of mobile mesonets and tornado pods are involved.

The most recent organized large-scale tornado field project ongoing at the time of this writing is VORTEX-Southeast (VORTEX-SE), a version of VORTEX focusing on tornadoes in the Southeastern United States across 2016 and 2017. The VORTEX-SE domain is considered unique due to the prevalence of nocturnal tornadoes, the timing of tornadoes relative to spring (compared to the plains), and the exceptionally dynamic terrain in the region. Due to the difficulty of deploying mobile radars in such an environment, most of the mobile radar assets in VORTEX-SE are situated in stationary locations before storms develop (Fig. 9), resulting in a large "net" of radar coverage during storm initiation and maintenance. VORTEX-SE also seeks to understand the social issues associated with understanding and heeding of tornado warnings, especially at night and when visibility is poor.

4 Scientific Advancements from Ground-Based Radar

This section introduces the achievements from ground-based radar observations since the advent of Doppler radar in the 1970s. Topics discussed in this section include tornado detection, radar analysis techniques, tornadogenesis and tornado dynamics, and dual-polarization and rapid-scan radar observations of tornadoes.

4.1 Tornado Detection

The first tornado detection studies were conducted with the NSSL Doppler radar observations in the 1970s. The first tornado vortex signature (TVS) observed by the NSSL Doppler radar was on 24 May 1973 near Union City, Oklahoma [38, 39]. They noted that the TVS was continuous with height and time and corresponded closely with the surface damage track. Most importantly, they noted that the TVS was evident 23 min before tornado formation, and thus the opportunity to improve tornado warning lead-time was documented. The TVS occurred first at mid-levels and descended to the lowest elevation angle prior to tornadogenesis. Between 1973 and 1976, the NSSL Doppler radar observed 10 TVSs associated with eight tornadoes and two cases without tornadoes [33]. The TVSs frequently appeared over 10 min before tornado formation, providing greater evidence of increased lead-time that could result from TVS detection prior to tornadogenesis. These early studies also noted several challenges, which complicate tornado detection even for current radar systems. First, the tornado's diameter must be large compared to the azimuthal and range resolution of the radar. Thus, for small tornado diameters or tornadoes at long ranges, radar resolution may not be sufficient to observe a TVS. At more distant ranges, the radar beam height is also too high to observe the tornado.

After these initial studies revealed that TVSs were often collocated with tornadoes, tornado detection research focused on developing automated techniques to detect mesocyclones and tornadoes. NSSL researchers developed an automated Mesocyclone Detection Algorithm (MDA) in the 1980s which computed the size and azimuthal shear of mesocyclones [223]. This algorithm formed the basis for the first MDA implemented on the NEXRAD radar network [48]. An upgrade to the MDA was made in the late 1990s that eliminated thresholds included in the original algorithm and provided more robust detection of mesocyclones of different scales throughout their lifecycle [178]. The upgraded MDA provided a probability of detection (POD) of mesocyclones of over 99% compared to 50–75% for the original MDA.

Early efforts at tornado detection focused on the TVS and assessing gate-to-gate shear. The first TVS algorithm implemented on the NEXRAD network employed stringent criteria resulting in a low POD [131]. To address this, the NSSL Tornado Detection Algorithm (TDA) was developed in the 1990s [131]. The TDA focused on examining the three-dimensional characteristics of the TVS, and developed new methods to track tornado vortices and characterize their intensity. The NSSL TDA resulted in a lower false alarm rate and a higher probability of detection compared to the original algorithm.

Some limitations exist to detecting and characterizing tornadoes based on gate-to-gate shear. The magnitude and location of the maximum gate-to-gate shear depends on the radar beam's location relative to the center of the vortex [201]. If the tornado is poorly sampled, then changes in the position of the radar beam relative to the tornado cause fluctuations in TVS intensity or location. Typically, the rotational velocity measured by the radar decreases as the range to the tornado or mesocyclone increases as a consequence of coarser spatial sampling of peak velocities and increased height above the ground.

4.2 Radar Analysis Techniques

In this section, radar analysis techniques critical to analyzing tornadoes and their parent thunderstorms are discussed. These techniques include dual-Doppler and ground-based velocity tracking display analyses for three-dimensional wind retrievals.

4.2.1 Dual-Doppler

Dual- or multi-Doppler analyses employ different techniques to combine Doppler velocity data from two or more radars to obtain three-dimensional winds. The Doppler or radial velocity observed by a radar is a function of the radars' azimuth angle θ, elevation angle ϕ, range r, the Cartesian wind components (u,v,w), and the scatterers' terminal fall speeds w_t.

$$v_r = u \cos\phi \sin\theta + v \cos\phi \cos\theta + (w - w_t) \sin\phi \quad (1)$$

From (1), it is apparent that the radar observes the component of the wind along the radar beam. When two or more radars observe sufficiently different components of the wind (e.g., usually when the angle between the radar beams is between 30° and 150°), the three-dimensional wind components can be retrieved by employing the continuity equation as an additional constraint.

Dual-Doppler analyses require careful data post-processing techniques to address several error sources while preserving the resolvable scales of motion. Doppler velocities are often aliased in severe thunderstorms because a pulsed Doppler radar system can only resolve Doppler velocities within the Nyquist interval or maximum unambiguous velocity v_a (2),

$$v_a = \frac{\lambda}{4T_s}, \tag{2}$$

where the radar's wavelength is λ and the pulse repetition time is T_s. To obtain Doppler velocities outside of the Nyquist interval, manual or automated dealiasing or unfolding of Doppler velocity data is required.[1] After dealiasing, Doppler velocity data are objectively analyzed (i.e., spatially filtered) to a common grid. Objective analysis parameters are selected to obtain the smallest resolvable scale while mitigating random errors [116, 183]. A two-pass Barnes objective analysis technique is often implemented and effectively suppresses noise while preserving resolved wavelengths, and is the most commonly applied method [116]. Finally, since the storm moves and evolves during a volume scan, the movement of the storm (advection) between observation times must be corrected. A constant storm motion and linear evolution is often assumed [69], but supercells and tornadoes can contain substantial non-linear advection and evolution, leading to large errors after advection correction. To mitigate the non-linear motion errors, a spatially variable advection correction technique was developed [167, 168] that reduces non-linear advection errors [144]. Dual-Doppler analyses using rapid-scan radars can also help reduce non-linear advection errors.

4.2.2 GBVTD

Dual-Doppler analyses require two or more Doppler radars located in a specific geometric configuration to obtain three-dimensional winds, which can be difficult to obtain for tornado measurements at close range. Moreover, the spatial resolution of dual-Doppler analyses is limited by objective analyses required to remove noisy data. Thus, scientists have explored single-Doppler techniques for retrieving three-dimensional winds. In the late 1990s, the Ground-Based Velocity Tracking Display (GBVTD) [107] was developed to retrieve mean radial and tangential velocities and their components at different wavenumbers in tropical cyclones. The GBVTD performs a Fourier decomposition of Doppler velocity data under the assumption of axisymmetry. This assumption is not required for dual-Doppler analyses, and thus flow asymmetries can introduce errors into GBVTD analyses. The GBVTD has been extended to ground-based radar studies of tornadoes [18, 90, 106, 179], including

[1] Staggered pulse repetition times (PRTs), which use two different PRTs between pairs of pulses, can be used to provide a higher Nyquist interval [54].

retrievals of vertical velocity by integrating the continuity equation. Simplified versions of the GBVTD have also been created to obtain the zeroth-order flow [58] or calculate angular momentum budgets [146]. An advantage of the GBVTD is that it can be applied to raw (dealiased) Doppler velocity data, and thus provides higher resolution than dual-Doppler analyses.

Several challenges exist to obtaining accurate three-dimensional wind retrievals from both GBVTD and dual-Doppler analyses. Doppler radar wind analyses assume that radar velocity measurements are measuring wind speeds; however, radars measure the reflectivity-weighted average of scatterers' velocities rather than air velocities. In tornadoes, scatterers are comprised of debris and hydrometeors which move at different velocities compared to the air as a consequence of debris centrifuging [58]. As the size or density of the scatterers increase, larger differences between air and debris velocities result and can lead to errors of tens of $\mathrm{m\,s^{-1}}$ [25, 58]. Since debris are centrifuged outward relative to radial air motion, Doppler radar measurements with debris are excessively divergent and thus can result in anomalously strong downdrafts [25, 135]. For shorter wavelength radars (e.g., W or Ka band), however, debris centrifuging errors are mitigated because smaller scatterers dominate the backscattered radar signal even when large debris are present [25].

4.3 Tornadogenesis and Tornado Dynamics

Although scientists have formulated compelling theories for tornadogenesis, the formation process of tornadoes is still not fully understood. In particular, scientists have been unable to identify why some thunderstorms produce tornadoes while other thunderstorms do not produce tornadoes. For example, numerous supercells will form during tornado outbreaks in regions where forecast parameters favor tornadic storms, yet some supercells will produce tornadoes and others will not. Tornadogenesis is difficult to observe since tornadoes are relatively infrequent events, and a large suite of observations must be collected in order to synthesize the dynamic, thermodynamic, and microphysical processes occurring in the parent storm [214]. Doppler radars have enabled detailed observations of dynamic processes occurring during tornadogenesis, and more recently, dual-polarization radars have provided information about microphysical processes occurring within the parent storm [97, 138].

Single- and dual-Doppler analyses provided a basic conceptual understanding of tornadogenesis in the 1970s and 1980s. Early studies identified a descending TVS, forming at mid-levels and descending to the surface [29, 33]. Early dual-Doppler analyses helped establish a conceptual model for supercells [109], identifying that supercell tornadoes develop at the interface between the updraft and the rear flank downdraft and established the importance of the rear-flank gust front in occlusions [29, 148]. Tilting of horizontal vorticity and subsequent stretching was identified as a mechanism for intensifying low-level vortices, occurring after the rear-flank downdraft develops and intensifies horizontal gradients of vertical velocity [30].

These dual-Doppler measurements provided valuable analyses to compare to the first three-dimensional supercell simulations [85–87, 152]. Combined with radar analyses, these observational and modeling studies elucidated the fundamental dynamics of supercell storms.

Early radar studies also noted a cyclic process of tornadogenesis that occurred in some supercells often during major tornado outbreaks. Cyclic mesocyclogenesis was first analyzed by Don Burgess using Doppler radar observations in the 1980s [40]. During mesocyclogenesis, vorticity maxima develop along the rear-flank gust front and a vorticity maximum approaching the updraft is stretched, resulting in more intense low-level rotation. Toward dissipation, the mesocyclone advects rearward, and a new vorticity maximum forms along the rear-flank gust front and the process repeats [56, 57]. During the occlusion phase, the mesocyclone's updraft becomes cut off from warm, moist air driving the buoyant updraft and the tornado dissipates. The most intense tornadoes result when the tornado remains beneath the updraft of the parent storm for a long period of time (mode 2 tornadogenesis), whereas weaker tornadoes result when a tornado moves rearward more quickly (mode 1 tornadogenesis) [56, 57, 62, 180]. For cyclic tornadic supercells (CTSs), mode 2 tornadogenesis often occurs later in the CTS's lifecycle resulting in long-track tornadoes after weaker, brief tornadoes. During occlusion phases, tornado diameter usually decreases; however, the tornado can remain as intense as the mature phase [62, 102].

Although many of these early studies established how supercell structure evolves during different phases of the tornado's lifecycle, the mechanisms by which low-level rotation intensifies and a tornado develops are not fully understood. Baroclinic generation of horizontal vorticity along storm-induced temperature gradients has been thought to provide an important source of vorticity [85, 120], although weak surface temperature gradients in the rear-flank downdraft and forward-flank downdraft have been observed in many tornado cases[2] [119, 166]. Numerical simulations have found that tornadogenesis can be initiated without baroclinic generation of vorticity by disrupting low-level inflow and transporting higher angular momentum air from aloft [112] or by transporting high angular momentum from aloft to the surface through descending rain curtains [52]. Recently, high-resolution simulations of the 8 May 2003 tornadic supercell identified that frictional generation of horizontal vorticity during internal momentum surges, with subsequent tilting, contributed to intensification of low-level rotation and produced a more intense tornado compared to simulations with no surface friction [161]. Thus, the 8 May 2003 simulations showed that baroclinic generation of vorticity was secondary to frictionally generated vorticity. In contrast, others have found that frictional vorticity generation was secondary to baroclinic generation in idealized and real supercell simulations [117, 127].

[2]Lack of large surface temperature gradients, however, does not preclude large surface temperature gradients aloft.

Exploring tornadogenesis processes from observations alone is challenging due to the lack of three-dimensional thermodynamic observations to complement three-dimensional wind retrievals from radar. Nonetheless, substantial progress has been made and recent tornadogenesis research has focused on the role of the rear-flank gust front (RFGF) or internal momentum surges on tornado formation and dissipation [92, 108, 170], as well as understanding the vorticity and angular momentum evolution throughout a tornado's lifecycle [124, 213]. High-resolution dual-Doppler data have been used to document tilting of horizontal vorticity and stretching of vertical vorticity into tornadoes [211], confirming the expected behavior from numerical simulations [199]. Using radar data from the Goshen, Wyoming tornado case during VORTEX2, dual-Doppler analyses were used to identify that storm-generated vorticity (likely baroclinic) originated largely from the forward-flank downdraft rather than the rear-flank downdraft [122]. The formation of the Goshen, Wyoming tornado also coincided with secondary RFGFSs that modulated tilting and stretching of horizontal vorticity [92].

A classic presentation of tornadic supercell characteristics is observed in the 2223 UTC dual-Doppler analysis of the Moore, Oklahoma tornado, which uses data from OU-PRIME and KOUN near the time of tornadogenesis. The dual-Doppler analysis is shown in Fig. 10, including horizontal winds overlaid on OU-PRIME radar reflectivity factor (Fig. 10a) and horizontal winds overlaid on vertical velocity and contoured vertical vorticity (Fig. 10b). An intense RFGFS is evident just south of the tornado with strong easterly winds and an intense downdraft (>10 m s^{-1}). The southeast edge of the RFGF has vertical velocities exceeding 20 m s^{-1}, and an intense updraft-downdraft circulation is evident straddling the tornado (vertical vorticity maximum labelled by "T"). Accordingly, the large horizontal gradient in

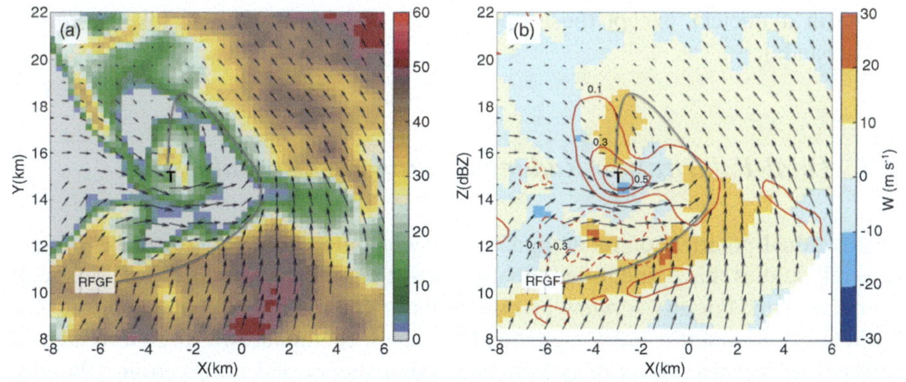

Fig. 10 Dual-Doppler analysis at 2223 UTC on 10 May 2010 at 750 m AGL with the shading indicating (**a**) radar reflectivity factor (dBZ) and (**b**) vertical velocity (m s^{-1}). The dual-Doppler analysis uses OU-PRIME and KOUN radar data. The location of the tornado is marked by the "T", and the rear-flank gust front (RFGF) is denoted by the solid gray line. The red solid lines show isocontours of vertical vorticity at 0.1, 0.3, and 0.5 s^{-1} while red dashed lines show vertical vorticity at −0.1 and −0.3 s^{-1}

vertical velocity allows horizontal vorticity to be tilted into vertical vorticity. In reflectivity, a thin hook echo is evident with intense precipitation wrapping into the northeast side of the tornado.

Detailed understanding of tornado dynamics and structure from Doppler radars became possible with mobile Doppler radar systems in the 1990s. Observations from the Doppler on Wheels (DOW) [209] and the UMass W-band radar [12] documented several characteristics of tornadoes. Tornado velocities exceeded 70 m s^{-1} and locally intense regions of azimuthal shear were observed [12, 209]. Radar reflectivity factor showed a maximum near the surface and decreased with height. Aloft, an annulus of high reflectivity was observed with a minimum in the center of the tornado called a WEH, which has been attributed to centrifuging of debris [25, 58]. The WEH radius expanded with height in some cases [206, 209], and had a maximum radius near the surface in other cases [19]. These observations also noted the lack of surface convergence [206], presumably due to biased Doppler velocities caused by debris centrifuging [25, 58].

Three-dimensional wind structure of tornadoes has been diagnosed using single- and dual-Doppler wind retrievals. Doppler radar studies have observed a range of tornado swirl ratios consistent with laboratory studies using GBVTD analyses [91, 106, 206] and indirectly by identifying tornado subvortices from locally intense regions of azimuthal shear [204, 207]. GBVTD analyses have also frequently revealed central downdrafts in tornadoes [106, 191] as well as low-level inflow [90, 106]. Compared to numerical simulations, however, inflow velocities retrieved from single-Doppler techniques are much lower. Multiple-Doppler analyses have also been used to resolve tornado-scale downdrafts and subvortices [74]. One noteworthy characteristic of tornadoes studied using GBVTDs is that they tend to be larger diameter tornadoes since most mobile radars do not have sufficient azimuthal resolution to adequately characterize smaller diameter tornadoes (e.g., less than 200 m in diameter), except those operating at high frequency at close range (W or Ka bands).

4.4 Dual-Polarization Radar Signatures of Tornadoes

Distinct dual-polarization radar signatures occur in tornadoes and tornadic storms. Several studies have identified dual-polarization radar signatures of supercell thunderstorms. For an overview on dual-polarization radar theory and applications, including polarimetric radar signatures of supercell thunderstorms, the reader is referred to recent reviews of polarimetric radar theory and observations [94–96]. Since the focus of this chapter is ground-based observations of tornadoes, this section emphasizes polarimetric radar signatures of tornadoes specifically rather than polarimetric signatures of the parent thunderstorm.

The first dual-polarization radar observations of tornadoes were conducted using KOUN, revealing the first tornado debris signature or TDS. Lofted tornado debris produces a unique polarimetric radar signature characterized by moderate to high radar reflectivity factor (Z_H), low co-polar cross-correlation coefficient (ρ_{HV}), and near-zero differential reflectivity (Z_{DR}) [156, 158]. The TDS should also be collocated with a TVS, but might exist without a TVS if the spatial resolution is not adequate to resolve the TVS [172]. Z_H is a function of the debris' size and concentration, and thus higher concentrations of debris or larger debris sizes tend to produce higher Z_H. Recent studies have observed a larger range of Z_H from 20 to 70 dBZ [22, 163]. ρ_{HV} measures the correlation between the horizontal and vertical backscattered radar returns, and low ρ_{HV} occurs for (1) large particle sizes due to Mie scattering, (2) varied particle types within a volume, and (3) irregular particle shapes. In tornadoes, lofted debris causes reduced correlation or low ρ_{HV} due to the debris' varied sizes, shapes, and compositions. Z_{DR} is the ratio of horizontal and vertical reflectivity factors, and tends to have positive (negative) values when a particle's major axis is aligned horizontally (vertically) for Rayleigh scattering. Since debris orientations have a low degree of common alignment, Z_{DR} tends to be near 0 dB. Nonetheless, coherent regions of positive and negative Z_{DR} have been observed suggesting some common degree of debris alignment in tornadoes [23, 45]. An example of a TDS observed by the OU PX-1000 radar during the 20 May 2013 Moore, Oklahoma EF-5 tornado is shown in Fig. 11, which shows a prominent circular region of low ρ_{HV} collocated with a TVS and two debris ejections to the south.

TDSs provide a valuable operational tool for remote tornado detection [20, 97, 158]. In particular, ρ_{HV} has been identified as the most useful polarimetric variable for detecting TDSs since Z_{DR} can be positive in areas of precipitation entrainment [20, 22]. Tornado detection capabilities using TDSs are particularly important at night or for rain-wrapped tornadoes when tornadoes are difficult to observe visually. Recent studies have developed automated fuzzy logic tornado detection algorithms to incorporate polarimetric TDSs, improving tornado detection statistics [175, 194]. The TDS can add additional detection capability for cases when a TVS is not detected or is not sufficiently resolved. For example, some tornadoes within a quasi-linear convective system (QLCS) on 31 August 2014 produced TDSs but not TVSs because the azimuthal resolution was not sufficient to resolve the TVS [172].

While TDSs can provide a valuable tool for tornado detection, several caveats are important to note. A TDS study using 744 reported tornadoes found that only 16% of tornadoes in storm reports produced a TDS [186]. However, the study noted that 100% of violent tornadoes and 58% of strong tornadoes produced a TDS, and these signatures were evident for longer periods of time compared to weak tornadoes. Thus, many weak tornadoes are not capable of lofting sufficient amounts of debris to the altitude of the radar beam to be detected by the radar, which becomes more challenging at distant observing ranges [97, 186]. Finally, tornado debris may remain lofted after a tornado dissipates, and thus the TDS is not necessarily a reliable indicator of a tornado during and after dissipation [22, 163]. In summary, the TDS can provide confirmation of a damaging tornado, however, the absence of a TDS does not preclude the possibility of an on-going tornado.

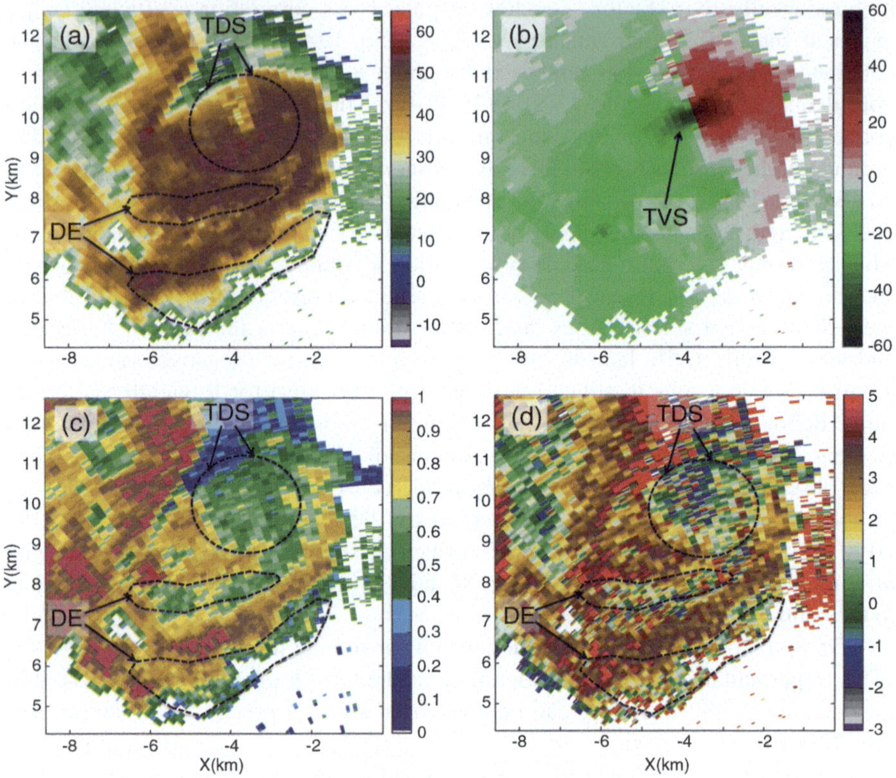

Fig. 11 Example of a TDS from the PX-1000 on 20 May 2013, showing (**a**) radar reflectivity factor (Z_{HH}), (**b**) Doppler velocity (v_r), (**c**) co-polar cross-correlation coefficient ρ_{HV}, and (**d**) differential reflectivity (Z_{DR}). The primary TDS is centered on the TVS with a minimum in Z_{HH} in the center of the tornado. Two debris ejections (labelled "DE") are located south of the primary TDS

Tornado debris signature studies have examined the relationship between TDSs and damage surveys, revealing that TDS parameters may provide information to estimate damage severity [22, 158, 162]. TDS parameters include TDS height or volume, 90th percentile reflectivity, or 10th percentile ρ_{HV}. The volume of the TDS can be tens of km^3 and extend to the top of the storm [22]. Using a large sample size of TDS cases, it has been found that tornadoes with higher TDS heights tend to have higher Enhanced Fujita (EF) ratings [185, 186]. These findings are consistent with long-distance debris transport studies which found that debris are transported farther in higher EF-rated tornadoes [73].

The dependence of TDS characteristics on EF-rating suggests that a relationship may exist between the TDS and tornado dynamics. High-resolution radar data from mobile and fixed platforms have been used to examine the relationship between TDSs and tornado dynamics. Tornado-scale TDS characteristics include lower ρ_{HV} in tornado subvortices with negative Z_{DR} on the periphery of subvortices [74]. In the center of the TDS, both low ρ_{HV} and higher ρ_{HV} have been observed. Low ρ_{HV}

may result from the presence of a small number of scatterers and lower signal-to-noise ratios [180, 192], while higher ρ_{HV} may result if raindrops or other Rayleigh scatterers have not been centrifuged from the center of the tornado or if rain drops are being recycled by the central downdraft [74]. The TDS also appears to be bounded by a wind speed of 35–40 $\mathrm{m\,s^{-1}}$ [74, 80].

Close-range, rapid-scan mobile radar observations of polarimetric radar signatures were obtained in 2011 and 2013 by Howie Bluestein's research group using the RaXpol. Prior to tornadogenesis of the 24 May 2011 El Reno, Oklahoma EF-5 tornado, a TDS was observed as a result of the developing tornadic circulation lofting debris from a previous tornado and/or from the surface [80]. The tornado and parent storm lofted debris often above 5 km (maximum height observed) during the mature stage. During a brief weakening period, the TDS height decreased and widespread debris fallout was noted, consistent with earlier predictions without tornado-scale velocity measurements [22]. From isosurface analyses, they identified wavelike structures within the TDS that propagated upward at over 30 $\mathrm{m\,s^{-1}}$ and hypothesized that these may be associated with centrifugal waves. From PPI scans, they identified "sawtooth" appendages of low ρ_{HV} similar to inflow bands of dust identified in observations. RaXpol radar data were also used to study the 2013 El Reno, Oklahoma EF-3 tornado [192]. As the tornado passed over a house causing EF-3 damage, they noted high radar reflectivity factors associated with lofted debris in a localized area. They also noted the presence of a "debris overhang" where low ρ_{HV} is collocated with the WEH at higher elevation angles, and suggested that this corresponded to debris being lofted into the storm-scale updraft.

The storm-scale flow impacts the three-dimensional structure of the TDS through strong updrafts and downdrafts. Debris are recycled into the storm-scale updraft [22, 192], which can be lofted to middle and upper levels of intense supercells. Strong downdrafts may also aid in the fallout of tornado debris [138]. The role of the rear flank downdraft (RFD) in "ejecting" debris from the tornadic circulation and main updraft base was investigated with respect to the 2013 Moore, Oklahoma EF-5 tornado [102]. A series of rear flank gust front surges (RFGFSs) were observed in the velocity data, along with co-located areas of significantly lowered cross-correlation coefficient (indicating debris). An example of two debris ejections is shown in Fig. 11c, evident by two bands of depressed ρ_{HV}. The study attributes these debris ejections to a combination of debris fallout within the RFD and debris re-lofting by the RFGF. Additionally, the RFD was assumed to play a major role in what the authors called a failed occlusion, resulting in a loop of the tornado track without the occurrence of a full occlusion. Similar debris ejections were seen at X band during the 2011 El Reno, Oklahoma tornado [79]. Photogrammetric analyses have also revealed areas of low ρ_{HV} associated with areas of lofted dust and debris along the rear-flank gust front [193] as well as other regions without apparent convergence.

An emerging application of dual-polarization TDSs is characterizing scatterers within the tornado. Such efforts are needed to correct debris centrifuging effects on Doppler velocity, as well as to provide information about hazards associated with tornadoes. Several studies have postulated that larger lofted particles cause higher radar reflectivity factor and low ρ_{HV} [22, 158]. Dual-wavelength radar observations and simulations also suggest that larger debris may be associated with large dual-wavelength ratios [23, 25]. To investigate how TDSs vary depending on tornado debris types, Boon Leng Cheong and a group of scientists and engineers developed a dual-polarization radar simulator that ingests radar cross section data for debris and an LES model to produce realistic dual-polarization TDS signatures, including low ρ_{HV} and negative Z_{DR} signatures [45]. The simulation has also been used to explore polarimetric spectral densities that represent dual-polarization radar variables as a function of Doppler velocity. Since hydrometeors and debris move at different speeds, it is expected that polarimetric spectral densities can be used to identify hydrometeors and debris within the Doppler spectrum, and subsequently filter the biased debris velocities caused by debris centrifuging [184].

4.5 Rapid-Scan Radar Applications

Before rapid-scan radar technology was developed, Howie Bluestein emulated a rapid-scan radar by scanning the lowest elevation angle every 10–15 s, and observed that the tornado wind field evolved in tens of seconds or less [18]. In their GBVTD analysis, they observed tangential wind speeds oscillating on time scales of 30–60 s, and the most intense winds occurring when the tornado diameter was smallest. These initial analyses helped demonstrate the need for the current suite of rapid-scan radars.

Howie Bluestein's research group has conducted several volumetric, rapid-scan studies of tornadogenesis using the MWR-05XP and the RaXpol. These studies have focused on the vertical development of TVSs to explore bottom-up or top-down tornadogenesis. Top-down tornadogenesis may occur through the dynamic pipe effect [110, 182] through radial convergence of angular momentum from the mesocyclone altitude to the surface. Four tornadogenesis cases were examined using MWR-05XP data, and analyses of these cases revealed that the mesocyclone-scale vortex signature developed upward from the surface [63]. Using simulated TVSs with similar temporal resolution to a NEXRAD radar, it became evident that inadequate temporal sampling of TVS evolution could create the appearance of a top-down tornadogenesis process even though bottom-up tornadogenesis was actually occurring. Using RaXpol data during tornadogenesis of an EF-5 tornado, it was observed that the low-level mesocyclone strengthened and its diameter decreased during the 3 min preceding tornadogenesis while at higher altitudes the mesocyclone intensity remained constant. Just prior to tornadogenesis, the upper levels intensified and the TVSs intensified rapidly through the vertical column in about 30 s [79]. Within an inversion layer at middle levels (between 1.5 and 2 km), they observed a lag in tornado intensification compared to heights above and below it. From this small sample size of rapid-scan observations, the lack of top-down tornadogenesis processes suggest the dynamic pipe effect may not be important to tornadogenesis.

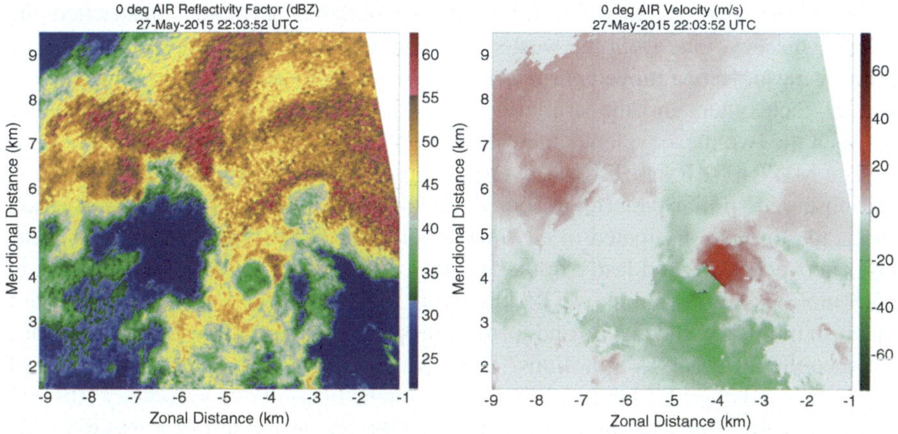

Fig. 12 Reflectivity factor (left; dBZ) and radial velocity (right; m s^{-1}) data from the Atmospheric Imaging Radar (AIR) during the 27 May 2015 Canadian, Texas tornado. The AIR collected 6-s volumes during this case

Rapid changes in tornado wind structure and radar reflectivity signatures have been identified from rapid-scan radar data. Rapid-scan radars have revealed changes in tornado structure on different scales [216], abrupt shifts in the tornado track and speed [64, 102], and evolution of subvortices [174, 193, 216]. During the Shawnee, Oklahoma EF-4 tornado, AIR data showed apparent, localized debris lofting on the right side of the tornado's track evident by reflectivity maxima that ascended with time. Using AIR data, a rapid transition from a large diameter tornado with a 1-km-wide WEH (Fig. 12) to a smaller diameter tornado with disorganized WEH was observed in just 20 s [103]. The Rapid-Scan DOW captured a complex track of a persistent subvortex including several loops and subvortex translation speeds exceeding 80 m s^{-1} [216]. Recently, the first rapid-scan, dual-Doppler analysis was conducted using rapid-scan DOW and RaXpol data [193], revealing a transition between a single-cell and a larger-swirl, multiple vortex structure. The analysis also provided sufficient spatial resolution to resolve 95 m s^{-1} winds within subvortices.

Rapid-scan radars have observed complex interactions between tornadoes and mesocyclones, resulting in tornado and mesocyclone intensity changes. Analyses of MWR-05XP data documented the merger of two tornadoes with similar TVS intensities [63, 66]. Prior to the merger, the original tornado weakened and moved away from the prior updraft while a second tornado formed in a region of strong convergence to its northwest near the intersection of the RFGF and forward-flank gust front. During the merger process, the TVSs rotated around each other in a helical manner (so-called Fujiwhara effect). After the merger, the single TVS was more intense than either of the two preceding TVSs. This interaction occurred during a period of less than 1 min, and its evolution could not have been documented without sub-minute temporal resolution radar data [66]. Using data from the AIR, the rapid development of a mesocyclone and its interaction with the primary mesocyclone were detailed [103]. As the second mesocyclone approached the primary mesocyclone from the west side, the primary mesocyclone weakened. The primary mesocyclone moved north as the two mesocyclones approached and rotated around each other, and the primary mesocyclone reintensified after the distance between the two mesocyclones increased.

The MWR-05XP and RaXpol were also used to document tornado dissipation and observed similar patterns in four tornado cases. For two MWR-05XP tornado cases, the tornado dissipated in an "inside out" process, first at 1.5 km, then in the layer above 1.5 km, and lastly in the lowest 1.5 km [64]. They noted differences in TVS motions below and above 1.5 km where the TVS moved to the right and left of storm motion, respectively, and tornado dissipation occurred first at this interface between the different TVS motions. They speculated that movement to the right of storm motion below 1.5 km resulted from an intensification of a secondary RFGFS, and it appeared that the initial dissipation height was at the top of a RFGFS. The upper part of the tornado then weakened as the inflow was cut off followed by a full occlusion near the surface. This dissipation process occurred over a period of about 3 min. Using higher-resolution RaXpol data from the 2011 El Reno tornado, the dissipation of an EF-3 tornado was documented [79]. They observed that the tornado weakened within a 3 min period in an "inside out" manner, with initial weakening occurring at 1–3 km. Tornado dissipation, however, occurred within about 30 s during which tornado intensity decreased at all heights.

5 Unsolved Tornado Research Questions and Future Radar Technology

This section discusses several unsolved tornado research areas and needed improvements in radar analysis techniques. The section concludes by presenting future radar systems that will lead to new scientific and engineering advancements to address these research areas and improve tornado forecasting.

5.1 Unsolved Tornado Research Questions

Although much progress has been made in tornadogenesis research to date and several compelling tornadogenesis theories have been developed from observational and numerical modeling studies, scientific understanding of tornadogenesis remains incomplete. Thus, it is challenging to assess why some supercells produce tornadoes while others do not, often within similar environments. Storm-scale processes leading to intensification of low-level rotation remain debated, and multiple processes may be involved at different spatiotemporal scales. In particular, the extent to which baroclinic generation of horizontal vorticity contributes to tornadogenesis remains contested, including which regions of the supercell produce substantial baroclinically generated vorticity. Other studies have noted that tornadogenesis can proceed without baroclinic generation of horizontal vorticity. Thus, to test these hypotheses, it is critical to combine the three-dimensional kinematic fields obtained from radar data with thermodynamic or microphysical information from complementary arrays of instrumentation (e.g., mobile mesonets, aircraft or UAVs, soundings, disdrometers) to understand how these dynamic, thermodynamic, and microphysical processes evolve during tornadogenesis.

The defining characteristics of tornadoes vary substantially among cases, including their diameters, velocities, and durations. Thus, it might be surmised that tornadogenesis processes differ among cases, and subsequent intensification and dissipation mechanisms may be case dependent. While a single, unifying theory of tornado processes might be desirable, a spectrum of processes are likely involved in tornadogenesis, intensification, maintenance, and dissipation. Thus, it is worthwhile to explore how tornado processes vary depending on these characteristics (e.g., weak vs. violent tornadoes). To date, a small sample size of rapid-scan radar observations have been collected, and thus more tornado cases are needed from rapid-scan radars to determine the tornadogenesis, intensification, and dissipation processes over the full spectrum of tornado characteristics. Rapid-scan, polarimetric radars can help examine how microphysical processes evolve during tornadogenesis, as well as the relationships between the polarimetric radar signatures and kinematic characteristics. Such studies could illuminate differences between tornadic and nontornadic storms, which may aid forecasters in providing more accurate and timely tornado warnings.

Advancements in numerical modeling have enabled high-resolution, storm-scale data assimilation of radar data and other measurements within supercells. Since polarimetric radar measurements are closely linked to the size distributions of hydrometeors, polarimetric data assimilation may improve microphysical analyses of supercells and assessments of microphysical evolution at different stages of supercell and tornado lifecycles. For such studies, microphysics schemes used in Numerical Weather Prediction (NWP) models must be considered carefully [99] and likely require two- or three-moment microphysics schemes or computationally intensive bin microphysics because more simplistic representations of hydrometeor distributions (e.g., single-moment) do not adequately represent unique hydrometeor size distributions in supercells. Data assimilation of other platforms, such as mobile mesonets, UAVs, soundings, etc., will also be helpful for providing more realistic supercell simulations needed to examine tornado-related processes. At present, state-of-the-art NWP models are capable of conducting nested grid simulations sufficient to coarsely resolve tornado-scale processes with 25–100 m resolution [161]. Continuing advancements in supercomputing capabilities will enable more frequent supercell simulations with sufficient resolution to resolve tornado-scale processes, and eventually these very high-resolution model resolutions may be available in operational forecast models.

The highest resolution tornado simulations are obtained from Large-Eddy Simulation models, resolving scales as small as 1–3 m. Although much has been learned through very high resolution tornado simulations and mobile radar data, the integration of the two could exploit the respective benefits of each. For example, Doppler radar data can not easily resolve the tornado's interaction with the surface or the corner flow region, but can characterize real cases and three-dimensional asymmetries. To date, tornado LES model experiments have been limited to idealized, axisymmetric boundary conditions for initialization, and thus these experiments do not account for a realistic, three-dimensional parent thunderstorm. Thus, methods are needed to incorporate high-resolution model radar data into LES models to enable "real" LES experiments. Such experiments could commence by using axisymmetric analyses (e.g., GBVTD) to initialize the lateral and upper boundary conditions of the LES model (Fig. 3). Then, data assimilation experiments could be conducted using mobile radar data.

Several improvements are still needed to improve the accuracy of radar analysis techniques used to study supercells and tornadoes. Very accurate three-dimensional wind fields are needed in supercells and tornadoes to understand tornadogenesis and tornado dynamics and to conduct trajectory analyses to assess vorticity generation. To date, dual-Doppler methods used to obtain three-dimensional winds in supercell thunderstorms (and occasionally tornadoes) have not been extensively validated. Therefore, their accuracy is poorly understood. Inadequate measurements of the low-level mass flux and non-linear advection are two challenging issues that can cause large errors, although dual-Doppler measurements between rapid-scan radars may help address the latter. Vertically pointing radar measurements may be helpful to measure vertical motions in supercells. In particular, vertically pointing measurements exploiting the large transmit beam of the AIR might be particularly helpful for obtaining volumetric measurements of vertical motion.[3] A network of vertically pointing frequency-modulated continuous wave (FMCW) radars might also provide a relatively inexpensive way to obtain multiple vertical velocity profiles in supercells, as well as improved microphysical retrievals.

5.2 *Upcoming Ground-Based Radar Technologies for Tornado Observations*

Although there is currently a plethora of tornado-scale and rapid-scanning radars in the United States, barriers are constantly being broken in order to develop the next generation of weather radars. There are three primary rapid-scanning radars that are currently under development at the time of this writing: the PX-10000 (or PX-10k; an updated version of the PX-1000), the Advanced Technology Demonstrator (ATD), and the Polarimetric Atmospheric Imaging Radar (PAIR). The first of these expected to be completed is the PX-10k, with completion expected in mid-2017.

[3]Such measurements would still likely require an additional radar sampling at a large cross beam angle.

PX-10k will be an X-band platform on a small trailer with an integrated generator for remote-area operations. With 100 MHz of available chirp bandwidth for pulse compression, along with 800-W solid-state transmitters for each polarization, PX-10k will be capable of impressive range resolutions for a weather radar while maintaining exceptional sensitivity of 10 dBZ at 50 km range. PX-10k will feature a slightly larger parabolic dish compared to PX-1000, resulting in a 3-dB beamwidth of 1.4° (at the nominal center frequency of 9.25 GHz). Additionally, the dish will be capable of a 180° s^{-1} or faster rotation rate, resulting in single-elevation updates in less than 2 s. PX-10k will also be able to alter its center frequency between 8.9–9.6 GHz. When also considering the chirp bandwidth, this capability can allow PX-10k to fit into a mobile radar field program dynamically based on its position relative to other X-band radars and their frequencies, making PX-10k exceptionally flexible.

The second expected system, the ATD (Fig. 13), is an NSSL-acquired fixed-site replacement for the NWRT in Norman, Oklahoma. The ATD was co-designed by MIT Lincoln Laboratory, NSSL, and General Dynamics, and will be delivered to NSSL from Lincoln Laboratory for installation completion by mid-2018. The ATD features an active phased array operating at S band, and is a dual-polarization proving ground for the future of phased-array weather radar. The ATD will be used to investigate polarimetry and multi-functionality in a combined weather and aviation radar, serving as one potential solution (MPAR) to the Spectrum Efficient National Surveillance Radar (SENSR) program. The ATD will feature a 1.7° beamwidth at broadside using 76 64-element panels to form a 4 m × 4 m array,

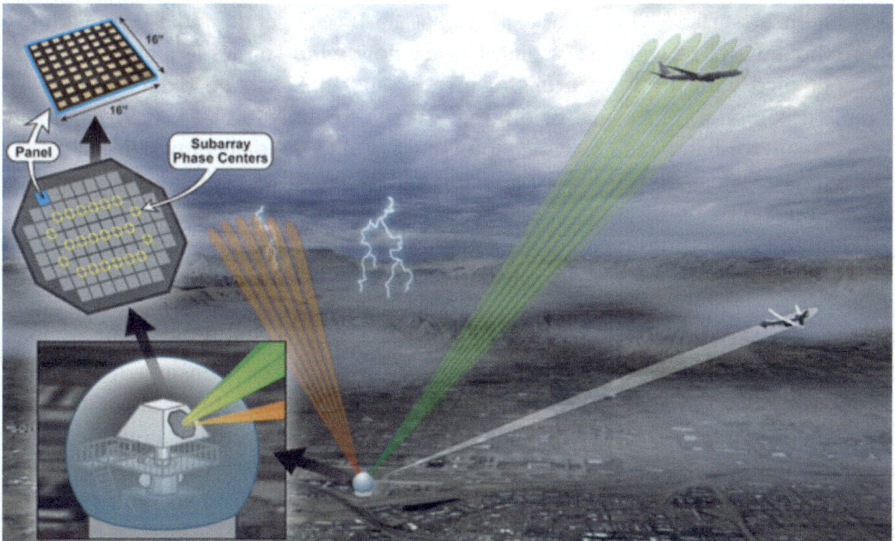

Fig. 13 Diagram of the Multifunction Phased Array Radar (MPAR) and Advanced Technology Demonstrator (ATD) concept. Figure provided by MIT Lincoln Laboratory

making it a relatively suitable tool for considering future full-scale weather radar in the United States. Four thousand eight hundred and sixty four total elements at 6 W of peak power each will result in a roughly 30 kW total peak power (without array weighting), and thus pulse compression will likely be needed to obtain the desired sensitivity and range resolution required for the ATD. Due to its phased-array design, a focus on rapid scanning and interleaved missions with adaptive capabilities will be possible.

Finally, the PAIR, an OU ARRC-led effort funded by the NSF and OU, will be a C-band mobile polarimetric imaging radar that takes lessons learned from the current AIR and applies polarimetric capabilities and a different frequency to a new design. The PAIR is expected to be completed in 2019 and will feature a 6.4-kW solid-state transmitter with the capability for simultaneous or alternating dual-polarimetry. The array will be large enough to accommodate a roughly 2° beamwidth at C band, but without much antenna motion, there will be little beam smearing. There will be enough chirp bandwidth to achieve range resolutions better than 10 m, and enough sensitivity to achieve −17 dBZ at 10 km range. The $90° \times 20°$ update rate will be on the order of 2.5 s. The C-band design of the PAIR will allow for significantly better attenuation characteristics than the existing X-band rapid-scan radars, and will also provide a second frequency for dual-Doppler polarimetric observations of tornadoes, leading to new scientific frontiers for understanding tornadic debris at various wavelengths. C band will also help supplement the SMART-R radars, only one of which is dual-polarization, and will allow for additional mission types including hurricane deployments.

Although rapid-scan X-band radars are often used to study tornadoes, the 1° beamwidths of these systems limit their ability to observe smaller diameter tornadoes and near-surface winds. For example, at a 5-km range, the radar beamwidth is approximately 87-m wide at X band. Corner flow winds in numerical simulations are contained with the lowest tens of meters, and thus X-band radars will rarely sample this region. Finer beamwidth radars exist, such as the UMass W-band and Texas Tech Ka-band radars, which provide 16- and 28-m beamwidths at 5 km, respectively. Thus, these finer beamwidth radars are likely capable of sampling corner flow regions of tornadoes. However, they currently lack rapid-scan and polarimetric capabilities, and are very susceptible to attenuation and even complete signal extinction. To address this need, the development of rapid-scan Ka- or W-band radars would provide a more complete understanding of tornado dynamics by meeting both the temporal and spatial resolution required for such studies. Shorter wavelengths are also less sensitive to debris and more sensitive to hydrometeors, providing smaller debris centrifuging errors and thus more accurate wind speed measurements [25]. Finally, new technological advancements could leverage dual- or multiple-frequency radar measurements to study tornadoes, providing more detailed information about scatterer characteristics [23] and mapping the three-dimensional structure of the debris centrifuging bias [25]. Dual- or multiple-frequency radar measurements would also provide additional information needed to obtain hydrometeor characteristics and distributions in supercells.

Although many of these technological advancements are currently limited to research radar systems, the eventual goal is to move toward a nationwide network of rapid-scan, polarimetric radars that can leverage the scientific and engineering advancements emerging from these systems. The scientific studies will improve understanding of tornado-related processes, such as tornadogenesis, and hopefully decrease tornado false alarm rates and increase tornado warning lead times. The scientific insight from these radars provides a glimpse at the future capabilities of an operational rapid-scan, polarimetric radar network.

Acknowledgements David Bodine is supported by National Science Foundation grant AGS-1303685. James Kurdzo is currently an MIT Lincoln Laboratory employee; no Laboratory funding or resources were used to produce the results/findings reported in this publication. The authors thank Casey Griffin for providing dual-Doppler data from 10 May 2010, and reviewing an early draft of the chapter.

References

1. Alexander, C.R., and J. Wurman. 2005. The 30 May 1998 Spencer, South Dakota, storm. Part I: The structural evolution and environment of the tornadoes. *Monthly Weather Review* 133 (1): 72–97.
2. Atkins, N.T., A. McGee, R. Ducharme, R.M. Wakimoto, and J. Wurman. 2012. The LaGrange tornado during VORTEX2. Part II: Photogrammetric analysis of the tornado combined with dual-Doppler radar data. *Monthly Weather Review* 140 (9): 2939–2958.
3. Beck, J.R., J.L. Schroeder, and J.M. Wurman. 2006. High-resolution dual-Doppler analyses of the 29 May 2001 Kress, Texas, cyclic supercell. *Monthly Weather Review* 134 (11): 3125–3148.
4. Bedard, A.J., and C. Ramzy. 1983. Surface meteorological observations in severe thunderstorms. Part I: Design details of TOTO. *Journal of Climate and Applied Meteorology* 22 (5): 911–918.
5. Benjamin, T.B. 1962. Theory of the vortex breakdown phenomenon. *Journal of Fluid Mechanics* 14: 593–629.
6. Bieringer, P., and P.S. Ray. 1996. A comparison of tornado warning lead times with and without NEXRAD Doppler radar. *Weather and Forecasting* 11: 47–52.
7. Biggerstaff, M.I., L.J. Wicker, J. Guynes, C. Ziegler, J.M. Straka, E.N. Rasmussen, A. Doggett, L.D. Carey, J.L. Schroeder, and C. Weiss. 2005. The shared mobile atmospheric research and teaching radar: a collaboration to enhance research and teaching. *Bulletin of the American Meteorological Society* 86 (9): 1263–1274.
8. Biglerm, S.G. 1956. A note on the successful identification and tracking of a tornado by radar. *Weatherwise* 9 (6): 198–201.
9. Bluestein, H.B. 1983. Surface meteorological observations in severe thunderstorms. Part II: Field experiments with TOTO. *Journal of Climate and Applied Meteorology* 22 (5): 919–930.
10. Bluestein, H.B. 1999. A history of severe-storm-intercept field programs. *Weather and Forecasting* 14 (4): 558–577.
11. Bluestein, H.B. 2013. *Severe Convective Storms and Tornadoes: Observations and Dynamics*. Berlin: Springer.
12. Bluestein, H.B., and A.L. Pazmany. 2000. Observations of tornadoes and other convective phenomena with a mobile, 3-mm wavelength, Doppler radar: the spring 1999 field experiment. *Bulletin of the American Meteorological Society* 81 (12): 2939–2951.

13. Bluestein, H.B., and W.P. Unruh. 1989. Observations of the wind field in tornadoes, funnel clouds, and wall clouds with a portable Doppler radar. *Bulletin of the American Meteorological Society* 70 (12): 1514–1525.
14. Bluestein, H.B., and G.R. Woodall. 1990. Doppler-radar analysis of a low-precipitation severe storm. *Monthly Weather Review* 118 (8): 1640–1665.
15. Bluestein, H.B., E.W. McCaul, G.P. Byrd, and G.R. Woodall. 1988. Mobile sounding observations of a tornadic storm near the dryline: the Canadian, Texas storm of 7 May 1986. *Monthly Weather Review* 116 (9): 1790–1804.
16. Bluestein, H.B., J.G. Ladue, H. Stein, D. Speheger, and W.F. Unruh. 1993. Doppler radar wind spectra of supercell tornadoes. *Monthly Weather Review* 121 (8): 2200–2222.
17. Bluestein, H.B., C.C. Weiss, and A.L. Pazmany. 2003. Mobile Doppler radar observations of a tornado in a supercell near Bassett, Nebraska, on 5 June 1999. Part I: Tornadogenesis. *Monthly Weather Review* 131 (12): 2954–2967.
18. Bluestein, H.B., W.C. Lee, M. Bell, C.C. Weiss, and A.L. Pazmany. 2003. Mobile Doppler radar observations of a tornado in a supercell near Bassett, Nebraska, on 5 June 1999. Part II: Tornado-vortex structure. *Monthly Weather Review* 131: 2968–2984.
19. Bluestein, H.B., C.C. Weiss, and A.L. Pazmany. 2004. The vertical structure of a tornado near Happy, Texas, on 5 May 2002: high-resolution, mobile, W-band, Doppler radar observations. *Monthly Weather Review* 132 (10): 2325–2337.
20. Bluestein, H.B., M.M. French, R.L. Tanamachi, S. Frasier, K. Hardwick, F. Junyent, and A.L. Pazmany. 2007. Close-range observations of tornadoes in supercells made with a dual-polarization, X-band, mobile Doppler radar. *Monthly Weather Review* 135 (4): 1522–1543.
21. Bluestein, H.B., M.M. French, I. PopStefanija, R.T. Bluth, and J.B. Knorr. 2010. A mobile, phased-array Doppler radar for the study of severe convective storms. *Bulletin of the American Meteorological Society* 91 (5): 579–600.
22. Bodine, D.J., M.R. Kumjian, R.D. Palmer, P.L. Heinselman, and A.V. Ryzhkov. 2013. Tornado damage estimation using polarimetric radar. *Weather and Forecasting* 28 (1): 139–158.
23. Bodine, D.J., R.D. Palmer, and G. Zhang. 2014. Dual-wavelength polarimetric radar analyses of tornadic debris signatures. *Journal of Applied Meteorology and Climatology* 53: 242–261.
24. Bodine, D.J., T. Maruyama, R.D. Palmer, C.J. Fulton, H.B. Bluestein, and D.C. Lewellen. 2016. Sensitivity of tornado dynamics to debris loading. *Journal of the Atmospheric Sciences* 73: 2783–2801.
25. Bodine, D.J., R.D. Palmer, T. Maruyama, C.J. Fulton, Y. Zhu, and B.L. Cheong. 2016. Simulated frequency dependence of radar observations of tornadoes. *Journal of Atmospheric and Oceanic Technology* 33 (9): 1825–1842.
26. Bowden, K.A., and P.L. Heinselman. 2016. A qualitative analysis of NWS forecasters' use of phased-array radar data during severe hail and wind events. *Weather and Forecasting* 31: 43–55.
27. Bowden, K.A., P.L. Heinselman, D.M. Kingfield, and R.P. 2015. Thomas. Impacts of phased-array radar data on forecaster performance during severe hail and wind events. *Weather and Forecasting* 30: 389–404.
28. Bradford, M. 2001. *Scanning the Skies: A History of Tornado Forecasting*. Norman: University of Oklahoma Press.
29. Brandes, E.A. 1977. Flow in severe thunderstorms observed by dual-Doppler radar. *Monthly Weather Review* 105 (1): 113–120.
30. Brandes, E.A. 1984. Vertical vorticity generation and mesocyclone sustenance in tornadic thunderstorms: the observational evidence. *Monthly Weather Review* 112: 2253–2269.
31. Brotzge, J., and Donner, W. 2013. The tornado warning process: a review of current research, challenges, and opportunities. *Bulletin of the American Meteorological Society* 94 (11): 1715–1733.
32. Brown, R.A., D.W. Burgess, J.K. Carter, L.R. Lemon, and D. Sirmans. 1975. NSSL dual-Doppler radar measurements in tornadic storms: a preview. *Bulletin of the American Meteorological Society* 56 (5): 524–526.

33. Brown, R.A., L.R. Lemon, and D.W. Burgess. 1978. Tornado detection by pulsed Doppler radar. *Monthly Weather Review* 106: 29–38.
34. Brown, R.A., J.M. Janish, and V.T. Wood. 2000. Impact of WSR-88D scanning strategies on severe storm algorithms. *Weather and Forecasting* 15 (1): 90–102.
35. Brown, R.A., V.T. Wood, and D. Sirmans. 2000. Improved WSR-88D scanning strategies for convective storms. *Weather and Forecasting* 15 (2): 208–220.
36. Brown, R.A., V.T. Wood, and D. Sirmans. 2002. Improved tornado detection using simulated and actual WSR-88D data with enhanced resolution. *Journal of Atmospheric and Oceanic Technology* 19 (11): 1759–1771.
37. Brown, R.A., V.T. Wood, R.M. Steadham, R.R. Lee, B.A. Flickinger, and D. Sirmans. 2005. New WSR-88D volume coverage pattern 12: results of field tests. *Weather and Forecasting* 20: 385–393.
38. Burgess, D.W. 1976. Single Doppler radar vortex recognition: Part I. Mesocyclone signatures. In *17th Radar Meteorology Conference*, Seattle, WA, 97–103. Boston: American Meteorological Society (Preprints).
39. Burgess, D.W., L.R. Lemon, and R.A. Brown. 1975. Tornado characteristics revealed by Doppler radar. *Geophysical Research Letters* 2 (5): 183–184.
40. Burgess, D.W., V.T. Wood, and R.A. Brown. 1982. Mesocyclone evolution statistics. In *12th Conference on Severe Local Storms*, San Antonio, TX, 422–424. Boston: American Meteorological Society (Preprints).
41. Burgess, D.W., M.A. Magsig, J. Wurman, D.C. Dowell, and Y. Richardson. 2002. Radar observations of the 3 May 1999 Oklahoma City tornado. *Weather and Forecasting* 17 (3): 456–471.
42. Burgess, D.W., E.R. Mansell, C.M. Schwarz, and B.J. Allen. 2010. Tornado and tornadogenesis events seen by the NOXP, X-band, dual-polarization radar during VORTEX2 2010. In *25th Conference on Severe Local Storms*. Boston: American Meteorological Society.
43. Cheong, B.L., R.D. Palmer, C.D. Curtis, T.Y. Yu, and D.S. Zrnić, and D. Forsyth. 2008. Refractivity retrieval using the phased-array radar: first results and potential for multimission operation. *IEEE Transactions on Geoscience and Remote Sensing* 46 (9): 2527–2537.
44. Cheong, B.L., R. Kelley, R.D. Palmer, Y. Zhang, M. Yeary, and T.Y. Yu. 2013. PX-1000: a solid-state polarimetric X-band weather radar and time-frequency multiplexed waveform for blind range mitigation. *IEEE Transactions on Instrumentation and Measurement* 62 (11): 3064–3072.
45. Cheong, B.L., D.J. Bodine, C.J. Fulton, S.M. Torres, T. Maruyama, and R.D. Palmer. 2017. SimRadar: a polarimetric radar time-series simulator for tornadic debris studies. *IEEE Transactions on Geoscience and Remote Sensing* 55: 2858–2870.
46. Church, C.R., J.T. Snow, G.L. Baker, and E.M. Agee. 1979. Characteristics of tornado-like vortices as a function of swirl ratio: a laboratory investigation. *Journal of the Atmospheric Sciences* 36: 1755–1776.
47. Collier, C.G., C.A. Fair, and D.H. Newsome. 1988. International weather-radar networking in western Europe. *Bulletin of the American Meteorological Society* 69: 16–21.
48. Crum, T., and Alberty, R. 1993. The WSR-88D and the WSR-88D operational support facility. *Bulletin of the American Meteorological Society* 74 (9): 1669–1687.
49. Crum, T., S. Smith, J. Chrisman, R. Vogt, M. Istok, R. Hall, B. Saffle. 2013. WSR-88D radar projects 2013 update. In *29th Conference on EIPT*, Austin, TX
50. Daniel, A.E., J.N. Chrisman, S.D. Smith, and M.W. Miller. 2014. New WSR-88D operational techniques: responding to recent weather events. In *30th Conference on Environmental Information Processing Technologies*, Atlanta, GA. Boston: American Meteorological Society.
51. Davies-Jones, R.P. 1973. The dependence of core radius on swirl ratio in a tornado simulator. *Journal of the Atmospheric Sciences* 30: 1427–1430.
52. Davies-Jones, R. 2008. Can a descending rain curtain in a supercell instigate tornadogenesis barotropically? *Journal of the Atmospheric Sciences* 65: 2469–2497.
53. Davies-Jones, R., R.J. Trapp, H.B. Bluestein. 2001. Tornadoes and tornadic storms. In *Severe Convective Storms*, ed. C.A. Doswell, 167–222. Boston: American Meteorological Society.

54. Doviak, R.J., and D.S. Zrnić. 1993. *Doppler Radar and Weather Observations*. 2nd ed. New York: Dover Publications.
55. Doviak, R.J., V. Bringi, A. Ryzhkov, A. Zahrai, and D. Zrnić. 2000. Considerations for polarimetric upgrades to operational WSR-88D radars. *Journal of Atmospheric and Oceanic Technology* 17: 257–278.
56. Dowell, D.C., and H.B. Bluestein. 2002. The 8 June 1995 Mclean, Texas, storm. Part I: Observations of cyclic tornadogenesis. *Monthly Weather Review* 130 (11): 2626–2648.
57. Dowell, D.C., and H.B. Bluestein. 2002. The 8 June 1995 Mclean, Texas, storm. Part II: Cyclic tornado formation, maintenance, and dissipation. *Monthly Weather Review* 130 (11): 2649–2670.
58. Dowell, D.C., C.R. Alexander, J.M. Wurman, and L.J. Wicker. 2005. Centrifuging of hydrometeors and debris in tornadoes: radar-reflectivity patterns and wind-measurement errors. *Monthly Weather Review* 133 (6): 1501–1524.
59. Evans, J., and D. Turnbull. 1989. Development of an automated windshear detection system using Doppler weather radar. *Proceedings of the IEEE* 77: 1661–1673.
60. Fiedler, B.H. 1998. Wind-speed limits in numerically simulated tornadoes with suction vortices. *Quarterly Journal of the Royal Meteorological Society* 124: 2377–2392
61. Fiedler, B.H., and R. Rotunno. 1986. A theory for the maximum windspeeds in tornado-like vortices. *Journal of the Atmospheric Sciences* 43: 2328–2340.
62. French, M.M., H.B. Bluestein, D.C. Dowell, L.J. Wicker, M.R. Kramar, and A.L. Pazmany. 2008. High-resolution, mobile Doppler radar observations of cyclic mesocyclogenesis in a supercell. *Monthly Weather Review* 136 (12): 4997–5016.
63. French, M.M., H.B. Bluestein, I. PopStefanija, C.A. Baldi, and R.T. Bluth. 2013. Reexamining the vertical development of tornadic vortex signatures in supercells. *Monthly Weather Review* 141 (12), 4576–4601.
64. French, M.M., H.B. Bluestein, I. PopStefanija, C.A. Baldi, and R.T. Bluth. 2014. Mobile, phased-array, Doppler radar observations of tornadoes at X band. *Monthly Weather Review* 142 (3): 1010–1036.
65. French, M.M., D.W. Burgess, E.R. Mansell, and L.J. Wicker. 2015. Bulk hook echo raindrop sizes retrieved using mobile, polarimetric Doppler radar observations. *Journal of Applied Meteorology and Climatology* 54 (2): 423–450.
66. French, M.M., P.S. Skinner, L.J. Wicker, and H.B. Bluestein. 2015. Documenting a rare tornado merger observed in the 24 May 2011 El Reno–Piedmont, Oklahoma, supercell. *Monthly Weather Review* 143 (8): 3025–3043.
67. Fujita, T.T. 1974. Jumbo tornado outbreak of 3 April 1974. *Weatherwise* 27 (3), 116–126.
68. Fujita, T.T., D.L. Bradbury, and C.F.V. Thullenar. 1970. Palm Sunday tornadoes of April 11, 1965. *Monthly Weather Review* 98: 29–69.
69. Gal-Chen, T. 1982. Errors in fixed and moving frame of reference: application for conventional and Doppler radar analysis. *Journal of the Atmospheric Sciences* 39: 2279–2300.
70. Geerts, B., D. Parsons, C.L. Ziegler, T.M. Weckwerth, M.I. Biggerstaff, R.D. Clark, M.C. Coniglio, B.B. Demoz, R.A. Ferrare, W.A. Gallus, K. Haghi, J.M. Hanesiak, P.M. Klein, K.R. Knupp, K. Kosiba, G.M. McFarquhar, J.A. Moore, A.R. Nehrir, M.D. Parker, J.O. Pinto, R.M. Rauber, R.S. Schumacher, D.D. Turner, Q. Wang, X. Wang, Z. Wang, J. Wurman. 2017. The 2015 plains elevated convection at night field project. *Bulletin of the American Meteorological Society* 98 (4): 767–786.
71. Golden, J.H., and B.J. Morgan. 1972. NSSL Notre Dame tornado intercept program, Spring 1972. *Bulletin of the American Meteorological Society* 53 (12): 1178–1180.
72. Goliger, A.M., and R.V. Milford. 1998. A review of worldwide occurrences of tornadoes. *Journal of Wind Engineering and Industrial Aerodynamics* 74–76: 111–121.
73. Grazilus, T. 1993. Significant Tornadoes 1680–1991. The Tornado Project of Environmental Films.
74. Griffin, C.B., D.J. Bodine, and R.D. Palmer. 2017. Kinematic and polarimetric radar observations of the 10 May 2010, Moore-Choctaw, Oklahoma tornadic debris signature. *Monthly Weather Review* 145: 2723–2741.

75. Groenemeijer, P., and T. Kühne. 2014. A climatology of tornadoes in Europe: results from the European severe weather database. *Monthly Weather Review* 142 (12): 4775–4790.
76. Heinselman, P.L., and S.M. Torres. 2011. High-temporal-resolution capabilities of the national weather radar testbed phased-array radar. *Journal of Applied Meteorology and Climatology* 50 (3): 579–593.
77. Heinselman, P.L., D.L. Priegnitz, K.L. Manross, T.M. Smith, and R.W. Adams. 2008. Rapid sampling of severe storms by the national weather radar testbed phased array radar. *Weather and Forecasting* 23 (5): 808–824.
78. Hirth, B.D., J.L. Schroeder, S.W. Gunter, and J.G. Guynes. 2012. Measuring a utility scale turbine wake using the TTUKa mobile research radars. *Journal of Atmospheric and Oceanic Technology* 29 (6): 765–771.
79. Houser, J.B., H.B. Bluestein, and J.C. Snyder. 2015. Rapid-scan, polarimetric, Doppler-radar observations of tornadogenesis and tornado dissipation in a tornadic supercell: the "El Reno, Oklahoma" storm of 24 May 2011. *Monthly Weather Review* 143 (7): 2685–2710.
80. Houser, J.L., H.B. Bluestein, and J.C. Snyder. 2017. A fine-scale radar examination of the tornadic debris signature and weak-echo reflectivity band associated with a large, violent tornado. *Monthly Weather Review* 144: 4101–4130.
81. Isom, B., R. Palmer, R. Kelley, J. Meier, D. Bodine, M. Yeary, B.L. Cheong, Y. Zhang, T.Y. Yu, and M.I. Biggerstaff. 2013. The atmospheric imaging radar: simultaneous volumetric observations using a phased array weather radar. *Journal of Atmospheric and Oceanic Technology* 30 (4): 655–675.
82. Joe, P., and S. Lapczak. 2002. Evolution of the Canadian operational radar network. In *Proceedings of ERAD*, Delft, Netherlands, 370–382.
83. Johnson, J.T., P.L. MacKeen, A. Witt, E.D.W. Mitchell, G.J. Stumpf, M.D. Eilts, and K.W. Thomas. 1998. The storm cell identification and tracking algorithm: an enhanced WSR-88D algorithm. *Weather and Forecasting* 13 (2): 263–276.
84. Karstens, C.D., T.M. Samaras, B.D. Lee Jr., W.A. Gallus, C.A. Finley. 2010. Near-ground pressure and wind measurements in tornadoes. *Monthly Weather Review* 138 (7): 2570–2588.
85. Klemp, J.B., and R. Rotunno. 1983. A study of the tornadic region within a supercell thunderstorm. *Journal of the Atmospheric Sciences* 40: 359–377.
86. Klemp, J.B., and R.B. Wilhelmson. 1978. The simulation of three-dimensional convective storms dynamics. *Journal of the Atmospheric Sciences* 35: 1070–1096.
87. Klemp, J.B., R.B. Wilhelmson, and P.S. Ray. 1981. Observed and numerically simulated structure of a mature supercell thunderstorm. *Journal of the Atmospheric Sciences* 38: 1558–1580.
88. Knupp, K.R., J. Walters, and M. Biggerstaff. 2006. Doppler profiler and radar observations of boundary layer variability during the landfall of Tropical Storm Gabrielle. *Journal of the Atmospheric Sciences* 63 (1): 234–251.
89. Knupp, K.R., T.A. Murphy, T.A. Coleman, R.A. Wade, S.A. Mullins, C.J. Schultz, E.V. Schultz, L. Carey, A. Sherrer, E.W. McCaul, B. Carcione, S. Latimer, A. Kula, K. Laws, P.T. Marsh, K. Klockow. 2014. Meteorological overview of the devastating 27 April 2011 tornado outbreak. *Bulletin of the American Meteorological Society* 95 (7): 1041–1062.
90. Kosiba, K., and J. Wurman. 2010. The three-dimensional axisymmetric wind field structure of the Spencer, South Dakota, 1998 tornado. *Journal of the Atmospheric Sciences* 67 (9): 3074–3083.
91. Kosiba, K.A., and J. Wurman. 2013. The three-dimensional structure and evolution of a tornado boundary layer. *Weather and Forecasting* 28 (6): 1552–1561.
92. Kosiba, K., J. Wurman, Y. Richardson, P. Markowski, P. Robinson, and J. Marquis. 2013. Genesis of the Goshen County, Wyoming, tornado on 5 June 2009 during VORTEX2. *Monthly Weather Review* 141 (4): 1157–1181.
93. Kumjian, M.R. 2011. Precipitation properties of supercell hook echoes. *Electronic Journal of Severe Storms Meteorology* 6 (5): 1–21.
94. Kumjian, M.R. 2013. Principles and applications of dual-polarization weather radar. Part I: Description of the polarimetric radar variables. *Journal of Operational Meteorology* 1: 226–242.

95. Kumjian, M.R. 2013. Principles and applications of dual-polarization weather radar. Part II: Warm- and cold-season applications. *Journal of Operational Meteorology* 1: 243–264.
96. Kumjian, M.R. 2013. Principles and application of dual-polarization weather radar. Part III: Artifacts. *Journal of Operational Meteorology* 1: 265–274.
97. Kumjian, M.R., and A.V. Ryzhkov. 2008. Polarimetric signatures in supercell thunderstorms. *Journal of Applied Meteorology and Climatology* 48: 1940–1961.
98. Kumjian, M.R., A.V. Ryzhkov, V.M. Melnikov, and T.J. Schuur. 2010. Rapid-scan super-resolution observations of a cyclic supercell with a dual-polarization WSR-88D. *Monthly Weather Review* 138 (10): 3762–3786.
99. Kumjian, M.R., Z.J. Lebo, and H.C. Morrison. 2015. On the mechanisms of rain formation in an idealized supercell storm. *Monthly Weather Review* 143: 2754–2773
100. Kurdzo, J.M., and R.D. Palmer. 2012. Objective optimization of weather radar networks for low-level coverage using a genetic algorithm. *Journal of Atmospheric and Oceanic Technology* 29 (6): 807–821.
101. Kurdzo, J.M., B.L. Cheong, R.D. Palmer, G. Zhang, and J.B. Meier. 2014. A pulse compression waveform for improved-sensitivity weather radar observations. *Journal of Atmospheric and Oceanic Technology* 31 (12): 2713–2731.
102. Kurdzo, J.M., D.J. Bodine, B.L. Cheong, and R.D. Palmer. 2015. High-temporal resolution polarimetric X-band Doppler radar observations of the 20 May 2013 Moore, Oklahoma, tornado. *Monthly Weather Review* 143 (7): 2711–2735.
103. Kurdzo, J.M., F. Nai, D.J. Bodine, T.A. Bonin, R.D. Palmer, B.L. Cheong, J. Lujan, A. Mahre, A.D. Byrd. 2017. Observations of severe local storms and tornadoes with the atmospheric imaging radar. *Bulletin of the American Meteorological Society* 98 (5): 915–935.
104. Kuster, C.M., P.L. Heinselman, and T.J. Schuur. 2016. Rapid-update radar observations of downbursts occurring within an intense multicell thunderstorm on 14 June 2011. *Weather and Forecasting* 31 (3): 827–851.
105. Lee, R.R., and A. White. 1998. Improvement of the WSR-88D mesocyclone algorithm. *Weather and Forecasting* 13 (2): 341–351.
106. Lee, W.C., and J. Wurman. 2005. Diagnosed three-dimensional axisymmetric structure of the Mulhall tornado on 3 May 1999. *Journal of the Atmospheric Sciences* 62 (7): 2373–2393.
107. Lee, W.C., B.J.D. Jou, P.L. Chang, and S.M. Deng. 1999. Tropical cyclone kinematic structure retrieved from single-Doppler radar observations. Part I: Doppler velocity patterns and the GBVTD technique. *Monthly Weather Review* 127: 2419–2439.
108. Lee, B.D., C.A. Finley, and C.D. Karstens. 2012. The Bowdle, South Dakota, cyclic tornadic supercell of 22 May 2010: surface analysis of rear-flank downdraft evolution and multiple internal surges. *Monthly Weather Review* 140 (11): 3419–3441.
109. Lemon, L.R., and C.A. Doswell. 1979. Severe thunderstorm evolution and mesocyclone structure as related to tornadogenesis. *Monthly Weather Review* 107 (9): 1184–1197.
110. Leslie, L.M. 1971. The development of concentrated vortices: a numerical study. *Journal of Fluid Mechanics* 48 (1): 1–21.
111. Lewellen, D.C., and W.S. Lewellen. 2007. Near-surface intensification of tornado vortices. *Journal of the Atmospheric Sciences* 64: 2176–2194.
112. Lewellen, D.C., and W.S. Lewellen. 2007. Near-surface vortex intensification through corner flow collapse. *Journal of the Atmospheric Sciences* 64: 2195–2209.
113. Lewellen, D.C., W.S. Lewellen, and J. Xia. 2000. The influence of a local swirl ratio on tornado intensification near the surface. *Journal of the Atmospheric Sciences* 57: 527–544.
114. Mahale, V.N., J.A. Brotzge, and H.B. Bluestein. 2014. The advantages of a mixed-band radar network for severe weather operations: a case study of 13 May 2009. *Weather and Forecasting* 29 (1): 78–98.
115. Mahre, A., T.Y. Yu, R. Palmer, and J. Kurdzo. 2017. Observations of a cold front at high spatiotemporal resolution using an X-band phased array imaging radar. *Atmosphere* 8 (2): 30.
116. Majcen, M., P. Markowski, Y. Richardson, D. Dowell, and J. Wurman. 2008. Multipass objective analyses of Doppler radar data. *Journal of Atmospheric and Oceanic Technology* 25 (10): 1845–1858.

117. Markowski, P.M. 2016. An idealized numerical simulation investigation of the effects of surface drag on the development of near-surface vertical vorticity in supercell thunderstorms. *Journal of the Atmospheric Sciences* 73: 4349–4385.
118. Markowski, P.M., and Y.P. Richardson. 2014. What we know and don't know about tornado formation. *Physics Today* 67 (9): 26–31.
119. Markowski, P.M., J.M. Straka, and E.N. Rasmussen. 2002. Direct surface thermodynamic observations within the rear-flank downdrafts of nontornadic and tornadic supercells. *Monthly Weather Review* 130 (7): 1692–1721.
120. Markowski, P., E. Rasmussen, J. Straka, R. Davies-Jones, Y. Richardson, and R.J. Trapp. 2008. Vortex lines within low-level mesocyclones obtained from Psuedo-Dual-Doppler radar observations. *Monthly Weather Review* 136: 3513–3535.
121. Markowski, P., Y. Richardson, J. Marquis, J. Wurman, K. Kosiba, P. Robinson, D. Dowell, E. Rasmussen, and R. Davies-Jones. 2012. The pretornadic phase of the Goshen County, Wyoming, supercell of 5 June 2009 intercepted by VORTEX2. Part I: Evolution of kinematic and surface thermodynamic fields. *Monthly Weather Review* 140 (9): 2887–2915.
122. Markowski, P.M., Y. Richardson, J. Marquis, R. Davies-Jones, J. Wurman, K. Kosiba, P. Robinson, E. Rasmussen, and D. Dowell. 2012. The pretornadic phase of the Goshen County, Wyoming, supercell of 5 June 2009 intercepted by VORTEX2. Part II: Intensification of Low-Level Rotation. *Monthly Weather Review* 140: 2916–2938.
123. Marquis, J., Y. Richardson, J. Wurman, and P. Markowski. 2008. Single- and dual-Doppler analysis of a tornadic vortex and surrounding storm-scale flow in the Crowell, Texas, supercell of 30 April 2000. *Monthly Weather Review* 136 (12): 5017–5043.
124. Marquis, J., Y. Richardson, P. Markowski, D. Dowell, and J. Wurman. 2012. Tornado maintenance investigated with high-resolution dual-Doppler and EnKF analysis. *Monthly Weather Review* 140 (1): 3–27
125. Marquis, J., Y. Richardson, P. Markowski, D. Dowell, J. Wurman, K. Kosiba, P. Robinson, G. Romine. 2014. An investigation of the Goshen County, Wyoming, tornadic supercell of 5 June 2009 using EnKF assimilation of mobile mesonet and radar observations collected during VORTEX2. Part I: Experiment design and verification of the EnKF analyses. *Monthly Weather Review* 142 (2): 530–554.
126. Maruyama, T. 2011. Simulation of flying debris using a numerically generated tornado-like vortex. *Journal of Wind Engineering and Industrial Aerodynamics* 99: 249–256.
127. Mashiko, W., H. Niino, and T. Kato. 2009. Numerical simulation of tornadogenesis in an outer-rainband minisupercell of Typhoon Shanshan on 17 September 2006. *Monthly Weather Review* 137: 4238–4260.
128. McLaughlin, D., D. Pepyne, B. Philips, J. Kurose, M. Zink, D. Westbrook, E. Lyons, E. Knapp, A. Hopf, A. Defonzo, R. Contreras, T. Djaferis, E. Insanic, S. Frasier, V. Chandrasekar, F. Junyent, N. Bharadwaj, Y. Wang, Y. Liu, B. Dolan, K. Droegemeier, J. Brotzge, M. Xue, K. Kloesel, K. Brewster, F. Carr, S. Cruz-Pol, K. Hondl, and P. Kollias. 2009. Short-wavelength technology and the potential for distributed networks of small radar systems. *Bulletin of the American Meteorological Society* 90 (12): 1797–1817.
129. Mead, J.B., G. Hopcraft, S.J. Frasier, B.D. Pollard, C.D. Cherry, D.H. Schaubert, and R.E. McIntosh. 1998. A volume-imaging radar wind profiler for atmospheric boundary layer turbulence studies. *Journal of Atmospheric and Oceanic Technology* 15 (4): 849–859.
130. Michelson, M., W. Shrader, and J. Wieler. 1990. Terminal Doppler Weather Radar. *Microwave Journal* 33: 139–148.
131. Mitchell, E.D.W., S.V. Vasiloff, G.J. Stumpf, A. Witt, M.D. Eilts, J.T. Johnson, and K.W. Thomas. 1998. The national severe storms laboratory tornado detection algorithm. *Weather and Forecasting* 13 (2): 352–366.
132. Moller, A.R.: The improved NWS storm spotters' training program at Ft. Worth, Tex. 1978. *Bulletin of the American Meteorological Society* 59 (12): 1574–1582.
133. Moller, A., C. Doswell, J. McGinley, S. Tegtmeier, and R. Zipser. 1974. Field observations of the Union City tornado in Oklahoma. *Weatherwise* 27 (2): 68–79.

134. Newman, J.F., and P.L. Heinselman. 2012. Evolution of a quasi-linear convective system sampled by phased array radar. *Monthly Weather Review* 140 (11): 3467–3486.
135. Nolan, D.S. 2013. On the use of Doppler radar-derived wind fields to diagnose the secondary circulations of tornadoes. *Journal of the Atmospheric Sciences* 70: 1160–1171.
136. Nolan, D.S., N.A. Dahl, G.H. Bryan, and R. Rotunno. 2017. Tornado vortex structure, intensity, and surface wind gusts in large-Eddy simulations with fully developed turbulence. *Journal of the Atmospheric Sciences* 74 (5): 1573–1597.
137. Palmer, R.D., B.L. Cheong, M.W. Hoffman, S.J. Frasier, and F.J. López-Dekker. 2005. Observations of the small-scale variability of precipitation using an imaging radar. *Journal of Atmospheric and Oceanic Technology* 22 (8): 1122–1137.
138. Palmer, R.D., D. Bodine, M. Kumjian, B. Cheong, G. Zhang, Q. Cao, H.B. Bluestein, A. Ryzhkov, T.Y. Yu, and Y. Wang. 2011. Observations of the 10 May 2010 tornado outbreak using OU-PRIME: potential for new science with high-resolution polarimetric radar. *Bulletin of the American Meteorological Society* 92 (7): 871–891.
139. Park, H.S., A.V. Ryzhkov, D.S. Zrnić, and K.E. Kim. 2009. The hydrometeor classification algorithm for the polarimetric WSR-88D: description and application to an MCS. *Weather and Forecasting* 24 (3): 730–748.
140. Payne, C.D., T.J. Schuur, D.R. MacGorman, M.I. Biggerstaff, K.M. Kuhlman, and W.D. Rust. 2009. Polarimetric and electrical characteristics of a lightning ring in a supercell storm. *Monthly Weather Review* 138 (6): 2405–2425.
141. Pazmany, A.L., J.B. Mead, H.B. Bluestein, J.C. Snyder, and J.B. Houser. 2013. A mobile rapid-scanning X-band polarimetric (RaXPol) Doppler radar system. *Journal of Atmospheric and Oceanic Technology* 30 (7): 1398–1413.
142. Petersen, W.A., K.R. Knupp, D.J. Cecil, J.R. Mecikalski, C. Darden, and J. Burks. 2007. The university of Alabama Huntsville thor center instrumentation: research and operational collaboration. In *33rd International Conference on Radar Meteorology*.
143. Polger, P.D., B.S. Goldsmith, R.C. Przywarty, and J.R. Bocchieri. 1994. National Weather Service warning performance based on the WSR-88D. *Bulletin of the American Meteorological Society* 75 (2): 203–214.
144. Potvin, C.K., L.J. Wicker, and A. Shapiro. 2012. Assessing errors in variational dual-Doppler wind syntheses of supercell thunderstorms observed by storm-scale mobile radars. *Journal of Atmospheric and Oceanic Technology* 29 (8): 1009–1025.
145. Putnam, B.J., M. Xue, Y. Jung, N. Snook, and G. Zhang. 2014. The analysis and prediction of microphysical states and polarimetric radar variables in a mesoscale convective system using double-moment microphysics, multinetwork radar data, and the ensemble Kalman filter. *Monthly Weather Review* 142 (1): 141–162.
146. Rasmussen, E.N., and J.M. Straka. 2007. Evolution of low-level angular momentum in the 2 June 1995 Dimmitt, Texas tornado cyclone. *Journal of the Atmospheric Sciences* 64: 1365–1378.
147. Rasmussen, E.N., J.M. Straka, R. Davies-Jones, C.A. Doswell, F.H. Carr, M.D. Eilts, and D.R. MacGorman. 1994. Verification of the origins of rotation in tornadoes experiment: VORTEX. *Bulletin of the American Meteorological Society* 75 (6): 995–1006.
148. Ray, P.S., R.J. Doviak, G.B. Walker, D. Sirmans, J. Carter, and B. Bumgarner. 1975. Dual-Doppler observations of a tornadic storm. *Journal of Applied Meteorology* 14: 1521–1530.
149. Rockney, V. 1958. The WSR-57 radar. In *Proceedings of Seventh Conference on Weather Radar*, Miami Beach, FL, F14–F20. Boston: American Meteorological Society.
150. Rotunno, R. 1979. A study in tornado-like vortex dynamics. *Journal of the Atmospheric Sciences* 36: 140–155.
151. Rotunno, R. 2013. The fluid dynamics of tornadoes. *Annual Review of Fluid Mechanics* 45: 59–84.
152. Rotunno, R., and J.B. Klemp. 1982. The influence of the shear-induced pressure gradient on thunderstorm motion. *Monthly Weather Review* 110: 136–151.
153. Rust, W.D. 1989. Utilization of a mobile laboratory for storm electricity measurements. *Journal of Geophysical Research* 94 (D11): 13305–13311.

154. Rust, W.D., and T.C. Marshall. 1989. Mobile, high-wind, balloon-launching apparatus. *Journal of Atmospheric and Oceanic Technology* 6 (1): 215–217.
155. Rust, W.D., D.W. Burgess, R.A. Maddox, L.C. Showell, T.C. Marshall, and D.K. Lauritsen. 1990. Testing a mobile version of a cross-chain LORAN atmospheric (M-CLASS) sounding system. *Bulletin of the American Meteorological Society* 71 (2): 173–180.
156. Ryzhkov, A., D. Burgess, D. Zrnić, T. Smith, and S. Giangrande. 2002. Polarimetric analysis of a 3 May 1999 tornado. In *21th Conference on Severe Local Storms*, San Antonio, TX. Boston: American Meteorological Society (Preprints).
157. Ryzhkov, A.V., S.E. Giangrande, and T.J. Schuur. 2005. Rainfall estimation with a polarimetric prototype of WSR-88D. *Journal of Applied Meteorology and Climatology* 44: 502–515.
158. Ryzhkov, A.V., T.J. Schurr, D.W. Burgess, and D. Zrnić. 2005. Polarimetric tornado detection. *Journal of Applied Meteorology and Climatology* 44: 557–570.
159. Ryzhkov, A.V., T.J. Schuur, D.W. Burgess, P.L. Heinselman, S.E. Giangrande, and D.S. Zrnić. 2005. The joint polarization experiment: Polarimetric rainfall measurements and hydrometeor classification. *Bulletin of the American Meteorological Society* 86 (6): 809–824.
160. Schenkman, A.D., M. Xue, A. Shapiro, K. Brewster, and J. Gao. 2011. Impact of CASA radar and Oklahoma mesonet data assimilation on the analysis and prediction of tornadic mesovortices in an MCS. *Monthly Weather Review* 139 (11), 3422–3445.
161. Schenkman, A.D., M. Xue, and M. Hu. 2014. Tornadogenesis in a high-resolution simulation of the 8 May 2003 Oklahoma City supercell. *Journal of the Atmospheric Sciences* 71: 130–154.
162. Schultz, C.J., L.D. Carey, E.V. Schultz, B.C. Carcione, C.B. Darden, C.C. Crowe, P.N. Gatlin, D.J. Nadler, W.A. Petersen, and K.R. Knupp. 2012. Dual-polarization tornadic debris signatures Part I: Examples and utility in an operational setting. *Electronic Journal of Operational Meteorology* 13: 120–137.
163. Schultz, C.J., S.E. Nelson, L.D. Carey, L. Belanger, B.C. Carcione, C.B. Darden, T. Johnstone, A.L. Molthan, G.J. Jedlovec, E.V. Schultz, C.C. Crowe, and K.R. Knupp. 2012. Dual-polarization tornadic debris signatures Part II: Comparisons and caveats. *Electronic Journal of Operational Meteorology* 13: 138–158.
164. Schwarz, C.M., and D.W. Burgess. 2010. Verification of the origins of rotation in tornadoes experiment, part 2 (VORTEX2): data from the NOAA (NSSL) X-band dual-polarized radar. In *25th Conference on Severe Local Storms*. Boston: American Meteorological Society.
165. Seo, B.C., and W.F. Krajewski. 2010. Scale dependence of radar rainfall uncertainty: initial evaluation of NEXRAD's new super-resolution data for hydrologic applications. *Journal of Hydrometeorology* 11 (5): 1191–1198.
166. Shabbott, C.J., and P.M. Markowski. 2006. Surface in situ observations within the outflow of forward-flank downdrafts of supercell thunderstorms. *Monthly Weather Review* 134: 1422–1441.
167. Shapiro, A., K.M. Willingham, and C.K. Potvin. 2010. Spatially variable advection correction of radar data. Part I: Theoretical considerations. *Journal of the Atmospheric Sciences* 67 (11): 3445–3456
168. Shapiro, A., K.M. Willingham, and C.K. Potvin. 2010. Spatially variable advection correction of radar data. Part II: Test results. *Journal of the Atmospheric Sciences* 67 (11): 3457–3470.
169. Simmons, K.M., and D. Sutter. 2005. WSR-88D radar, tornado warnings, and tornado casualties. *Weather and Forecasting* 20 (3): 301–310.
170. Skinner, P.S., C.C. Weiss, M.M. French, H.B. Bluestein, P.M. Markowski, and Y.P. Richardson. 2014. VORTEX2 observations of a low-level mesocyclone with multiple internal rear-flank downdraft momentum surges in the 18 May 2010, Dumas, Texas supercell. *Monthly Weather Review* 142 (8): 2935–2960.
171. Skinner, P.S., C.C. Weiss, L.J. Wicker, C.K. Potvin, and D.C. Dowell. 2015. Forcing mechanisms for an internal rear-flank downdraft momentum surge in the 18 May 2010 Dumas, Texas supercell. *Monthly Weather Review* 143 (11): 4305–4330.
172. Skow, K.D., and C. Cogil. 2017. A high-resolution aerial survey and radar analysis of quasi-linear convective system surface vortex damage paths from 31 August 2014. *Weather and Forecasting* 32: 441–467.

173. Smull, B.F., and R.A. Houze. 1987. Dual-Doppler radar analysis of a midlatitude squall line with a trailing region of stratiform rain. *Journal of the Atmospheric Sciences* 44 (15): 2128–2149.
174. Snyder, J.C., and H.B. Bluestein. 2014. Some considerations for the use of high-resolution mobile radar data in tornado intensity determination. *Weather and Forecasting* 29 (4): 799–827.
175. Snyder, J.C., and A.V. Ryzhkov. 2015. Automated detection of polarimetric tornadic debris signatures using a hydrometeor classification algorithm. *Journal of Applied Meteorology and Climatology* 54: 1861–1870.
176. Snyder, J.C., H.B. Bluestein, V. Venkatesh, and S.J. Frasier. 2013. Observations of polarimetric signatures in supercells by an X-band mobile Doppler radar. *Monthly Weather Review* 141 (1): 3–29.
177. Stensrud, D.J., L.J. Wicker, K.E. Kelleher, M. Xue, M.P. Foster, J.T. Schaefer, R.S. Schneider, S.G. Benjamin, S.S. Weygandt, J.T. Ferree, and J.P. Tuell. 2009. Convective-scale warn-on-forecast system. *Bulletin of the American Meteorological Society* 90: 1487–1499.
178. Stumpf, G.J., A. Witt, E.D. Mitchell, P.L. Spencer, J.T. Johnson, M.D. Eilts, K.W. Thomas, and D.W. Burgess. 1998. The National Severe Storms Laboratory mesocyclone detection algorithm for the WSR-88D. *Weather and Forecasting* 13 (2): 304–326.
179. Tanamachi, R.L., H.B. Bluestein, W.C. Lee, M. Bell, A. Pazmany. 2007. Ground-based velocity track display (GBVTD) analysis of W-band Doppler radar data in a tornado near Stockton, Kansas, on 15 May 1999. *Monthly Weather Review* 135 (3): 783–800.
180. Tanamachi, R.L., H.B. Bluestein, J.B. Houser, S.J. Frasier, and K.M. Hardwick. 2012. Mobile, X-band, polarimetric Doppler radar observations of the 4 May 2007 Greensburg, Kansas, tornadic supercell. *Monthly Weather Review* 140 (7): 2103–2125
181. Tanamachi, R.L., P.L. Heinselman, and L.J. Wicker. 2015. Impacts of a storm merger on the 24 May 2011 El Reno, Oklahoma, tornadic supercell. *Weather and Forecasting* 30 (3): 501–524.
182. Trapp, R.J., and R. Davies-Jones. 1997. Tornadogenesis with and without a dynamic pipe effect. *Journal of the Atmospheric Sciences* 54 (1): 113–133.
183. Trapp, R., and C.A. Doswell. 2000. Radar data objective analysis. *Journal of Atmospheric and Oceanic Technology* 17: 105–120.
184. Umeyama, A.Y., S.M. Torres, and B.L. Cheong. 2017. Bootstrap dual-polarimetric spectral density estimator. *IEEE Transactions on Geoscience and Remote Sensing* 55: 2299–2312.
185. Van Den Broeke, M.S. 2015. Polarimetric tornadic debris signature variability and debris fallout signatures. *Journal of Applied Meteorology and Climatology* 54: 2389–2405.
186. Van Den Broeke, M.S., and S.T. Jauernic. 2014. Spatial and temporal characteristics of polarimetric tornadic debris signatures. *Journal of Applied Meteorology and Climatology* 53 (10): 2217–2231.
187. Vasiloff, S.V. 2001. Improving tornado warnings with the Federal Aviation Administration's terminal Doppler weather radar. *Bulletin of the American Meteorological Society* 82 (5): 861–874.
188. Vasiloff, S.V., and K.W. Howard. 2009. Investigation of a severe downburst storm near Phoenix, Arizona, as seen by a mobile Doppler radar and the KIWA WSR-88D. *Weather and Forecasting* 24 (3): 856–867.
189. Wakimoto, R.M., and J.W. Wilson. 1989. Non-supercell tornadoes. *Monthly Weather Review* 117 (6): 1113–1140.
190. Wakimoto, R.M., N.T. Atkins, and J. Wurman. 2011. The LaGrange tornado during VORTEX2. Part I: Photogrammetric analysis of the tornado combined with single-Doppler radar data. *Monthly Weather Review* 139 (7): 2233–2258.
191. Wakimoto, R.M., P. Stauffer, W.C. Lee, N.T. Atkins, and J. Wurman. 2012. Finescale structure of the LaGrange, Wyoming, tornado during VORTEX2: GBVTD and photogrammetric analyses. *Monthly Weather Review* 140 (11): 3397–3418.
192. Wakimoto, R.M., N.T. Atkins, K.M. Butler, H.B. Bluestein, K. Thiem, J. Snyder, and J. Houser. 2015. Photogrammetric analysis of the 2013 El Reno tornado combined with mobile X-band polarimetric radar data. *Monthly Weather Review* 143 (7): 2657–2683.

193. Wakimoto, R.M., N.T. Atkins, K.M. Butler, H.B. Bluestein, K. Thiem, J.C. Snyder, J. Houser, K. Kosiba, and J. Wurman. 2016. Aerial damage survey of the 2013 El Reno tornado combined with mobile radar data. *Monthly Weather Review* 144 (5): 1749–1776.
194. Wang, Y., and T.Y. Yu. 2015. Novel tornado detection using an adaptive neuro-fuzzy system with S-band polarimetric weather radar. *Journal of Atmospheric and Oceanic Technology* 32: 195–208.
195. Wang, Y., T.Y. Yu, M. Yeary, A. Shapiro, S. Nemati, M. Foster, D.L. Andra, and M. Jain. 2008. Tornado detection using a neuro-fuzzy system to integrate shear and spectral signatures. *Journal of Atmospheric and Oceanic Technology* 25 (7): 1136–1148.
196. Weber, M.E., J.Y.N. Cho, J.S. Herd, Flavin, J.M., Benner, W.E., Torok, G.S. 2007. The next-generation multimission U.S. surveillance radar network. *Bulletin of the American Meteorological Society* 88: 1739–1751.
197. Weiss, C.C., J.L. Schroeder, J. Guynes, P.S. Skinner, and J. Beck. 2009. The TTUKa mobile Doppler radar: Coordinated radar and in situ measurements of supercell thunderstorms during Project VORTEX2. In *34th Conference on Radar Meteorology*, Williamsburg, VA. 11B.2. Boston: American Meteorological Society (Preprints).
198. Whiton, R.C., P.L. Smith, S.G. Bigler, K.E. Wilk, and A.C. Harbuck. 1998. History of operational use of weather radar by U.S. weather services. Part I: The pre-NEXRAD era. *Weather and Forecasting* 13: 219–243.
199. Wicker, L.J., and R.B. Wilhelmson. 1995. Simulation and analysis of tornado development and decay within a three-dimensional thunderstorm. *Journal of the Atmospheric Sciences* 52: 2675–2703.
200. Wilson, J.W., G.B. Foote, N.A. Crook, J.C. Fankhauser, C.G. Wade, J.D. Tuttle, C.K. Mueller, and S.K. Krueger. 1992. The role of boundary-layer convergence zones and horizontal rolls in the initiation of thunderstorms: a case study. *Monthly Weather Review* 120 (9): 1785–1815.
201. Wood, V.T., and R.A. Brown. 1997. Effects of radar sampling on single-Doppler velocity signatures of mesocyclones and tornadoes. *Weather and Forecasting* 12: 928–938.
202. Wood, V.T., R.A. Brown, and D. Sirmans. 2001. Technique for improving detection of WSR-88D mesocyclone signatures by increasing angular sampling mesocyclone signatures by increasing angular sampling. *Weather and Forecasting* 16 (1): 177–184.
203. Wurman, J. 2001. The DOW mobile multiple-Doppler network. In *30th International Conference on Radar Meteorology*.
204. Wurman, J. 2002. The multiple-vortex structure of a tornado. *Weather and Forecasting* 17 (3): 473–505.
205. Wurman, J., and C.R. Alexander. 2005. The 30 May 1998 Spencer, South Dakota, storm. Part II: Comparison of observed damage and radar-derived winds in the tornadoes. *Monthly Weather Review* 133 (1): 97–119.
206. Wurman, J., and S. Gill. 2000. Finescale radar observations of the Dimmitt, Texas (2 June 1995), tornado. *Monthly Weather Review* 128 (7): 2135–2164.
207. Wurman, J., and K. Kosiba. 2013. Finescale radar observations of tornado and mesocyclone structures. *Weather and Forecasting* 28 (5): 1157–1174.
208. Wurman, J., and M. Randall. 2001. An inexpensive, mobile, rapid-scan radar. In *30th International Conference on Radar Meteorology*, Munich, Germany. 3.4. Boston: American Meteorological Society (Preprints).
209. Wurman, J., J.M. Straka, and E.N. Rasmussen. 1996. Fine-scale Doppler radar observations of tornadoes. *Science* 272: 1774–1777.
210. Wurman, J., J. Straka, E. Rasmussen, M. Randall, and A. Zahrai. 1997. Design and deployment of a portable, pencil-beam, pulsed, 3-cm Doppler radar. *Journal of Atmospheric and Oceanic Technology* 14 (6): 1502–1512.
211. Wurman, J., Y. Richardson, C. Alexander, S. Weygandt, and P.F. Zhang. 2007. Dual-Doppler analysis of winds and vorticity budget terms near a tornado. *Monthly Weather Review* 135 (6): 2392–2405.

212. Wurman, J., Y. Richardson, C. Alexander, S. Weygandt, and P.F. Zhang, 2007. Dual-Doppler and single-Doppler analysis of a tornadic storm undergoing mergers and repeated tornadogenesis. *Monthly Weather Review* 135 (3): 736–758.
213. Wurman, J., K. Kosiba, P. Markowski, Y. Richardson, D. Dowell, and P. Robinson. 2010. Finescale single- and dual-Doppler analysis of tornado intensification, maintenance, and dissipation in the Orleans, Nebraska, supercell. *Monthly Weather Review* 138 (12): 4439–4455.
214. Wurman, J., D. Dowell, Y. Richardson, P. Markowski, E. Rasmussen, D. Burgess, L. Wicker, and H.B. Bluestein. 2012. The second verification of the origins of rotation in tornadoes experiment: VORTEX2. *Bulletin of the American Meteorological Society* 93 (8): 1147–1170.
215. Wurman, J., K. Kosiba, and P. Robinson. 2013. In situ, Doppler radar, and video observations of the interior structure of a tornado and the wind–damage relationship. *Bulletin of the American Meteorological Society* 94 (6): 835–846.
216. Wurman, J., K. Kosiba, P. Robinson, and T. Marshall. 2014. The role of multiple-vortex tornado structure in causing storm researcher fatalities. *Bulletin of the American Meteorological Society* 95 (1): 31–45.
217. Xue, M., S. Liu, and T.Y. Yu. 2007. Variational analysis of oversampled dual-Doppler radial velocity data and application to the analysis of tornado circulations. *Journal of Atmospheric and Oceanic Technology* 24 (3): 403–414.
218. Yu, T.Y., Y. Wang, A. Shapiro, M.B. Yeary, D.S. Zrnić, and R.J. Doviak. 2007. Characterization of tornado spectral signatures using higher-order spectra. *Journal of Atmospheric and Oceanic Technology* 24 (12): 1997–2013.
219. Yussouf, N., and D.J. Stensrud. 2008. Impact of high temporal frequency radar data assimilation on storm-scale NWP model simulations. In *24th Conference on Severe Local Storms*, 9B.1. Boston: American Meteorological Society.
220. Zrnić, D., and M. Istok. 1979. Wind speeds in two tornadic storms and a tornado, deduced from Doppler spectra. *Journal of Applied Meteorology* 19 (12): 1405–1415.
221. Zrnić, D.S., and R.J. Doviak. 1975. Velocity spectra of vortices scanned with a pulse-Doppler radar. *Journal of Applied Meteorology* 14 (8): 1531–1539.
222. Zrnić, D.S., R.J. Doviak, D.W. Burgess, and R.A. Brown. 1977. Probing tornadoes with a pulse Doppler radar. *Quarterly Journal of the Royal Meteorological Society* 103 (438): 707–720.
223. Zrnić, D.S., D.W. Burgess, and L.D. Hennington. 1985. Automatic detection of mesocyclonic shear with Doppler radar. *Journal of Atmospheric and Oceanic Technology* 2: 425–438.
224. Zrnić, D.S., J.F. Kimpel, D.E. Forsyth, A. Shapiro, G. Crain, R. Ferek, J. Heimmer, W. Benner, T.J. McNellis, and R.J. Vogt. 2007. Agile-beam phased array radar for weather observations. *Bulletin of the American Meteorological Society* 88: 1753–1766.

Ground-Based Weather Radar to Investigate Thunderstorms

Luca Baldini, Nicoletta Roberto, Mario Montopoli, and Elisa Adirosi

1 Introduction

Thunderstorms are short-duration local storms produced by cumulonimbus clouds. They are accompanied by electrification phenomena, and, but not always, strong winds, heavy rain, and hail. In their early phases, a distinctive characteristic is the strong updraft. Instead, a strong downdraft in the column of precipitation is the signature of dissipating stages.

The unique quality of thunderstorm is the electrical activity. Therefore, the study of thunderstorms includes investigating lightning phenomena and charging mechanisms in relation with the evolution in time of the properties of clouds.

Thunderstorms often reach altitudes of 15 km or more, higher than altitudes reached by civil aircraft during cruise. Therefore, they are linked with several specific risks for air traffic and their presence along the flight route must be promptly detected for the safety of flight and passengers. Thunderstorms are frequently associated with several hydrogeological threats caused by high precipitation rates that can produce landslides and flash floods especially in small catchment basins characterized by a very fast hydrological response.

Because of the short-life and localized nature of thunderstorms, remote sensing instruments and weather radars in particular, are necessary for observing, investigating, and, possibly, predicting such phenomena. Radar (RAdio Detection And Ranging) is a device that uses radio waves to detect and locate "targets" that, for weather radar, are clouds and precipitation particles. Physical properties of

L. Baldini (✉) · N. Roberto · M. Montopoli · E. Adirosi
National Research Council of Italy (CNR), Institute of Atmospheric Sciences and Climate ISAC, Rome, Italy
e-mail: l.baldini@isac.cnr.it; n.roberto@isac.cnr.it; m.montopoli@isac.cnr.it; elisa.adirosi@artov.isac.cnr.it

C. Andronache (ed.), *Remote Sensing of Clouds and Precipitation*, Springer Remote Sensing/Photogrammetry,
https://doi.org/10.1007/978-3-319-72583-3_4

such particles are estimated from some characteristics of the waves they scatter back to the radar, such as the amplitude, phase, and polarization. Weather radar measurements are located in space through a time delay with respect to the transmit pulse and the antenna pointing angle. To achieve an adequate spatial resolution, considering physical and technological constrains, antenna beamwidth is limited to 1° in most of cases. Many national weather services manage networks made of strategically located S-band or C-band radars to provide seamless measurements. Small and sometimes mobile X-band systems are increasingly used, also in a networked manner, for intensive monitoring of relatively small areas ranging from small catchment basins to highly populated metropolitan areas, particularly prone to the effects of severe weather [43]. In general, for same average transmission power and antenna beam width, the higher is the frequency, the systems become more sensitive and smaller in size. This implies less expensive and more manageable systems at higher frequencies. On the other hand, frequencies above the S band (at this band attenuation effects can be neglected) imply enhanced attenuation of meteorological returns caused by the propagation through precipitation paths. However, since dual-polarization techniques also provide a way for the mitigation of the path attenuation effects, thunderstorms associated with high precipitation rates that determine significant attenuation effects have been studied also using C and X band dual-polarization systems. This chapter focuses on observations collected by systems at S and C band with characteristics similar to the systems used by operational services.

Early weather radar provided reflectivity factor of precipitating clouds, which is still the basic measurement for every weather radar (see articles in [1] for an overview of early days of radar in meteorology). Systems measuring radial Doppler velocity of storms started to be available in 1970s. In 1988, US deployed a country-wide network based on Doppler S-band systems called WSR-88D and C-band Terminal Doppler Weather Radar (TDWR) at major airports. Later, radars with Doppler capabilities became standard for operational weather services. The use of polarimetric information of weather radar returns was investigated since 1960s. Among the possible schemes, the linear orthogonal scheme operating at the horizontal (h) and vertical (v) states (eigen-polarization states of the rain medium) proposed in the 1970s was considered to be a convenient solution also for operational implementations [57]. Apart from some exceptions, the adoption of this approach (now referred to as "dual-polarization") in operational systems took place mostly after year 2000. In 2017, a significant part of operational weather radars worldwide are dual-polarization Doppler systems (for example, all the WSR-88D system are dual-polarized since 2013) and plans for dual-polarization upgrades are ongoing almost everywhere. Dual-polarization technology has led to the definition of new radar measurements that not only improve the accuracy of the quantitative estimation of precipitation but also help to reveal additional properties of precipitating clouds. For more details on measurement principles and technology of weather radar, including Doppler and dual-polarization systems, the reader can refer to popular textbooks like those of Doviak and Zrnić [16] and Bringi and Chandrasekar [10].

When starting analyzing properties of thunderstorms in terms of radar observations, the identification of convective areas is usually performed at the first stage. Convective cells are identified in horizontal sections of radar reflectivity data based on relative peaks of reflectivity [59]. They are generally short-lived (20–30 min) and the rapid update rate of radar measurements (typically from less than 5–10 min) allows to infer their evolution in space and time [67] and to investigate their morphology in relation with synoptic conditions [2]. This is a key advantage of weather radar networks with respect to other weather monitoring systems, such as meteorological satellites. In addition, weather radar scans atmosphere using different elevation angles. A 3D picture of a storm is thus obtained that can be used to obtain information like height extension and, more in general, the vertical structure of a precipitating cloud. Doppler weather radar estimates an average speed of masses associated to precipitation along the radial direction that is referred to as mean Doppler velocity. To obtain the other two components of velocity, synchronous Doppler measurements from different radars in overlapping coverage areas must be combined allowing to point out 3D fluxes within a thunderstorm [33]. With the affirmation of dual-polarization as the standard technology for providing advanced radar-based weather products, new measurements have been available that can allow to identify hydrometeors and locate their spatial distribution in a precipitating cloud. In convective storms, such measurements provide clearer signatures related to specific phases of the life of a thunderstorm allowing more precise identification of local features, such as the regions of updraft and hail shaft. Advanced analysis methods aim at identifying, from radar measurements, conditions leading to lightning. It should be noted that weather radar does not allow catching all the prestorm activity from which convection triggering mechanism could be pointed out. However, recent research based on dual polarization measurements aims at detecting some prestorm features using returns from non-meteorological scatterers like insects whose movements are sensitive to air motion [45].

This chapter is organized as follows. The section following this introduction focuses on methods based single polarization and Doppler radar data to investigate the inner structure of convection and to understand its evolution in time. Then, a Sect. 3 introduces the dual-polarization radar measurements and explains, through specific examples, their characteristics that are directly related to different phases of a thunderstorm. Finally, some advances in establishing relationships between cloud microphysics as estimated by radar (especially dual-polarized systems) and cloud electrification processes are presented in Sect. 4. Section 5 completes the chapter by highlighting current efforts of radar research on storm electrifications.

2 Single Polarization, Doppler and Dual Doppler Methods

Methods for extracting properties of convective storms, and particularly, to detect the presence of hail have been proposed for single polarization weather radar since the end of 1950s [15]. Single polarization hail detection methods are based on analysis of vertical profile of reflectivity: the height of a contour of high reflectivity (e.g. 45 dBZ) above the freezing level or the height of radar echo tops have some relations with hail occurrence at ground [62]. Later, a relatively simple algorithm for extracting probability of hail and hail size based on height of 0 °C isothermal, radar echo top, and temperature-derived quantities that has been operational in WSR-88D radars was devised [69]. The more recent algorithms (see next section of this article) for hail identification embedded in Hydrometeor Classification Systems (HCS) based on dual-polarization radar measurements have clearly outperformed earlier reflectivity based methods [29].

Another important advantage when observing convective systems is represented by velocity information. Indeed, the availability of Doppler velocity in weather radar systems allowed to reveal average motion along with the radar beam (i.e. the radial direction along the line of sight of the antenna pointing). This is because weather radars only send and receive energy in one direction per pulse and, consequently, the strength of the motion detected is, necessarily, only the one in the direction of the energy emitted. The term "average motion" refers to the fact that the various velocity components of all the targets within the radar beam are averaged to produce the output velocity. Note that weather radar needs the presence of hydrometeors or even other scatterers, such as insects [46, 51] to track wind trajectories. Radar can measure the speed of hydrometeors thanks to the Doppler effect (or the Doppler shift) that is the change in frequency or wavelength of a wave (or other periodic event) for an observer moving relative to the source of the wave. It is named after the Austrian physicist Christian Doppler, who described it in 1842. In the weather radar case, a target (e.g. a bunch of hydrometeors in the atmosphere) moving with respect to the emission source of the wave (i.e. the radar) will produce continuous time variations of the target-source distance making the time of arrival of the backscatter wave fronts progressively longer or shorter, depending on the movement direction of the target. Weather radars measure the time variation of the target-source distance (i.e. velocity) through measurements of the time variation of the received signal phase (i.e. angular frequency). Usually, weather radar targets are liquid and/or solid particles whose movement is determined by several factors, including gravity, buoyancy, and local wind. In a thunderstorm, updraft/downdraft and turbulent motions govern the transportation of these particles. Thus, in the velocity domain, the received radar signal can be reasonably seen as a distribution (usually referred to as spectrum) of various contributions having different velocities and centered at the value of the average speed of the whole target motion. The spread of the spectrum of velocities (i.e. the spectrum width) is a measure of turbulence effects of the target within the radar beam. The coherent sum of all the spectrum velocities contributions is the signal received by weather radars.

Since radars estimate velocity by measuring time variations of the phase of the received signal, we can have a phase folding when phase variations exceed 180 deg for each direction leading to velocity ambiguity (or aliased velocity). In other words, velocity ambiguity appears when the target under observation moves fast enough to produce a radial velocity larger than the maximum value measurable (V_{max}) by the radar. In that case, the radar will still register a velocity within the interval ($-V_{max}$, V_{max}) which is ambiguous (i.e. erroneous). Such maximum velocity is given by $V_{max} = PRF \times \lambda/4$, where λ is the radar wavelength and *PRF* is the frequency of repetition of the pulses transmitted. The dependency on λ suggests that, for example, X band radars are more likely prone to velocity ambiguity than S band systems using the same *PRF*. Increasing the *PRF* increases V_{max} but at the cost of a reduced range coverage of radar observations, i.e. decreasing the maximum unambiguous range $r_{max} = c/(2 \times PRF)$, where c is the speed of light. A trade-off between range and velocity aliasing is then evident, since $r_{max} V_{max} = c\lambda/8$. Velocity unfolding techniques are needed also at S-band to recover the unambiguous velocity fields while maintaining range capabilities unchanged. They can be of two types: based on modifications of the transmitted sequence of pulses (also refereed as transmission coding techniques) or based on post processing approaches. The first class includes methods like dual *PRF* [14] and staggered *PRF* [75] as well as polarization [26] and phase coding [7]. In the dual *PRF* mode, a group of pulses is sent at one *PRF* followed by a second group of pulses at the other *PRF*. Two different velocities, one for each *PRF*, are estimated and compared to solve for the unambiguous velocity. A drawback is that for an equal amount of time spent in observing the same scene, the number of samples collected in each subparts of the signal (i.e. those parts associated to the two *PRF*s) is smaller than when using a single *PRF*. Consequently, dual *PRF* velocity estimates have higher standard deviation. To overcome this problem, a staggered *PRF* can be used. N sequences of pulses having the same *PRF* are transmitted with different time shifts producing a transmitted sequence of consecutive pulses separated each other by T_i (with $T_1 < T_2 < T_N$). The goal of the method is to find the best arrangements for T_1, T_2, T_N, given maximum unambiguous velocity and range. Large N improves the unfolding success rate but increases the complexity of an eventual filtering procedure to remove clutter from precipitation data. The unambiguous velocity is obtained by a combination of the velocity spectra obtained by pulse pair analysis.

Post-processing techniques are based on the identification of unrealistic gradients in the ambiguous velocity field measured by Doppler radars whatever is the transmission strategy. The correction of ambiguities is accomplished based on: (1) the continuity of the reconstructed velocity field using external information of the environmental wind [16]; (2) self-consistent approaches relying on image processing techniques [74]; (3) the velocity azimuth display (VAD) allowing to check velocity continuity along azimuths [12], or its numerous modifications such as extended VAD [58], gradient VAD [23], or AR VAD [72].

Techniques for three dimensional wind vector reconstruction from measurements of radial components of velocity collected by single [36, 61] or multiple Doppler radar systems [8, 17] have been proposed. Such techniques are based on simplified assumptions of the structure of the wind field under study and often an incomplete description of the local air circulation in the final reconstructed wind product have to be accepted. In proximity of airports, Doppler radars are used to detect strong variation of wind in terms of intensity and direction with altitude, usually referred to as wind shear. Downdrafts from small-scale single cell thunderstorms are the most hazardous in terms of wind shear and pose an important challenge for the accuracy of air security warning systems [68].

In the framework of the analysis of severe events, the quantification of the amount of rotation present in thunderstorms is a relevant application of Doppler weather radars. The Doppler radar ability to detect rotation has improved the meteorologists' ability to look inside thunderstorms and determine the presence of a rotational component, which is often a precursor to the development of tornadoes [11, 70]. A notable example of rotation is given in Fig. 1 that shows a supercell thunderstorm detected by the WSR-88D in Raleigh/Durham, NC, on April 16th, 2011. From this figure, some interesting features can be pointed out. In terms of radar reflectivity (panels a, c) a clear hook-shaped echo is evident in the middle lower left side of panel a. Hook echoes are a sign of a strong rear flank downdraft coupled with a strong inflow region (annotated in panel c). It should be noticed that low reflectivity values are measured in the inflow region because air is rising in that region, and no significant rain is falling there. Contrarily, higher values of reflectivity are expected everywhere the downdraft is dominant. In terms of Doppler signature (panels b, d),

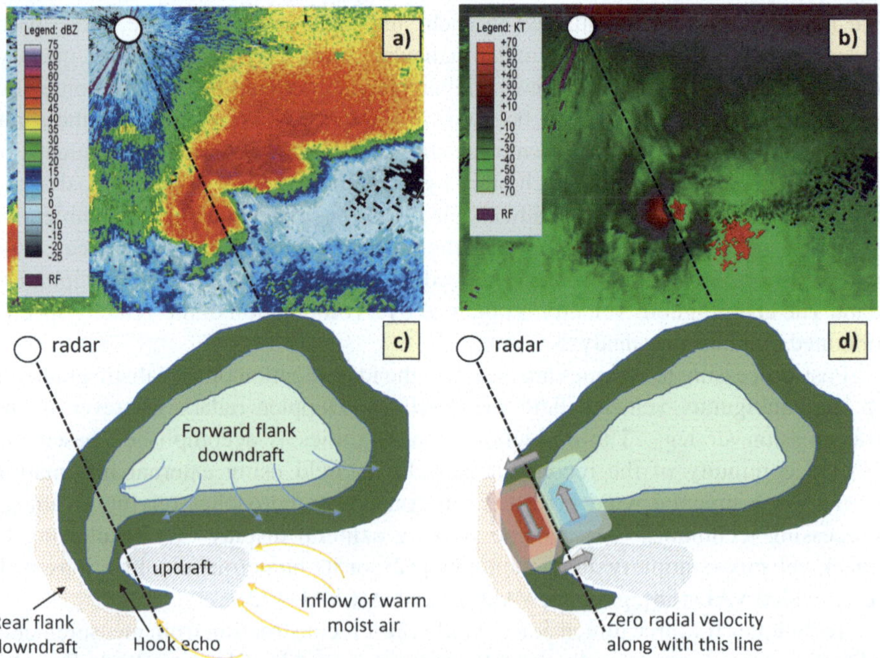

Fig. 1 Examples of radar base reflectivity image collected at 0.5° elevation (**a**) and corresponding Doppler radial velocity image (**b**) by the KRAX WSR-88D (Raleigh, NC), April 16th, 2011 at 20:40UTC. Scales are in dB and knot for panels (**a**) and (**b**) respectively. Panels (**c**) and (**d**) show a schematization of main characteristic supercell features and expected radar Doppler signature, respectively. Supplemental material on this event can be found in [40]

it can be seen that to the right of the dotted line, winds in the mesocyclone are generally moving away from the radar (warmer colors indicate a higher values of radial velocities moving away from the radar) while the opposite happens at the left side of the dotted line where winds are generally moving toward the radar (colder colors indicate a higher values of radial velocities moving toward the radar).

Actually, the reddish blur East of the radial in panel (b) is an example of Doppler ambiguities that is shown without corrections. Thus, we have a compact spatial domain divided into two sides where winds suddenly change. When these two sides are right next to each other, which implies that the winds are changing very rapidly across a small distance, we can detect a mesocyclone signature (counter-clockwise in our example). It is worth remarking that we cannot state anything on what is going on when winds are blowing perpendicular to the radar beam (dotted line) unless we have information from another radar located at different position.

Another important meteorological application where Doppler radar data are fundamental is the assimilation of Doppler wind fields in weather forecast models to increase prediction skills especially for severe weather [71]. Finally, among the less popular applications of Doppler radar, bird tracking [24] and volcanic clouds characterization [47] worth to be mentioned.

3 Dual-Polarization Radar Signatures of Convection

Dual-polarization technique was proposed in late 1970s as a result of research aiming at demonstrating that useful information is embedded in polarization of echoes backscattered by clouds and precipitation particles. Polarization of echoes are mainly determined by particles' properties, such as shape (e.g. raindrops are oblate during fall), size distribution, density, habit, falling behavior (as in the case of tumbling hailstones), and temperature. The dual polarization (linear orthogonal) scheme uses horizontal and vertical polarization states, corresponding to the eigen-polarization states of the rain medium that do not change through propagation path.

Such scheme has become the reference standard for advanced operational weather radar only after year 2000. Nowadays, most of operational systems implement dual-polarization using the so-called STAR (simultaneous transmit and receive) mode: they simultaneously transmit horizontally and vertically polarized pulses and simultaneously receive h and v states using two receivers. Actually, in the STAR scheme a linear 45° polarization state is transmitted at each pulse as the result of the simultaneous combination of h and v polarization states. As explained later on, this poses some limitations in the availability of some measurements provided by dual polarization radars.

Other radar system schemes such as the dual polarization alternate pulse transmission, or the "fully polarimetric" scheme that provides a larger set of measurements, are now used mostly in research systems. Other radar system schemes using foresee alternate pulse transmission where h and v pulses are switched during transmission following a specific pulse sequence, or the "fully polarimetric" scheme that provides a larger set of measurements and where h and v pulses are transmitted using two separate transmission channels are mostly used in research systems.

The basic set of measurements provided by a dual polarization system includes the equivalent reflectivity factor Z_h, usually at horizontal polarization ($mm^6\,m^{-3}$—more often ten times the common logarithm is taken and the unit is referred to as dBZ), the differential reflectivity Z_{dr} (often expressed in decibels) defined as the ratio between co-polar returns from h and v polarized pulses (it is actually obtained as the ratio between average power of returns at h and v receiving channels); the copolar correlation coefficient (ρ_{hv}), defined as the amplitude of the complex correlation coefficient between co-polar returns from h and v polarized pulses and the propagation differential phase shift which is the phase of the complex co-polar correlation coefficient. The differential phase shift is actually a propagation variable that expresses the propagation phase difference between h and v polarized radar waves. The half of the range derivative of the range profile of the differential propagation phase shift is known as specific differential phase shift (K_{dp}, deg km^{-1}). A further measurement is the linear depolarization ratio (LDR, dB), defined as the ratio between cross-polar and co-polar average power level. However, LDR is not directly available in the STAR scheme because cross polar components are not available in that scheme. In addition Z_{dr} from STAR may be incorrect in case of cross coupling effects (i.e. back scattered signal power moves from one polarization state to the orthogonal one generating cross-polar components) as described in [63] and [30].

Each measurement is sensitive to some specific physical properties of the particles being sensed so that it is possible use dual-polarization measurements to identify the prevailing hydrometeor type in each radar resolution volume. Identification of hydrometeors has found reasonably good success as applications called Hydrometeor Classification Systems (HCS) used in most of the operational networks making use of dual-polarization radars. Most of them are based on fuzzy logic (FL) approach that incorporates in an automated classificator the existing knowledge (either empirical or theory-based) about hydrometeor discrimination capability of the different dual-polarization measurement

Fundamentally, based on the approaches described [37] and [76], many fuzzy logic HCS have been proposed along the years. They uses different input measurement sets, different hydrometeor classes as output, and different membership functions. The latter are functions that expresses the degree of truth that a measured variable (or variable vectors) has with respect to hydrometeors belonging to a given class. Recent improvements have been obtained by using methods based on supervised, semi-supervised, and unsupervised approaches, such as clustering [4, 28], neural networks, support vector machine [53] In the case of convective storms, HCS are fundamental to support meteorologists to quickly and effectively analyze the vertical structure of precipitation or to highlight areas where hail shaft is occurring. An exception to the FL scheme is the Bayesian classificator described in [41].

The possibility to detect the hail shaft below the melting level, that is the layer below the 0 °C isothermal, thick a few hundred meters, in which ice particles melt into raindrops, in contrast to the surrounding rain areas, was among the early applications of dual polarization technique [3]. Above the melting layer (ML), a distinctive feature of dual-polarization measurements are the columns of positive Z_{dr} whose tops can extend to several kilometers above the 0 °C level. Columns of Z_{dr} reveal areas of strong updraft where raindrops are lifted into colder heights then freezing and losing stability thus leading, at top heights, to a rapid decrease in Z_{dr}. Observations using probes showed that warm rain processes involving collision/coalescence of drops are dominant within the positive Z_{dr} column [32]. Size sorting due to growth of some drops as they fall through the updraft (i.e. those drops whose terminal velocity exceeds the updraft speed and, thus, are exposed to high liquid water contents) are largely responsible for the observed vertical structure of Z_h and Z_{dr} within the column. Raindrops can grow to large sizes in such columns although they are found in very low concentrations. A 9 mm diameter can be considered an upper limit aloft, but measurements in [25] report a maximum drop diameter of 9.7 mm at ground. An example of Z_{dr} column signature is shown in Fig. 2, showing two convective cells at around 20 and 50 km from radar observed by a C-band dual polarization system in Italy. Data are collected in the RHI (Range Height Indicator) scanning mode in which elevation angles vary while azimuth is kept constant. Although this mode is not common in operational radars, data collected in this way are more consistent and representative of instantaneous development of the rain cells and for this reason they can better reveal characteristics in a vertical section. Note that attenuation correction algorithm is applied to Z_h and Z_{dr} for a fair interpretation of such measurements. The typical Z_h convective signature is shown (upper left panel) with a high reflectivity core that extend above the ML (a dash line in the figure marks the 0 °C isothermal determined from a nearby radiosounding). In the cell closest to the radar, below the melting layer, the region with Z_{dr} greater than 0 dB (upper right panel) is nearly coincident with the region of high reflectivity (upper left panel). This region coincides with downdraft and the high values of Z_{dr} indicate relatively large raindrops. Positive Z_{dr} values above the ML due to updraft effect on supercooled water droplets are present in both cells. Actually, in the far rain cell the narrow Z_{dr} column above the ML is shifted 2 km left with respect to larger values of Z_{dr} below ML. A similar behavior is also noted in terms of Z_h for the same rain cell. This is suggesting the coexistence of cycling updraft and downdraft regime characterized by a Z_{dr} column and large Z_h and Z_{dr} below ML, respectively. Lower values of Z_{dr} below ML of the far rain cell with respect to the closest one could be due to differences in the strengths of updrafts and downdrafts (supposed C-band attenuation to be properly compensated). Higher values of K_{dp} (up to 2.5 deg km^{-1}) are found in the core of the convective cells and above the ML (around 1.0 deg km^{-1}) indicating the presence of liquid water (lower left panel).

By applying a hydrometeor classification algorithm the location of hydrometeors involved in this thunderstorm structure become easily detectable as shown in the lower right panel. The possible hydrometeor classes set are those of rain, hail, hail-rain mix, dry snow, wet snow, and graupel-small hail and the algorithm used

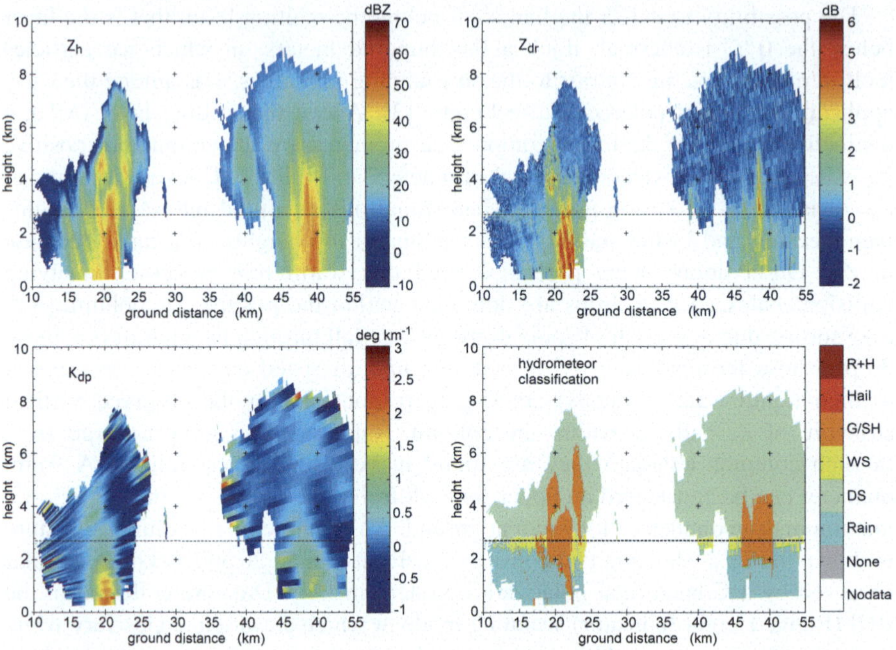

Fig. 2 An RHI presentation of a scanning at fixed azimuth performed by the Polar 55 C radar of the Institute of Atmospheric Sciences and Climate of the National Research Council of Italy located in Rome (Italy), 41.84°N, 12.65°E, and 130 m above sea level. Data were collected at 27° azimuth on October 12th, 2012. Shown are Z_h (top left) and Z_{dr} (top right, *Kdp* (bottom left), and hydrometeor classification results (bottom right)). In the colorbar of hydrometeor classification, "DS", "WS", "G/SH", and "R+H" mean dry snow, wet snow, graupel/small hail, hail mix, respectively; "Nodata" and "none" indicate absence of measurements and inability of HCS to provide an acceptable result. Black dashed line identifies the 0 °C level derived from the closest radiosounding performed at the Pratica di Mare airport (41.66°N, 12.48°E)

to assign a class to each radar triplet of Z_h, Z_{dr} and K_{dp} is based on the support vector machine algorithm described in [53]. Regions in the cell core are classified as composed of graupel that are the typical hydrometeor involved in the cloud electrification mechanisms. In general, the early detection of insurgence of Z_{dr} columns can reveal the strength of up-draft associated with convection and is an important tool for very short-term weather forecast [34]. Also somehow related to the mechanism determining Z_{dr} columns, the mixed phase region extending from 0 °C to 10 °C consisting of supercooled raindrops, partially freezing raindrops, and other forms of wet ice hydrometeors, can be detected by enhanced values of LDR which often form a "cap" on the positive Z_{dr} column. They are also associated with lower values of ρ_{hv} that, in this case, can be used as a substitute of LDR for systems that do not allow such measurement.

To further point out the behavior of the set of dual polarization measurements in convection, Fig. 3 is discussed. It shows different dual-polarization measurements

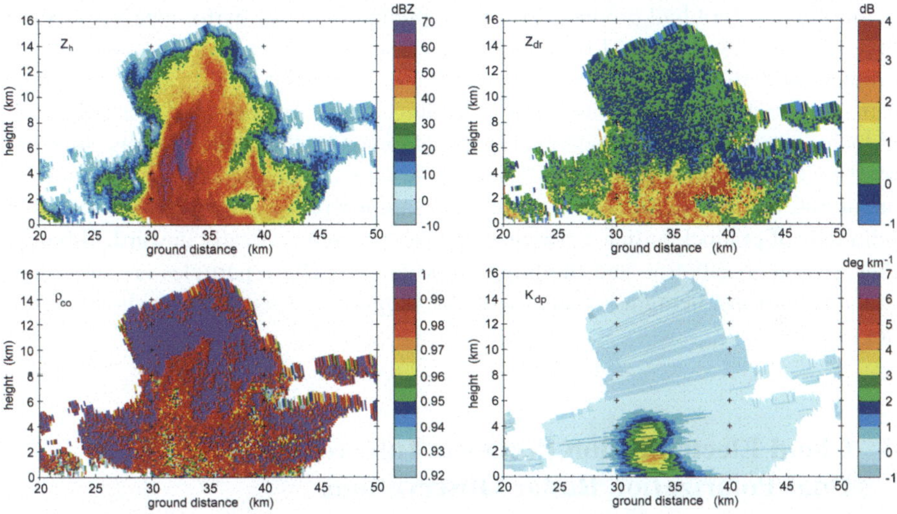

Fig. 3 RHI plot of radar measurements collected at by the NASA S-band NPOL radar along the 130.4° azimuth angle during the IFloodS campaign on May 20th, 2013 at 02:16UTC. Radar was located at 42.27°N, 92.51°E. From top left, clockwise, shown are reflectivity factor, differential reflectivity, co-polar correlation coefficient and specific differential phase shift. Height is referred to sea level

collected, again, in the RHI antenna mode by the NASA NPOL S-band radar during the Ground Validation Iowa Flood Studies (IFloodS) field experiment as a part of the Ground Validation program of NASA/JAXA Global Precipitation Measurement mission [50]. Measurements of reflectivity (top left panel) highlight a convective cell that extends up to more than 14 km above the radar site. Higher values of reflectivity are found here above the 0 °C isothermal (estimated to be closer to 3900-m height above sea level) and are associated both to nearly 0 dB values of Z_{dr} (top right) and low values of the co-polar correlation coefficient (bottom left). These values indicates the presence of mixed tumbling particles. High values of positive K_{dp} (this measurement is insensitive to tumbling particles) are measured where liquid particles are present (bottom right). In general, regions of positive K_{dp} near 0 °C are indicative of water loading up to these levels. Such water contributes to the initiation of the downdraft phase. Ice crystals in the upper levels of convective storms can be vertically aligned by strong electric field forces and are associated to negative K_{dp}. Measurements collected in Florida in 1991 during the Convection and Precipitation Electrification project (CaPE) and analyzed in [31] showcase this behavior. For dual-polarization radar employing the STAR mode, propagation through non-spherical hydrometeors falling with a mean canting angle that differs from 0° determines coupling between the returns of the h and v polarized waves. Radially-oriented stripes of enhanced positive and negative Z_{dr} values can be found and are due to rotation of orientations of ice crystals with respect to the h and v polarization planes that results in coupling effects.

In ordinary (not hail producing) storms, predominant ice hydrometeor types are graupel particles originated by a large ice crystal that accretes supercooled cloud droplets as it falls. Supercooled droplets freeze upon impact with the ice crystals (this deposit is referred to as "rime") leading to the formation of graupel. Such particles can typically take a conical shape as shown by the measurements collected by the Colorado State University CHILL take a radar and the images collected by an aircraft equipped with a volume particle sampler probe shown in [18]. A stable fall mode of graupel particles presents the apex pointed vertically upward, although they exhibits oscillation and tumbling while falling. These particles are important in thunderstorms, being associated with the triggering of lightning as shown in the next section.

4 Cloud Electrification Processes Inferred by Dual-Polarization Radar Observations

Formation of lightning is strongly dependent on the microphysics of a precipitating cloud. Charges in clouds are typically vertically stratified on a tripolar structure [66]: positive charges are accumulated in the upper part of cloud, negative charges in the middle, and lastly, at the bottom of cloud a low intense center of positive charges is found (see Fig. 4, right panel). At the state-of-the art, the most accepted theory used to explain the cloud electrification mechanism is that of non-inductive charging (NIC). The NIC theory explains charging of ice hydrometeors as the result of collisions between ice crystals and graupel in an environment with supercooled water droplets and ice crystals [55] then fostering the electric charge separation. The charge separation obtained by the NIC process can produce an in-tense electrification field (hundreds of $\mathrm{kV\,m^{-2}}$) able to overcome the dielectric strength of air (i.e. the minimum value of electric field able to produce the formation of an electrically conductive path in an insulating material also referred as dielectric breakdown level) and generate the current flow responsible of the observed electrical discharges. Laboratory experiments have confirmed the NIC theory documenting that the factors that mainly influence charge intensity and its polarity are cloud liquid water content and environmental temperature [60]. In fact, the relative growth rate of ice hydrometeors is affected by the increase of supersaturation conditions in clouds [55]. In synthesis, this mechanism requires the presence of large ice hydrometeors (i.e. graupel or hail pellets) that collide with ice crystals in a suspension of supercooled water droplets. While the strength of the updraft is responsible of the transport of air mass in the upper parts of cloud and then of the vertical distribution of the charges involved, the quantity of charges separated, when sufficient enough, triggers the electric breakdown of the atmosphere producing lightning. A typical configuration of hydrometeor distribution fostering lighting formation in thunderstorms is sketched in Fig. 4 (left panel). It can be summarized as follows:

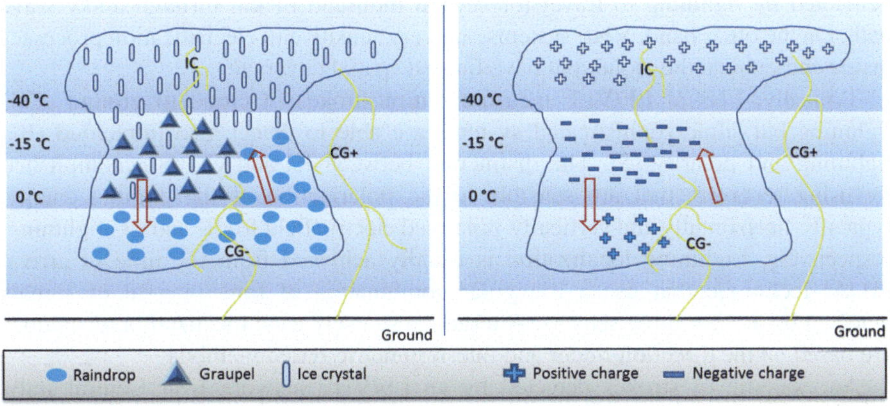

Fig. 4 Conceptual map of hydrometeor distribution (left panel) and charge distribution (right panel) in a thunderstorm. Positive and negative cloud to ground, (CG+ and CG− respectively) and intracloud (IC) lightning are represented as yellow curved lines, while red arrows indicate updraft and downdraft. Note that ice crystals are vertically aligned

- in *glaciated phase* (at ∼−40 °C) ice crystals that bring positive charges, in the upper part of clouds are found;
- in *mixed phase I* (from ∼−15 to −10 °C) graupel particles that bring negative charges in the middle of cloud are found;
- in *mixed phase II* (below −10 °C) there are some graupel and hail that precipitate and bring positive charge at bottom of cloud since at temperature greater than −10 °C, graupel is charged positive [60].

In terms of radar signatures, the following subsections shows how dual-polarization radars can identify such phases.

4.1 Dual Polarization Radar Measurements Collected During Lightings

Dual-polarization weather radars are able to provide information on ice hydrometeors responsible of cloud charging mechanisms with the proper spatial and temporal resolution. In order to get insights on cloud electrical activity, radar measurements has been conveniently combined with lightning measurements since 1980s [27]. Different lightning networks (LN) with local and global coverage are available to detect position, intensity and polarity (positive or negative) of lightning. Networks are based on ELF (extremely low frequency) to VLF (very low frequency) or VHF (very high frequency) systems. Those networks working at frequencies below VHF have usually large spatial coverage. In fact, they can detect lightings beyond the line of sight exploiting the ionospheric propagation effects that allow the signal generated by lightning to travel for several thousand of km through a sky wave path. On the other hand, VHF systems, are very sensitive to electrification processes inside and outside the cloud, but have limited spatial coverage.

Typically, LNs at LF/VLF measure return strokes of cloud to ground (CG) lightning but some sophisticated systems are able to detect also intracloud (IC) lightning and polarity [5]. The distinction between IC and GC is accomplished by using receivers that are sensible to the polarization of the emitted energy, typically, horizontally and vertically-polarized electric field for IC and CG lightning, respectively. Lightning localization is usually achieved from the time of arrival (TOA) technique that use a triangular combination of time interval of signals detected by at least three sensors. Alternative sparsely used localization techniques are based on the direction finder and interferometric-based methods.

An example of strokes detected by an LF/VHF network over Central Italy, superimposed on the radar reflectivity observed by a C-band weather radar located in Rome is shown on top panel of Fig. 5. It is possible to note the correlation between lightning and radar reflectivity values: high numbers of strokes are found over area with high radar reflectivity values (greater than 40 dBZ).

Several works that combine observations from ground based dual polarization radars and lightning network have demonstrated good correlations between lightning and microphysical properties [65]. For instance, linear relations were found between the amount of graupel and flash density by using global satellite observations [49] and using ground based observations collected by C-band dual-polarization radar [52]. These correlations can be exploited in different fields of application such as numerical weather prediction (NWP) models and now-casting techniques. Typically, NWP models use lightning flashes, which consist of one or more strokes measured in a very short time interval and close spatial proximity, rather than strokes. Some methods to convert stroke data into flashes need to be applied. Commonly, the National Lightning Detection Network (NLDN) thresholds of maximum temporal separation of 0.5 s and maximum lateral distance of 10 km radius between successive strokes is used. Scatter plot in Fig. 5 (bottom panel) shows the linear relations between the columnar ice water content (IWC) due to graupel and the spatial density of flashes for different case studies [52]. The linear relationships reported in [52] shows an average agreement with other literature results [22, 49]. However, different slopes of the IWC vs. flash density are obtained for each case study considered (see table shown in Fig. 5 bottom panels). The possible relation of such slopes with other characteristics of thunderstorm that can be estimated from dual-polarization radar measurements, such as up-draft intensity is at present, subject of investigation.

In other words, from Fig. 5 it is noticed that a same number of flash density is associated to a quite large variability of IWC. Therefore, not only the accurate calculation of IWC is important for the prediction of the flash density but also the efficiency of electrification mechanism plays an important role. In this respect a deepen investigation of updrafts can explain the case study-driven slope variability shown in Fig. 5. Actually, combining dual-polarization radar observations and VHF LN measurements, [65] found an increase in the number of VHF sources and CG

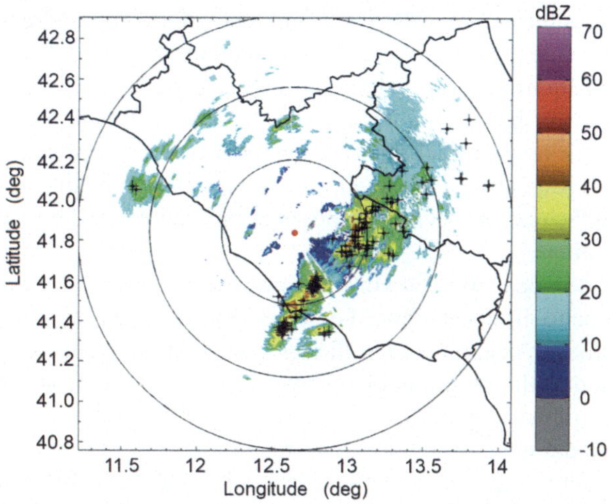

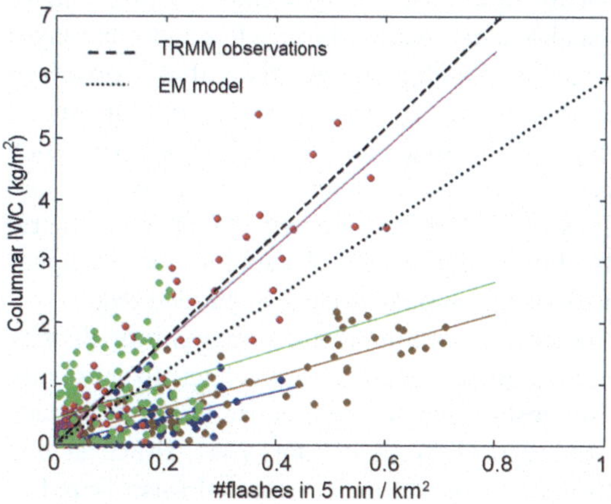

DATE	R^2	Slopes
13 Sep 2012	0.506	2.104
14 Oct 2012	0.824	2.547
15 Oct 2012	0.856	7.911
14 Oct 2012	0.506	2.782

Fig. 5 Radar reflectivity factor (top panel) with superimposed the strokes (black crosses) detected by LINET [6]. Data were collected on October 14th 2015 15:05 UTC with an elevation angle of 1.6°. In the bottom panel, shown is the scatterplot between the number of flashes recorded in 5 min over 1 km^2 surface from LINET and IWC ($kg\,m^{-2}$) estimated for dual polarization measurements collected by Polar 55C [52]. For reference, shown in black dot line corresponds to the linear relation between and a minimum threshold of columnar ice water con-tent of graupel from [22] and, in black dash line, is the linear relation between columnar precipitation ice mass and lightning flash density over land in [49]. Table shows for different cases (colors of rows correspond to those that identify the same case in the scatter plot) the coefficient of determination (R^2) and the slope of the linear relation found

when updraft strength is above 10 m s^{-1} and when the volume of graupel is greater than 2000 km^3, while [56] documented an increment of updraft strength and graupel volume prior to lighting jump that is the last step of discharge process. Another important aspect worth to mention is the fact that IWC estimation can suffer of large uncertainty. In this respect dual-polarization radar measurements aid to reduce such uncertainty with respect to single polarization systems being able to distinguish the ice region from the rest as previously shown in this chapter and then better constrain the IWC estimations.

During last decades, specific abilities have been revealed by dual-polarization radar measurements in determining key cloud microphysics characteristics related to lightning activity. As explained in the previous section, characteristics of Z_{dr} and K_{dp} measurements (the columns of positive Z_{dr} in particular) are able to identify important microphysical processes of convection. With reference to the schemes of Fig. 4, Z_{dr} signatures are dominant in mixed phase layers, while K_{dp} signatures become important in the glaciated phase layers [13, 38, 65]. Different signatures of Z_{dr} and K_{dp} are found in the layers that are significant for lightning initiation: a layer with positive Z_{dr} and K_{dp} values is just above the melting layer while in higher layer usually Z_{dr} and K_{dp} present negative values. Another characteristic that is observed in clouds that are involved in electrification processes is the presence of in the mixed phase layer of graupel particles having conical shape and that fall down with apex vertically oriented. Typical signatures of these hydrometeors are negative Z_{dr} and K_{dp} values associated with high values of Z_h [19, 52]. Specific radar polarimetric signatures related to lightning are also found in glaciated phase region in the upper part of clouds. Strongly negative values of K_{dp} and Z_{dr} are documented [38].

Decreasing K_{dp} and Z_{dr} values occurring simultaneously to lighting density increasing were observed [21]. These negative signatures are related to vertically aligned ice particles. Some studies suggested that strong electric field can induce instantaneous dipole in pristine ice that can align ice crystals [64]. However, the presence of graupel that falling with major axis vertically oriented (also found in glaciated layer) can provide negative values of K_{dp}.

4.2 Applications of Combined Radar and Lightning Measurements

Some important experimental field campaigns were devoted to explore the main characteristics of storms with electrification in relation to polarimetric radar signatures, such as STEPs (Severe Thunderstorm Electrification and Precipitation Study) conducted in US during May–July 2000 [35] and more recently the campaigns being carried out in Brazil since 2010 under the framework of the CHUVA (Cloud Processes of the Main Precipitation Systems in Brazil) project [39]. In particular, in the CHUVA project, specific experiments were conducted to study characteristics of cloud electrification: an ad hoc scanning strategy was designed for an X-band dual-polarization Doppler radar in order to detect vertical distribution of polarimetric radar measurements with adequate resolution. Radar data were combined with the lightning measurements detected from a dense network of VHF sensors installed in the South West radar sector. From observations of millions of vertical profiles of polarimetric radar variables, [42] showed the main features of storms with and without lightning activity at different heights. They identified the polarimetric radar signature in cloud vertical profile deducing the main hydrometeors and processes involved in cloud electrification. They found variations of K_{dp} and Z_{dr} at different cloud layers. Enhanced positive K_{dp} was found in the layer with temperature ranging from 0 to 15 °C likely due to lifted supercooled water droplets, while negative K_{dp} and Z_{dr} values were typically found in layers where temperatures ranges from 15 to 40 °C (mixed phase I). They ascribed these values to the presence of conical graupel. Finally, K_{dp} was found decreased in negative values by a factor of ten times in the glaciated phase layer where the lightning density increases rapidly (14 VHF sources per km in 4 min). Such K_{dp} low values are in agreement with the presence of aligned ice crystals, when lightning occurs. Results of this work support, through large statistics, many outcomes found in last decades on polarimetric radar signatures at different cloud layers, often obtained analyzing few case studies.

Lightning in convective thunderstorms impacts on several activities such as aviation and management of electric infrastructures. Therefore, important efforts were devoted to develop lighting forecast systems in last decades. Depending on time resolution and spatial scale, different forecast systems and prediction models have been made available. Those focused in short time (few hours) and limited areas are typically based on weather radar data [48]. Most of radar techniques to predict lightning are based on the identification of precursor signals of convection initiation, such as estimation of ice water content, identification and quantification of graupel and estimation of updraft intensity. Lightning prediction at higher temporal and spatial resolution are usually obtained by using techniques based on satellite data [44] or Numerical Weather Prediction models [20]. The latter are based on lightning simulation models that reproduce the charging mechanism in cloud using one-dimensional [22] or three-dimensional cloud structure associated with sophisticated electrification schemes. Lightning model simulations can obtain relevant benefits and important improvement from in-depth knowledge on cloud microphysical processes and their relation to the electrical activity revealed by studies from polarimetric radar measurements. Furthermore, this knowledge can be exploited in NWP models to improve microphysical schemes, cloud parametrization and model initialization.

5 Conclusion

Weather radar observations have been used since decades to detect and possibly predict severe weather. In the case of thunderstorms, weather radar are expected to reveal important characteristics, such as height of top of the storm, extension, wind features, rain rate, from which some storm severity indices can be obtained.

Doppler radars available since 1970s have played a fundamental role in identifying kinematic structures of convective systems. A fundamental contributions to obtain details about thunderstorm structure and its time evolution has been provided by dual-polarization radar systems. Analysis of specific signatures of dual-polarization measurements, eventually supported by automatic classification algorithms, allows to identify different phases of the life of a convective cell. A relevant role in relations to cloud electrification mechanisms is played by the identification of the Z_{dr} columns and of the K_{dp} signatures in the upper part of a cloud. Radar observations become more useful when used in synergy with measurements from other instruments. In fact, multisensors observations offer a chance to study thunderstorms from multiple point of views. Recent results reported in the literature highlight the capabilities of combining dual polarization radar and lightning networks observations for studying cloud structure and microphysical properties associated to cloud charging mechanisms. In spite of the advancements in understanding of cloud electrification mechanisms, critical issues remain open. One is the cloud transition from storm to thunderstorm, i.e. the comprehension of lightning formation processes ongoing from the first radar echo to the time of the first lightning that is the fundamental contribution to nowcasting of lightning occurrence. Rapid updating of radar measurements, eventually achieved by the used electronically steered antennas [73] and high spatial resolution measurements achievable through dense radar networks coupled with nowcasting tools [54] are expected to bring advancement in the identification, quantification and prediction of high density ice which is related to the triggering of cloud electrification mechanisms. Important contributions to detect lighting or convection precursors are expected from combining weather radar observations with observations from visible-infrared satellite-borne instruments, such as those that will be available from 2021 by the Meteosat Third Generation (MTG) able to detect clouds and electrification from the early stage of its development, or from water vapor characterization derived from Global Navigation Satellite System (GNSS) measurements that can provide alert on the initiation of a convective system [9].

Acknowledgements Authors acknowledge the two reviewers (Roberto Cremonini and an anonymous) whose suggestions have contributed to improving the quality and the significance of this article. Authors acknowledge the Department of Civil Protection of Italy for having supported along the years CNR-ISAC research on convective systems in the Mediterranean region.

References

1. Atlas, D., ed. 1990. *Radar in Meteorology*. Boston: American Meteorological Society. https://doi.org/10.1007/978-1-935704-15-7.
2. Austin, P.M., and R.A. Houze. 1972. Analysis of the structure of precipitation patterns in New England. *Journal of Applied Meteorology* 11: 926–935. https://doi.org/10.1175/1520-0450(1972)011<0926:AOTSOP>2.0.CO;2.
3. Aydin, K., T. Seliga, and V. Balaji. 1986. Remote sensing of hail with a dual linear polarization radar. *Journal of Climate and Applied Meteorology* 25: 1475–1484. https://doi.org/10.1175/1520-0450(1986)025%3C1475:RSOHWA%3E2.0.CO;2.

4. Bechini, R., and V. Chandrasekar. 2015. Semisupervised robust hydrometeor classification method for dual-polarization radar applications. *Journal of Atmospheric and Oceanic Technology* 32: 22–47. https://doi.org/10.1175/JTECH-D-14-00097.1.
5. Betz, H.-D., K. Schmidt, P. Oettinger, and M. Wirz. 2004. Lightning detection with 3-D discrimination of intracloud and cloud-to-ground discharges. *Geophysical Research Letters* 31: L11108. https://doi.org/10.1029/2004GL019821.
6. Betz, H.-D., K. Schmidt, and W.P. Oettinger. 2009. LINET-an international VLF/LF lightning detection network in Europe. In *Lightning: Principles, Instruments and Applications*, Chap. 5, ed. H.-D. Betz, U. Schumann, and P. Laroche. Dordrecht: Springer. https://doi.org/10.1007/978-1-4020-9079-05.
7. Bharadwaj, N., and V. Chandrasekar. 2007. Phase coding for range ambiguity mitigation in dual-polarized Doppler weather radars. *Journal of Atmospheric and Oceanic Technology* 24: 1351–1363. https://doi.org/10.1175/JTECH2061.1.
8. Bousquet, O., and P. Tabary. 2014. Development of a nationwide real-time 3 D wind and reflectivity radar composite in France. *Quarterly Journal of the Royal Meteorological Society* 140: 611–625. https://doi.org/10.1002/qj.2163.
9. Brenot, H., J. Neméghaire, L. Delobbe, N. Clerbaux, P. De Meutter, A. Deckmyn, A. Delcloo, L. Frappez, and M. Van Roozendael. 2013. Preliminary signs of the initiation of deep convection by GNSS. *Atmospheric Chemistry and Physics* 13: 5425–5449. https://doi.org/10.5194/acp-13-5425-2013.
10. Bringi, V.N., and V. Chandrasekar. 2001. *Polarimetric Doppler Weather Radar: Principles and Applications*. Cambridge: Cambridge University Press.
11. Brown, R.A., W.C. Bumgarner, K.C. Crawford, and D. Sirmans. 1971. Preliminary Doppler velocity measurements in a developing radar hook echo. *Bulletin of the American Meteorological Society* 52: 1186–1188. https://doi.org/10.1175/1520-0477(1971)052<1186:PDVMIA>2.0.CO;2.
12. Browning, K.A., and R. Wexler. 1968. The determination of kinematic properties of a wind field using Doppler radar. *Journal of Applied Meteorology* 7: 105–113. https://doi.org/10.1175/1520-0450(1968)007<0105:TDOKPO>2.0.CO;2.
13. Caylor, I., and V. Chandrasekar. 1996. Time-varying ice crystal orientation in thunderstorms observed with multiparameter radar. *IEEE Transactions on Geoscience and Remote Sensing* 34: 847–858. https://doi.org/10.1109/36.508402.
14. Dazhang, T., S.G. Geotis, R.E. Passarelli Jr., A.L. Hansen, and C.L. Frush. 1984. Evaluation of an alternating-PRF method for extending the range of unambiguous Doppler velocity. In *Proceedings of the 22nd Conference on Radar Meteorology*, Zurich, Switzerland, 10–13 Sept 1984, 523–527.
15. Donaldson, R.J. 1961. Radar reflectivity profiles in thunderstorms. *Journal of Meteorology* 18: 292–305. https://doi.org/10.1175/1520-0469(1961)018%3C0292:RRPIT%3E2.0.CO;2.
16. Doviak, R.J., and D.S. Zrnić. 1994. *Doppler Radar and Weather Observations*. 2nd ed. Washington: Academic.
17. Doviak, R.J., P.S. Ray, R.G. Strauch, and L.J. Miller. 1976. Error estimation in wind fields derived from dual-Doppler radar measurement. *Journal of Applied Meteorology* 15: 868–878. https://doi.org/10.1175/1520-0450(1976)015%3C0868:EEIWFD%3E2.0.CO;2.
18. El-Magd, A., V. Chandrasekar, V.N. Bringi, and W. Strapp. 2000. Multiparameter radar and in situ aircraft observation of graupel and hail. *IEEE Transactions on Geoscience and Remote Sensing* 38: 570–578. https://doi.org/10.1109/36.823951.
19. Evaristo, R., T.M. Bals-Elsholz, E.R. Williams, D.J. Smalley, M.F. Donovan, and A. Fenn. 2013. Relationship of graupel shape to differential reflectivity: theory and observations. Presented at 93rd American Meteorological Society Annual Meeting, Austin, TX, 5–10 Jan 2013.
20. Federico, S., E. Avolio, M. Petracca, G. Panegrossi, P. San, D. Casella, and S. Dietrich. 2014. Simulating lightning into the RAMS model: implementation and preliminary results. *Natural Hazards and Earth System Sciences* 14: 2933–2950. https://doi.org/10.5194/nhess-14-2933-2014.

21. Figueras i Ventura, F., F. Honor, and P. Tabary. 2013. X-band polarimetric weather radar observations of a hailstorm. *Journal of Atmospheric and Oceanic Technology* 30: 143–2151. https://doi.org/10.1175/JTECH-D-12-00243.1.
22. Formenton, M., G. Panegrossi, D. Casella, S. Dietrich, A. Mugnai, P. San, F. Di Paola, H.-D. Betz, C. Price, and Y. Yair. 2013. Using a cloud electrification model to study relationships between lightning activity and cloud microphysical structure. *Natural Hazards and Earth System Sciences* 13: 1085–1104. https://doi.org/10.5194/nhess-13-1085-2013.
23. Gao, J., K.K. Droegemeier, J. Gong, and Q. Xu. 2004. A method for retrieving mean horizontal wind profiles from single-Doppler radar observations contaminated by aliasing. *Monthly Weather Review* 132: 1399–1409. https://doi.org/10.1175/1520-0493-132.1.1399.
24. Gasteren, H.V., I. Holleman, W. Bouten, E.V. Loon, and J. Shamoun-Baranes. 2008. Extracting bird migration information from C-band Doppler weather radars. *Monthly Weather Review* 150: 674–686. https://doi.org/10.1111/j.1474-919X.2008.00832.x.
25. Gatlin, P.N., M. Thurai, V.N. Bringi, W. Petersen, D. Wolff, A. Tokay, L. Carey, and M. Wingo. 2015. Searching for large raindrops: a global summary of two-dimensional video disdrometer observations. *Journal of Applied Meteorology and Climatology* 54: 1069–1089. https://doi.org/10.1175/JAMC-D-14-0089.1.
26. Golestani, Y., V. Chandrasekar, and R.J. Keeler. 1995. Dual polarized staggered PRT scheme for weather radars: analysis and application. *IEEE Transactions on Geoscience and Remote Sensing* 33 (2): 239-246. https://doi.org/10.1109/36.377923.
27. Goodman, S.J., D.E. Buechler, P.D. Wright, and W.D. Rust. 1988. Lightning and precipitation history of a microburst-producing storm. *Geophysical Research Letters* 15: 1185–1188.
28. Grazioli, J., D. Tuia, and A. Berne. 2015. Hydrometeor classification from polarimetric radar measurements: a clustering approach. *Atmospheric Measurement Techniques* 8: 149–170. https://doi.org/10.5194/amt-8-149-2015.
29. Heinselman, P., and A. Ryzhkov. 2006. Validation of polarimetric hail detection. *Weather Forecast* 21: 839–850. https://doi.org/10.1175/WAF956.1.
30. Hubbert, J.C., S.M. Ellis, M.M. Dixon, and G.G. Meymaris. 2010. Modeling, error analysis, and evaluation of dual-polarization variables obtained from simultaneous horizontal and vertical polarization transmit radar. Part II: experimental data. *Journal of Atmospheric and Oceanic Technology* 27: 1599–1607. https://doi.org/10.1175/2010JTECHA1337.1.
31. Hubbert, J.C., S.M. Ellis, W.-Y. Chang, S. Rutledge, and M. Dixon. 2014. Modeling and interpretation of S-Band ice crystal depolarization signatures from data obtained by simultaneously transmitting horizontally and vertically polarized fields. *Journal of Climate and Applied Meteorology* 53: 1659–1677. https://doi.org/10.1175/JAMC-D-13-0158.1.
32. Illingworth, A.J., J.W.F. Goddard, and S.M. Cherry. 1987. Polarization radar studies of precipitation development in convective storms. *Quarterly Journal of the Royal Meteorological Society* 113: 469–489. https://doi.org/10.1002/qj.49711347604.
33. Kropfli, R.A., and L.J. Miller. 1975. Thunderstorm flow patterns in three dimensions. *Monthly Weather Review* 103: 70–71. https://doi.org/10.1175/1520-0493(1975)103<0070:TFPITD>2.0.CO;2.
34. Kumjian, M., A. Khain, N. Benmoshe, E. Ilotoviz, A. Ryzhkov, and V. Phillips. 2014. The anatomy and physics of ZDR columns: investigating a polarimetric radar signature with a spectral bin microphysical model. *Journal of Applied Meteorology and Climatology* 53: 1820–1843. https://doi.org/10.1175/JAMC-D-13-0354.1.
35. Lang, T.J., et al. 2004. The severe thunderstorm electrification and precipitation study. *Bulletin of the American Meteorological Society* 85: 1107–1125. https://doi.org/10.1175/BAMS-85-8-1107.
36. Lhermitte, R.M., and D. Atlas. 1961. Precipitation motion by pulse Doppler radar. In *Proceedings of 9th Weather Radar Conference*, Kansas City, MO, 23–26 Oct 1961, 218–223.
37. Liu, H., and V. Chandrasekar. 2000. Classification of hydrometeors based on polarimetric radar measurements: development of fuzzy logic and neuro-fuzzy systems, and in situ verification. *Journal of Atmospheric and Oceanic Technology* 17: 140–164. https://doi.org/10.1175/1520-0426(2000)017<0140:COHBOP>2.0.CO;2.

38. López, R.E., and J.-P. Aubagnac. 1997. The lightning activity of a hailstorm as a function of changes in its microphysical characteristics inferred from polarimetric radar observations. *Journal of Geophysical Research* 102: 16799–16813. https://doi.org/10.1029/97JD00645.
39. Machado, L.A., et al. 2014. The Chuva project: how does convection vary across Brazil? *Bulletin of the American Meteorological Society* 95: 1365–1380. https://doi.org/10.1175/BAMS-D-13-00084.1.
40. Madaus, L. 2011. How to recognize rotation on Doppler radar. http://lukemweather.blogspot.it/2011/04/how-to-recognize-rotation-on-doppler.html. Accessed May 20, 2017.
41. Marzano, F.S., D. Scaranari, M. Montopoli, and G. Vulpiani. 2008. Supervised classification and estimation of hydrometeors from C-band dual-polarized radars: a Bayesian approach. *IEEE Transactions on Geoscience and Remote Sensing* 46: 85–98. https://doi.org/10.1109/TGRS.2007.906476.
42. Mattos, E.V., L.A.T. Machado, E.R. Williams, and R.I. Albrecht. 2016. Polarimetric radar characteristics of storms with and without lightning activity. *Journal of Geophysical Research—Atmospheres* 121: 14201–14220. https://doi.org/10.1002/2016JD025142.
43. McLaughlin, D., et al. 2009. Short-wavelength technology and the potential for distributed networks of small radar systems. *Bulletin of the American Meteorological Society* 90: 1797–1817. https://doi.org/10.1175/2009BAMS2507.1.
44. Mecikalski, J.R., W.M. MacKenzie Jr., M. König, and S. Muller. 2010. Cloud top properties of growing cumulus prior to convective initiation as measured by Meteosat Second Generation. Part II: use of visible reflectance. *Journal of Applied Meteorology and Climatology* 49: 2544–2558. https://doi.org/10.1175/2010JAMC2480.1.
45. Melnikov, V.M., D.S. Zrnić. 2017. Observations of convective thermals with weather radar. *Journal of Atmospheric and Oceanic Technology*. https://doi.org/10.1175/JTECH-D-17-0068.1.
46. Melnikov, V.M., R.J. Doviak, D.S. Zrnić, and D.J. Stensrud. 2011. Mapping Bragg scatter with a polarimetric WSR-88D. *Journal of Atmospheric and Oceanic Technology* 28: 1273–1285. https://doi.org/10.1175/JTECH-D-10-05048.1.
47. Montopoli, M. 2016. Velocity profiles inside volcanic clouds from three-dimensional scanning micro-wave dual-polarization Doppler radars. *Journal of Geophysical Research—Atmospheres* 121: 7881–7900. https://doi.org/10.1002/2015JD023464.
48. Mosier, R.M., C. Schumacher, R.E. Orville, and L.D. Carey. 2011. Radar nowcasting of cloud-to-ground lightning over Houston, Texas. *Weather Forecast* 26: 199–212. https://doi.org/10.1175/2010WAF2222431.1.
49. Petersen, W.A., H.J. Christian Jr., and S.A. Rutledge. 2005. TRMM observations of the global relation-ship between ice water content and lightning. *Geophysical Research Letters* 32: L14819. https://doi.org/10.1029/2005GL023236.
50. Petersen, W.A., D. Marks, and D. Wolff. 2014. GPM ground validation NASA S-band dual polarimetric (NPOL) Doppler radar IFloodS. http://ghrc.nsstc.nasa.gov/ available from the NASA EOSDIS Global Hydrology Resource Center Distributed Active Archive Center, Huntsville, Alabama. http://dx.doi.org/10.5067/GPMGV/IFLOODS/NPOL/DATA101.
51. Rennie, S.J., A.J. Illingworth, S.L. Dance, and S.P. Ballard. 2010. The accuracy of Doppler radar wind retrievals using insects as targets. *Meteorological Applications* 17: 419–432. https://doi.org/10.1002/met.174.
52. Roberto, N., E. Adirosi, L. Baldini, D. Casella, S. Dietrich, P. Gatlin, G. Panegrossi, M. Petracca, P. San, and A. Tokay. 2016. Multi-sensor analysis of convective activity in central Italy during the HyMeX SOP 1.1. *Atmospheric Measurement Techniques* 9: 535–552. https://doi.org/10.5194/amt-9-535-2016.
53. Roberto, N., L. Baldini, E. Adirosi, L. Facheris, F. Cuccoli, A. Lupidi, and A. Garzelli. 2017. A support vector machine hydrometeor classification algorithm for dual-polarization radar. *Atmosphere* 8: 134. https://doi.org/10.3390/atmos8080134.
54. Ruzanski, E., and V. Chandrasekar. 2015. Short-term predictability of weather radar quantities and lightning activity. In *2015 IEEE International Geoscience and Remote Sensing Symposium (IGARSS)*, Milan, 2608–2611. https://doi.org/10.1109/IGARSS.2015.7326346.

55. Saunders, C.P.R., H. Bax-Norman, C. Emersic, E.E. Avila, and N.E. Castellano. 2006. Laboratory studies of the effect of cloud conditions on graupel/crystal charge transfer in thunderstorm electrification. *Quarterly Journal of the Royal Meteorological Society* 132: 2653–2673. https://doi.org/10.1256/qj.05.218.
56. Schultz, C.J., L.D. Carey, E.V. Schultz, and R.J. Blakeslee. 2015. Insight into the kinematic and microphysical processes that control lightning, jumps. *Weather Forecast* 30: 1591–1621. https://doi.org/10.1175/WAF-D-14-00147.1.
57. Seliga, T.A., and V.N. Bringi. 1976. Potential use of radar differential reflectivity measurements at orthogonal polarizations for measuring precipitation. *Journal of Applied Meteorology* 15: 69–76. https://doi.org/10.1175/1520-0450(1976)015<0069:PUORDR>2.0.CO;2.
58. Srivastava, R., T. Matejka, and T. Lorello. 1986. Doppler radar study of the trailing anvil region associated with a squall line. *Journal of the Atmospheric Sciences* 43: 356–377. https://doi.org/10.1175/1520-0469(1986)043<0356:DRSOTT>2.0.CO;2.
59. Steiner, M., R.A. Houze Jr., S.E. Yuter. 1995. Climatological characterization of three-dimensional storm structure from operational radar and rain gauge data. *Journal of Applied Meteorology* 34: 1978–2007. https://doi.org/10.1175/1520-0450(1995)034<1978:CCOTDS>2.0.CO;2.
60. Takahashi, T. 1978. Riming electrification as a charge generation mechanism in thunderstorms. *Journal of the Atmospheric Sciences* 35: 1536–1548. https://doi.org/10.1175/1520-0469(1978)035<1536:REAACG>2.0.CO;2.
61. Waldteufel, P., and H. Corbin. 1979. On the analysis of single-Doppler radar data. *Journal of Applied Meteorology* 18: 532–542. https://doi.org/10.1175/1520-0450(1979)018<0532:OTAOSD>2.0.CO;2.
62. Waldvogel, A., B. Federer, and P. Grimm. 1979. Criteria for the detection of hail cells. *Journal of Applied Meteorology* 18: 1521–1525. https://doi.org/10.1175/1520-0450(1979)018<1521:CFTDOH>2.0.CO;2.
63. Wang, Y., V. Chandrasekar, and V.N. Bringi. 2006. Characterization and evaluation of hybrid polarization observation of precipitation. *Journal of Atmospheric and Oceanic Technology* 23: 552–572. https://doi.org/10.1175/JTECH1869.1.
64. Weinheimer, A.J., and A.A. Few. 1987. The electric field alignment of ice particles in thunderstorms. *Journal of Geophysical Research* 92(D12): 14833–14844. https://doi.org/10.1029/JD092iD12p14833.
65. Wiens, K.C., S.A. Rutledge, and S.A. Tessendorf. 2005. The 29 June 2000 supercell observed during STEPS. Part II: lightning and charge structure. *Journal of the Atmospheric Sciences* 62: 4151–4177. https://doi.org/10.1175/JAS3615.1.
66. Williams, E.R. 1989. The tripole structure of thunderstorm. *Journal of Geophysical Research* 94(D11): 13:151–167. https://doi.org/10.1029/JD094iD11p13151.
67. Wilson, J.W. 1966. Movement and predictability of radar echoes. In *National Severe Storms Laboratory Technical Memorandum ERTM-NSSL-28, 1966*, 30.
68. Wilson, J.W., R.D. Roberts, C. Kessinger, and J. McCarthy. 1984. Microburst wind structure and evaluation of Doppler radar for airport wind shear detection. *Journal of Climate and Applied Meteorology* 23: 898–915. https://doi.org/10.1175/1520-0450(1984)023<0898:MWSAEO>2.0.CO;2.
69. Witt, A., M.D. Eilts, G.J. Stumpf, J.T. Johnson, E.D. Mitchell, and K.W. Thomas. 1998. An enhanced hail detection algorithm for the WSR-88D. *Weather Forecast* 13: 286–303. https://doi.org/0.1175/1520-0434(1998)013<0286:AEHDAF>2.0.CO;2.
70. Wood, V.T., and R.A. Brown. 1977. Effects of radar sampling on single-Doppler velocity signatures of mesocyclones and tornadoes. *Weather Forecast* 12: 928–938. https://doi.org/0.1175/1520-0434(1997)012<0928:EORSOS>2.0.CO;2.
71. Xiao, Q., Y. Kuo, J. Sun, W. Lee, E. Lim, Y. Guo, and D.M. Barker. 2005. Assimilation of Doppler radar observations with a regional 3DVAR system: impact of Doppler velocities on forecasts of a heavy rainfall case. *Journal of Applied Meteorology* 44: 768–788. https://doi.org/10.1175/JAM2248.1.

72. Xu, Q., K. Nai, and L. Wei. 2010. Fitting VAD winds to aliased Doppler radial-velocity observations: a global minimization problem in the presence of multiple local minima. *Quarterly Journal of the Royal Meteorological Society* 136: 451–461. https://doi.org/10.1002/qj.589.
73. Yoshida, S., T. Adachi, K. Kusunoki, S. Hayashi, T. Wu, T. Ushio, and E. Yoshikawa. 2017. Relationship between thunderstorm electrification and storm kinetics revealed by phased array weather radar. *Journal of Geophysical Research—Atmospheres* 122: 3821–3836. https://doi.org/10.1002/2016JD025947.
74. Zhang, J., and S. Wang. 2006. An automated 2D multipass Doppler radar velocity dealiasing scheme. *Journal of Atmospheric and Oceanic Technology* 23: 1239–1248. https://doi.org/10.1175/JTECH1910.1.
75. Zrnić, D.S., and P.R. Mahapatra. 1985. Two methods of ambiguity resolution in pulsed Doppler weather radars. *IEEE Transactions on Aerospace and Electronic Systems* 21: 470–483. https://doi.org/10.1109/TAES.1985.310635.
76. Zrnić, D.S., A. Ryzhkov, J. Straka, Y. Liu, and J. Vivekanandan. 2001. Testing a procedure for automatic classification of hydrometeor types. *Journal of Atmospheric and Oceanic Technology* 18: 892–913. https://doi.org/10.1175/1520-0426(2001)018<0892:TAPFAC>2.0.CO;2.

Cloud Radars

Takeshi Maesaka

1 Introduction

So far, many weather radars have been deployed around the world, and recently some of them have been upgraded with dual-polarization capabilities. Observation results by the operational radars are provided in real time to warn about impending disasters caused by heavy rainfall. Since historically the radar meteorology started from research to eliminate precipitation noise in military radar, the primary target of weather radar is precipitation. Now, most meteorologists recognize that the echo of weather radar comes from rain, snow or sleet, and not from cloud droplets.

On the other hand, the research of cloud microphysics is significant and is especially needed in weather and climate studies. Cloud processes are critical in numerical modeling, but cloud microphysics representation remains inadequate. One of the difficulties of the observational studies of clouds is that in-situ measurements cannot be performed on the ground, except for fog. To solve this problem, we must use instruments on an airplane or balloon-borne sounding system. Nevertheless, with these methods, it is quite difficult to observe a cloud for a long time. Thus, a remote sensing instrument, primarily a cloud radar, which is similar to a weather radar devised to detect clouds, is needed for steady and long-term observation of clouds.

A Quantitative Precipitation Estimation (QPE) is now maturing with a dual-polarization radar, and researchers' interests are shifting to a Quantitative Precipitation Forecast (QPF). In particular, the prediction of localized heavy rainfall is one of significant interest. To make the leading time of such prediction longer, data from observations of the early stage of cumulonimbus clouds, i.e. cumulus clouds,

T. Maesaka (✉)
National Research Institute for Earth Science and Disaster Resilience, Tsukuba, Ibaraki, Japan
e-mail: maesaka@bosai.go.jp

C. Andronache (ed.), *Remote Sensing of Clouds and Precipitation*,
Springer Remote Sensing/Photogrammetry,
https://doi.org/10.1007/978-3-319-72583-3_5

are needed as initial conditions for a numerical forecasting model. However, most operational weather radar can capture only precipitation, and cannot detect clouds. The cloud radar is also expected to issue a warning of the localized heavy rainfall.

A historical overview of cloud radar development is summarized in [4]. The U.S. Air Force started the development of cloud radar just after World War II and deployed the vertically pointing radar (AN/TPQ-11) to monitor the cloud-deck near airports. The cloud radar uses millimeter wavelength, shorter than the centimeter wavelength used by precipitation radar, for reasons described later. However, the reliability of microwave devices for such radio wave was not enough for stable operation. For this reason, the Air Force's system was retired, and the cloud radar was not prevalent. The recent advances in the millimeter wavelength devices and data processing allowed the development of stable cloud radar system. At first, the cloud radar started as a vertically pointing radar. Afterwards the radar evolved into airborne, spaceborne, and scanning radars.

This chapter describes some implementations and conditions needed to observe clouds by weather radar, with a focus on sensitivity and wavelength usage. Then, an example of recent observations, using a dual-polarization scanning cloud radar system, is described, and results are discussed.

2 Sensitivity of Cloud Radar

Weather radar transmits radio waves to clouds or precipitations, and receives radio waves scattered by them. These transmitted and received radio wave powers (P_t and P_r) satisfies the following radar equation (for distributed targets) as

$$P_r = \frac{G^2\lambda^2\theta^2 c\tau}{2^{10}\pi^2(\ln 2)} \cdot P_t \cdot \frac{1}{r^2} \cdot \eta \tag{1}$$

where G is the antenna gain, λ is the wavelength, θ is the beam width, c is the speed of light, τ is the pulse width, r is the distance from radar, and η is the radar reflectivity. Note that G and θ are almost proportional to λ^{-2} and λ, respectively, so the first fraction of the right side in Eq. (1) is not a function of wavelength. Under the Rayleigh approximation (when particle size is much smaller than wavelength), the reflectivity factor η is denoted as

$$\eta = \frac{\pi^5}{\lambda^4}|K|^2 Z \tag{2}$$

where $|K|^2$ is the refractive index factor for the observed hydrometeors, and Z is the radar reflectivity factor. The radar reflectivity factor Z is expressed as

$$Z = \sum_D N(D)D^6 \tag{3}$$

where D is the particle diameter, and $N(D)$ is the number density of the particle with the diameter D. Typically, the radar reflectivity factor Z for rainfall is larger than 10 dBZ, and Z for drizzle is about 0 dBZ.

When the hydrometeors are detectable by the radar, the received power P_r must be larger than a noise level (or a minimum sensitivity, almost -110 dBm, typically) of the radar system for a certain radar reflectivity factor (minimum detectable reflectivity factor). Although the minimum detectable reflectivity factor varies with the distance from the radar, the typical value is larger than 0 dBZ in case of the ordinal precipitation radar with centimeter wavelength. The value is enough for the precipitation.

Because the typical diameter of a cloud droplet is several tens of micrometers, the reflectivity factor of cloud is quite smaller than that of precipitation as expected from Eq. (3). Although the number density of cloud droplets is much larger than that of precipitations, the sixth-power of the diameter is more effective in Eq. (3). The reflective factor of clouds is less than -10 dBZ, which it is difficult to detect by the precipitation radar with centimeter wavelength. To make the cloud observation possible by weather radar, the following alternations are needed as suggested from Eqs. (1) and (2).

Using Longer Pulse Width The longer pulse width τ makes the received power P_r higher to improve the radar sensitivity; however, it also makes the range resolution coarse. Fortunately, now pulse compression technology, which is implemented by modulating the transmitted pulse (linear frequency modulation is typically used) and then calculating the correlation between the received and transmitted pulses, is available. The technique enables to make the range resolution compressed (improved) while preserving the sensitivity.

Using High Transmitted Power Equation (1) indicates that higher transmitted power P_t makes the received power P_r higher. This implementation needs high power microwave devices, which are more expensive than the low power devices.

Making the Observation Range Shorter The observation range of ordinal precipitation radar is several hundreds of kilometers. Because the received power P_r is inversely proportional to the second power of the distance from the radar, the high sensitivity is achieved only in near radar range ($\sim$10 km).

Using Shorter Wavelength Under the Rayleigh approximation, the received power P_r is inversely proportional to the fourth power of wavelength (Eq. (2)). The shorter wavelength makes the received power P_r higher. This implementation is more effective than the other ways described above because of the fourth power. If the wavelength changes from 3 cm to 3 mm, for example, the sensitivity improves 10^4 times (40 dB). This sensitivity improvement is why millimeter-wavelength is usually used for cloud observations.

3 Bragg Scattering

The weather radar sometimes captures a clear-air echo (an angel echo) in a non-precipitating area. This echo comes from the Bragg scattering, which is caused by refractive index inhomogeneity in the atmosphere. The actual condition of the inhomogeneity is a variation in water vapor associated with turbulence. This echo sometimes results in a misdetection of the precipitation, although its intensity is weaker than that of the precipitation echo. On the other hand, this echo sometimes works in our favor as enabling to acquire Doppler velocities in a non-precipitating area.

For the cloud observation, the cloud echo is contaminated with the Bragg scattering. Figure 1 illustrates conditions of the wavelength and radar reflectivity factor where the backscattered radar reflectivity from clouds equals that from the

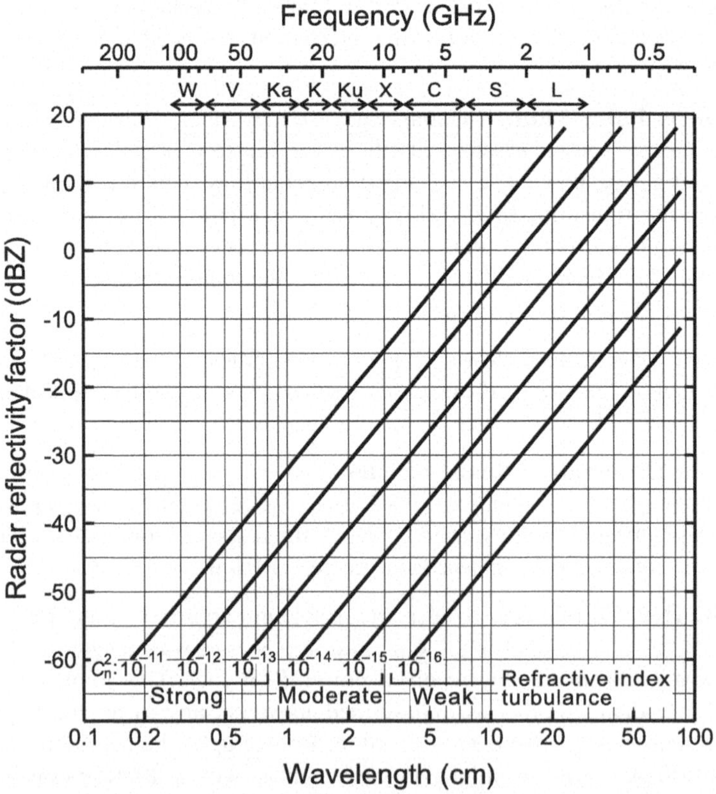

Fig. 1 Conditions of wavelength and radar reflectivity factor where the backscattered radar reflectivity from clouds equals that from the Bragg scattering for the variety of turbulences (based on [2]). C_n^2 is the refractive index structure parameter characterizing the turbulent fluctuation in refractive index

Bragg scattering for the variety of turbulences [2]. The radar reflectivity of Bragg scattering η_B is a function of wavelength, and is denoted as

$$\eta_B = 0.38\lambda^{-1/3}C_n^2 \quad (4)$$

where C_n^2 is the refractive index structure parameter characterizing the turbulent fluctuation in refractive index. The lines as shown in Fig. 1 are deduced by conjunction of Eqs. (2) and (4). When the backscattered radar reflectivity from clouds is larger than that from the Bragg scatter (the condition of the wavelength and radar reflectivity factor of cloud locates lower right to the lines in Fig. 1), the cloud signal is buried in the Bragg scattering. If the wavelength is 10 cm (S-band radio wave) and the target cloud radar reflectivity factor is −20 dBZ under the condition of $C_n^2 = 10^{(-13)}$, for example, we cannot detect the cloud by radar. Figure 1 suggests that centimeter-wavelength is unfavorable for observing the cloud whose radar reflectivity index is usually less than −10 dBZ, though the implementations by the pulse width, transmitted power and observation range, as described in the previous section, can make the radar sensitivity high enough to detect the cloud signal with the centimeter-wavelength. This is also why millimeter-wavelength is selected for this purpose.

4 Atmospheric Transmissivity

Atmospheric transmissivity for radio waves is also important. If the radio waves are attenuated between the radar and targeted cloud, the cloud signal becomes weak to be varied in noise. Figure 2 shows the specific attenuations by dry air and water vapor, which are the major components in atmosphere, for the various radio frequencies [5]. The attenuation by dry air, which is caused by molecular oxygen (O_2) absorption, has two peaks around 60 and 120 GHz in the frequency range in Fig. 2. The peak of the water vapor attenuation is around 22 GHz. Because frequencies associated with these peaks should be avoided, the Ka- and W-bands (27–40 GHz and 75–110 GHz in frequency, 7.5–11 mm and 2.7–4.0 mm in wavelength) are used for the millimeter-wavelength cloud radar. These bands are in atmospheric radio windows (ranges of radio frequency that the atmosphere lets through); however the values of specific attenuation for these bands (0.1–1 dB km^{-1}) are much larger than that for centimeter-wavelength radio band (0.01 dB km^{-1}, typically). This is why the observation range of millimeter-wavelength cloud radar cannot be extended as long as that of the centimeter-wavelength precipitation radar.

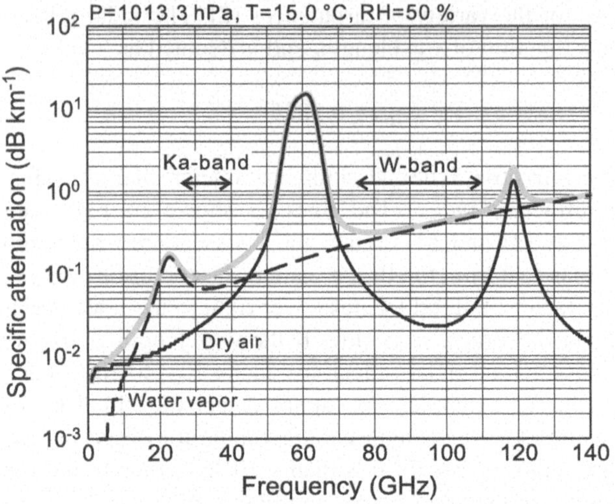

Fig. 2 Specific attenuations by atmosphere for the various radio frequencies when air pressure, temperature and relative humidity are 1013.6 hPa, 15 °C and 50%, respectively (based on [5]). Solid and dashed lines indicate the attenuation by dry air (oxygen O_2) and water vapor H_2O, respectively. Thick gray line indicates the total attenuation

5 Types of Cloud Radars

The previous sections described the implementations and conditions needed to observe the cloud by radar, and showed the millimeter-wavelength radio wave is suitable for the cloud radar. Now various kinds of weather radar with the millimeter-wavelength are developed to investigate the cloud microphysics. Here, the types of cloud radars are classified from the viewpoints of antenna scanning and capability, and these advantage and disadvantage are summarized.

Vertical Pointing Radar The history of cloud radar starting with the vertically pointing radar was already mentioned. Because the mechanics setup is so simple, this radar is suitable for long-term monitoring of clouds. The observation results are usually displayed and analyzed in a time-height cross section.

Scanning Radar This radar has a parabolic or Cassegrain antenna for which azimuth and elevation angles can be independently adjusted by an axial rotation mechanism. This radar usually performs PPI (Plan Position Indicator), and RHI (Range Height Indicator), scans (rotating the antenna horizontally with fixed elevation angle, and rotating the antenna vertically with fixed azimuth angle, respectively) to capture the three-dimensional structure of clouds. Comparing with the vertically pointing radar, an integration time (sample number of pulses) of the scanning radar tends to be shorter (smaller). This feature makes the weak cloud detectability of the scanning radar worse. For recovering the detectability, further implementations to improve the sensitivity are needed, such as using the pulse compression or high-power microwave amplifier.

Doppler Radar Doppler radar can observe a radial (a direction away from or toward the radar) component of target motion speed (Doppler velocity), which is approximately vector addition of wind and fall velocity of particles, by determining the frequency difference between the transmitted and received radio waves (Doppler effect). In pulse radar, the observed Doppler velocity is confined in Nyquist velocity denoted as

$$V_N = \frac{\lambda \cdot PRF}{4} \tag{5}$$

where *PRF* is a Pulse Repetition Frequency (PRF). The velocity exceeding the Nyquist velocity is folded by aliasing. If the PRF is 2000 Hz, for example, the Nyquist velocities for the wavelength of 3 cm and 3 mm are 15 m s^{-1} and 1.5 m s^{-1}, respectively. This difference indicates that the Doppler velocity observed by the millimeter-wavelength cloud radar is not as useful as that by centimeter-wavelength precipitation radar for wind observation.

On the other hand, the Doppler capability works well with the vertical pointing cloud radar. The shorter wavelength makes the Nyquist velocity smaller; however it makes the resolution of the Doppler velocity higher. (What is measured in radar system is a phase difference [−180° to +180°] between current and previous pulses, and the range of the phase difference corresponds the velocity range between $-V_N$ and $+V_N$.) The Doppler velocity observed by vertical pointing radar includes great information of a terminal fall velocity of the cloud particles. Because the terminal fall velocities of small ice crystals and cloud droplets are very small, it helps a better understanding of the cloud microphysics.

Dual-Polarization Radar Dual-polarization radar radiates both horizontally and vertically polarized waves simultaneously or alternately pulse by pulse to investigate the target shape information. The dual-polarization capability succeeded in the accurate QPE and attenuation correction by analyzing the differential phase shift between horizontal and vertical waves. When the size of raindrop becomes larger, its shape becomes distorted like a hamburger bun by an aero-dynamical effect. These QPE and attenuation corrections are based on the axis-ratio change with the drop size development. However, the shape of cloud droplet is almost sphere, because its size and terminal fall velocity are quite small. This situation indicates that the dual-polarization capability is worthless for the estimation of cloud water content and radio wave attenuation. For the ice-crystal cloud and mixed cloud, this ability provides useful information to analyze the particle types. Note that the dual-polarization capability greatly helps a data quality control to remove non-meteorological echoes.

Dual-Wavelength Radar Dual-wavelength radar radiates two wavelength radio waves simultaneously. Sometimes the two radio waves share one antenna, and sometimes two antennas are collocated to observe the same direction. Although the dual-polarization radar is worthless for water cloud observation, this radar provides Dual-Wavelength Ratio (DWR), which is a ratio of the radar reflectivity factors for two wavelengths, and which is well correlated with the cloud water content and the radio wave attenuation by the water cloud. The Ka- and W-band radio waves are usually selected for cloud observation because of the reasons already mentioned in this section.

6 Dual-Polarization Scanning Cloud Radar

Recently, municipal engineers showed that urban city is vulnerable to localized, and heavy rainfall. Thus, precipitation monitoring and forecasting are essential for water management. A weather radar is a powerful tool for this purpose; however what current operational radars (S-, C- and X-band) can detect is only precipitation particles (hydrometeors excluding cloud particles). This situation means that the rainfall has already started when the radar catches the echo. Meanwhile, the heavy rainfall is caused by a mature cumulonimbus, which develops from a cumulus. So, the detection of the cumulus is expected to make the lead time of heavy rainfall forecasting longer. However, the current operational weather radar cannot observe the cumulus which consists of the cloud particles. For an early warning of heavy rainfall, National Research Institute for Earth Science and Disaster Resilience (NIED), Japan started deploying the cloud radar network around Tokyo metropolitan area [6]. This network consists of five cloud radars and covers the most of the densely inhabited area around Tokyo (Fig. 3). In this section, the specifications of the NIED's cloud radar are introduced, and some observation results are presented.

6.1 Requirements Specification

A key microphysical process of the evolution from cumulus to cumulonimbus is a collision-coalescence process. To monitor the initiation of cumulonimbus cloud, the cloud radar must be able to observe the cloud in which the conversion from cloud to rain drops has just activated. This conversion process reminds us of the Kessler's warm rain parameterization [3], which is formulated as

$$\frac{dq_r}{dt} = \alpha(q_c - q_{c0})H(q_c - q_{c0}) \tag{6}$$

where q_r and q_c are mixing ratios of rain and cloud, respectively, α is a tuning constant, and $H(x)$ is the Heaviside step function. When the mixing ratio of cloud q_c exceeds the threshold q_{c0}, the conversion is activated. The values $\alpha \sim 10^{-3}\,\mathrm{s}^{-1}$ and $q_{c0} \sim 1.0\,\mathrm{g\,kg}^{-1}$ are usually selected for deep cumulus convection. This parameterization suggests the requirements specification of the cloud radar.

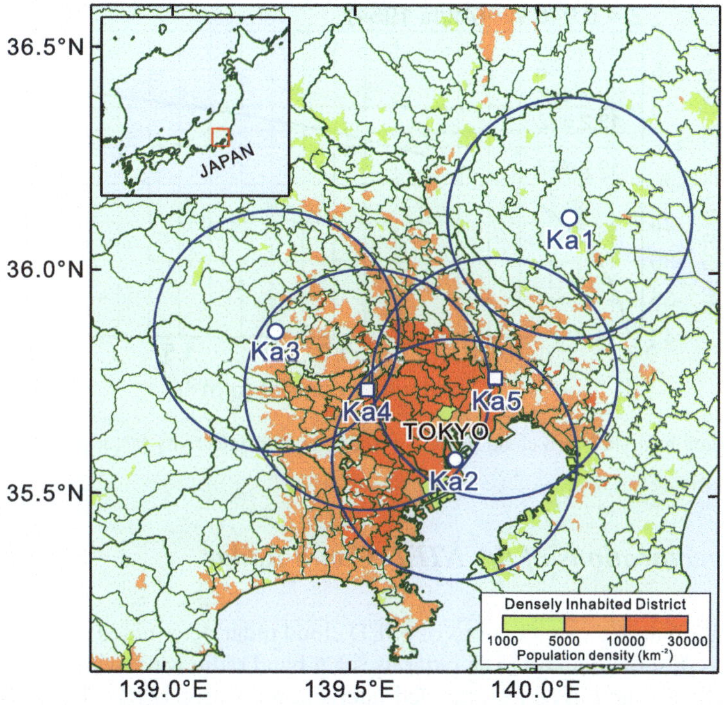

Fig. 3 Locations of NIED's scanning cloud radars. Open circles indicate the locations of Dual-polarization radars (Ka1–Ka3), while open squares indicate the locations of single-polarization radars (Ka4 and Ka5). Solid blue circles show the observation ranges ($r = 30$ km) of each radar. Colored areas illustrate densely inhabited districts provided by National Land Numerical Information, Japan

Figure 4 shows a relationship between cloud water content and radar reflectivity factor proposed by Atlas [1]. Because air density in lower troposphere is almost 1 kg m^{-3}, the mixing ratio of cloud and cloud water content are substitutable each other. This relationship indicates that the radar reflectivity factor corresponding to the mixing ratio of cloud of 1.0 g kg^{-1} is -13.2 dBZ. This means that the radar reflectivity factor of cumulus cloud is less than -13.2 dBZ. Considering that the Kessler's original value of q_{c0} was 0.5 g kg^{-1} (corresponding to the radar reflectivity factor of -19.2 dBZ), the requirements specification of the cloud radar is determined as "The radar reflectivity factor of -20 dBZ is detectable at the range of 20 km by single-polarization radar." In case of dual-polarization radar, the value is -17 dBZ, because radar transmit power is divided into the horizontal and vertical polarizations equally. By defining the detectable received power as 3 dB larger than noise level, the radar specifications needed for cumulus observation are derived from Eqs. (1) and (2). The attenuation by the atmosphere (0.15 dB km^{-1}) is also taken into account.

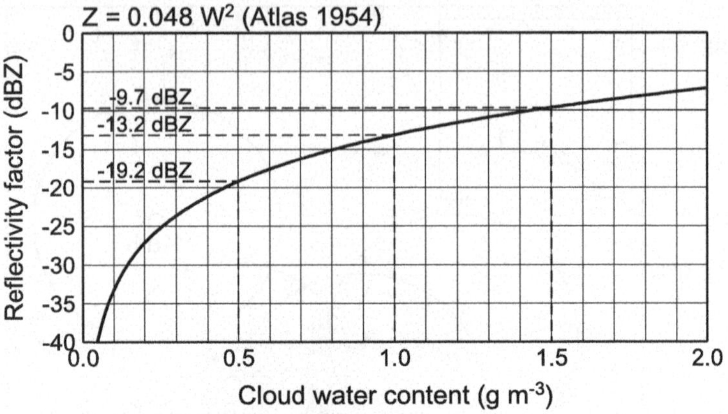

Fig. 4 Relationship between cloud water content and radar reflectivity factor proposed by Atlas [1]

6.2 *Specifications of the NIED Cloud Radar*

Table 1 summarizes specifications of NIED cloud radar determined as satisfying the conditions described above. The radar uses Ka-band radio wave with 3 kW transmit power, which is amplified by extended interaction klystron (EIK). The pulse width of 55 μs (long pulse) is usually used with pulse compression technology to improve the range resolution and sensitivity. While the pulse is transmitting, the radar system cannot receives the signal. This means that the radar cannot observe the echo in the range closer than $(\tau c)/2$ (8.25 km, with $\tau = 55\,\mu s$). This range is covered by short pulse (1.0 μs) observation.

The cloud radar mounts a Cassegrain antenna with the diameter of 2.2 m (Fig. 5). This antenna creates sharp pencil beam with the beam width of 0.3° to make the special resolution finer than the same size antennas with S-, C- and X-band radio waves. The antenna gain (54 dBi) is also larger than those with S-, C- and X-band radio waves.

As shown in Fig. 3, three of the cloud radars are dual-polarization radars, while two radar radars are single-polarization radar with 3 dB higher sensitivity than the dual-polarization radars. Both types of radar output received power, radar reflectivity factor, Doppler velocity, Doppler spectrum width, signal quality index (SQI, a.k.a. normalized coherent power) and signal to noise ratio (SNR). Furthermore, the dual-polarization radars output polarimetric parameters, differential reflectivity (Z_{DR}), differential phase (Φ_{DP}), specific differential phase (K_{DP}), correlation coefficient (ρ_{HV}) and linear depolarization ratio (LDR, available with single-polarization transmit).

Table 1 Specifications of NIED's scanning cloud radars

Features	Specifications
Frequency	34.815–34.905 GHz (Ka-band)
Occupied bandwidth	13 MHz
Microwave amplifier	Extended interaction klystron (EIK)
Transmit power	3 kW
Pulse compression	Linear frequency modulation (2 MHz)
Pulse width (short)	0.5 and 1.0 μs
Pulse width (long)	30, 45, 55, 80 and 100 μs
Pulse repetition frequency	Max. 2500 Hz @ pulse width (long) = 30 μs
IF digitizer	16 bit, 36 MHz
Antenna	Cassegrain antenna ($\Phi = 2.2$ m)
Antenna gain	54 dBi (Typically)
Beam width	0.3° (Typically)
Antenna Sidelobe level	−23 dB (Typically)
Polarization	H/V simultaneous (3 radars)
	Single H or V (2 radars)
Observation range	30 km
Output data resolution	75 or 150 m
Nyquist Velocity	5.38 m s^{-1} @ PRF=2500 Hz
Dual PRF	3:2, 4:3 and 5:4 for velocity de-aliasing
Clutter filters	IIR or spectrum interpolation
Output data	Received power (Pr)
	Radar reflectivity factor (Z)
	Doppler velocity (V)
	Doppler spectrum width (W)
	Signal quality index (SQI)
	Signal to noise ratio (SNR)
Output data (dual-pol.)	Differential reflectivity (Z_{DR})
	Differential phase (Φ_{DP})
	Specific differential phase (K_{DP})
	Correlation coefficient (ρ_{HV})
	Linear depolarization ratio (LDR, available with single polarization transmit)

6.3 *Cloud Radar Observations*

Figure 6 shows the rainfall intensity observed by NIED's and Ministry of Land, Infrastructure, Transport and Tourism's (MLIT's) X-band dual-polarization radars at 1925 JST (Japan Standard Time) on 7 September 2015. The stationary front located to the south of Tokyo, and had been stagnant for several hours. The moderate rainfall were analyzed near the front, and weak rainfall (<1 mm h^{-1}) were also analyzed to

Fig. 5 Photo of NIED's scanning cloud radar (Ka1)

the north of frontal zone. At this time, the Ka1 radar made a Range Height Indicator (RHI) scan to observe the vertical cross-section as shown in Fig. 6.

Figure 7 indicates the RHI vertical cross-section observed by the Ka1 radar at 1921 JST 7 September 2015. In the radar reflectivity factor field (Fig. 7a), the noise is also illustrated, and the noise gap at the range of 8.25 km corresponds to the boundary of short and long pulses observation. The shallow weak rainfall (15 dBZ_e echo) was observed 20 km far from the radar, and this rainfall corresponded to the weak rainfall on the ticked line in Fig. 6. Shallow stratus cloud (fog) with the reflectivity factor less than −20 dBZ_e are also captured below 2 km ASL. The Ka-band cloud radar succeeded in observing the stratus cloud, which it is impossible for X-, C- and S- bands weather radar to detect. The cloud radar well captured cirrostratus in the layer between 5.5 and 9.0 km ASL. Because 0 °C level was 4650 m around the Ka1 radar at that time, ice particles prevailed in the cirrostratus. The Ka-band cloud radar is useful for ice cloud observation.

Figure 7b shows de-aliased Doppler velocity observed by the Ka1 radar. The RHI scan was observed with dual-PRF (Pulse Repetition Frequency) of 1980 and 1584 Hz (5:4). These PRFs correspond to the Nyquist velocities of 4.257 and 3.4056 $m\,s^{-1}$ (from Eq. (5) with $\lambda = 8.6$ mm). By applying the dual-PRF de-aliasing method, the maximum Doppler velocity is improved to the least common multiple

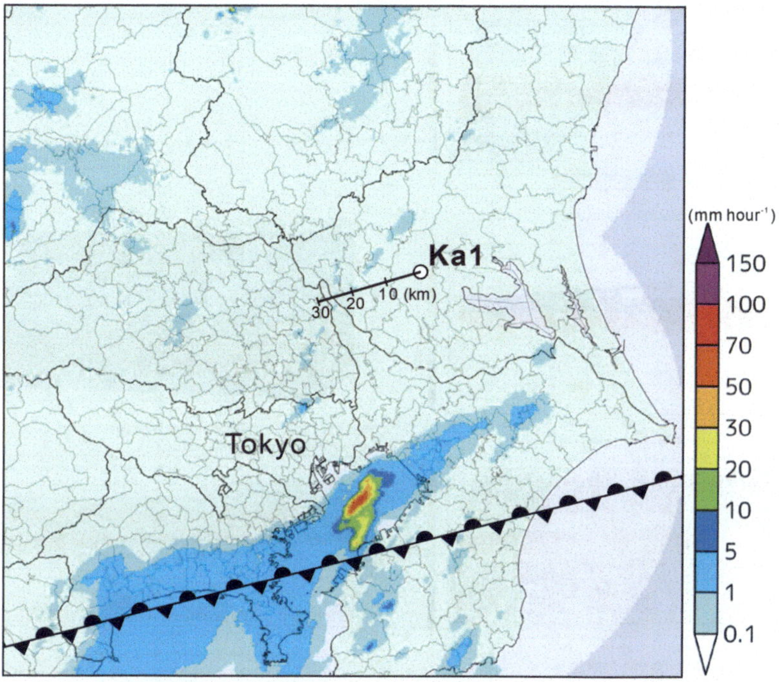

Fig. 6 Rainfall intensity observed by NIED's and Ministry of Land Infrastructure, Transport and Tourism's (MLIT's) X-band dual-polarization radars at 1925 JST (Japan Standard Time) 7 September 2015. Open circle indicates the location of NIED Ka1 radar, and ticked line indicates a vertical cross-section of RHI (Range Height Indicator) scan as shown in Fig. 7. Stationary front is subjectively drawn by considering Japan Meteorological Agency's surface observation data

of the Nyquist velocities (17.028 m s^{-1}). However a part of the Doppler velocity above 7 km ASL exceeded the maximum, and aliased (negative Doppler velocity area embedded in the positive velocity). Some parts of the Doppler velocities were also discontinuous due to de-aliasing failure. Because the Nyquist velocity of the Ka-band radar is smaller than those of X-, C- and S-bands, the de-aliasing failure sometimes occurs in the area where spatial variation of wind is large. Figure 7c shows differential reflectivity (Z_{DR}) observed by the Ka1 radar. The Z_{DR} is a difference between horizontal and vertical radar reflectivity factors, and is usually expressed in decibel scale. The definition is

$$10 \log Z_{DR} = 10 \log \frac{Z_H}{Z_V} = 10 \log Z_H - 10 \log Z_V \tag{7}$$

where subscriptions H and V indicate the horizontal and vertical polarizations, respectively. The positive Z_{DR} indicates the existence of the particle which has longer horizontal axis than vertical axis. In the weak rainfall (drizzle) and stratus area, the Z_{DR} value was small, because the shapes of drizzle and cloud drops

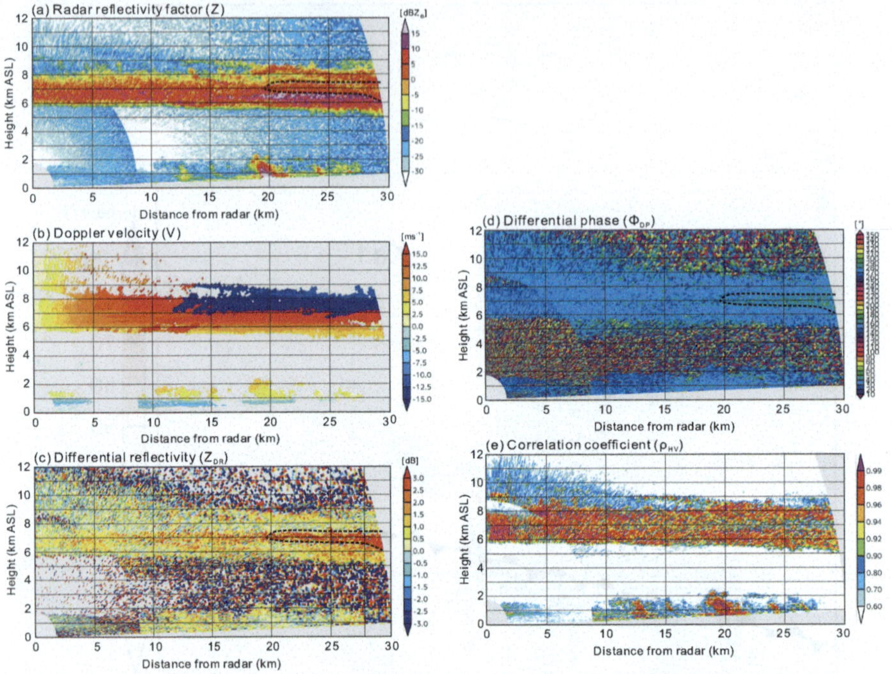

Fig. 7 RHI (Range Height Indicator) scan observed by NIED Ka1 radar at 1921 JST 7 September 2015. Azimuth angle of RHI scan was 256.6°. (**a**) Reflectivity factor, (**b**) Doppler velocity (de-aliased), (**c**) differential reflectivity, (**d**) differential phase and (**e**) correlation coefficient. Broken lines in (**a**), (**c**) and (**d**) indicate the high differential reflectivity area

are almost sphere. On the other hand, the large positive Z_{DR} was observed in the cirrostratus (around the height of 7 km ASL). This indicates the existence of flat ice particles, like plates or dendrites.

Figure 7d shows differential phase (Φ_{DP}) observed by the Ka1 radar. When the radio wave travels through the cloud and precipitation, the propagation speed is slightly delayed. If the cloud and precipitation particles has longer horizontal axis than vertical axis, the horizontal polarization is more delayed than the vertical polarization. The difference of the delay is detected as differential phase between horizontal and vertical polarizations. The distant delay of one wavelength corresponds to the Φ_{DP} of 360°. Note that the Φ_{DP} monotonically increases with range from radar in the precipitation and cloud which consist of the particles with longer horizontal axis, because the effect of the delay is accumulated with the radio wave propagation (forward scattering phase shift). In the lower layer of Fig. 7d, the Φ_{DP} was slightly increasing with the distance from radar. This indicates the existence of slightly flatten raindrops, but the number of such a drop may not be so large. In the cirrostratus, a peak of Φ_{DP} was found in the high Z_{DR} area (denoted by broken lines in Fig. 7), and the Φ_{DP} was not increasing monotonically around the peak. This peak might be created by back scattering phase shift, which is sometimes observed in the area where big and flatten particles exist. Just below the large values of Z_{DR} and Φ_{DP} denoted by broken lines in Fig. 7, a peak of radar reflectivity factor was found in Fig. 7a. This might be a result of the aggregation growth of plates and dendrites.

Figure 7e shows correlation coefficient between horizontal and vertical polarization (ρ_{HV}) observed by the Ka1 radar. This value is in the range between 0 and 1. When horizontal/vertical axis ratio of the particle fluctuates pulse by pulse (e.g., tumbling of non-spherical particle), the ρ_{HV} has small value. The ρ_{HV} of pure rainfall is almost one. The observed ρ_{HV} also depends on signal to noise ratio (SNR). When the SNR is very small in weak echo region, the ρ_{HV} has small value. This is why the ρ_{HV} is small in the lower stratus cloud.

7 Conclusion

Because the radar reflectivity of a cloud is quite smaller compared to that of precipitation, the cloud observation by weather radar needed various adaptations to improve the sensitivity. This chapter describes how the dual-polarization cloud radar provides valuable information about cloud microphysics. Observations from cloud radars can be used for advanced warning of heavy rainfall events. Recent progress in radar technology enables us to produce a stable and reliable dual-polarization cloud radar, which gradually will become prevalent in the world. This type of radar is expected to aid the progress of cloud microphysics studies and provide much-needed data for a better characterization of cloud systems.

Acknowledgements I am very grateful to Dr. Constantin Andronache for the chance to contribute this chapter and encouragement. I also thank all my colleagues in the National Research Institute for Earth Science and Disaster Resilience for installation, operation and maintenance of the cloud radar network. Some X-band radar data used in this chapter are provided by Ministry of Land, Infrastructure, Transport and Tourism, Japan via Data Integration and Analysis System (DIAS), Japan.

References

1. Atlas, D. 1954. The estimation of cloud parameters by radar. *Journal of Meteorology* 11: 309–317.
2. Gossard, E.E. 1988. Measuring drop-size distributions in clouds with a clear-air-sensing Doppler radar. *Journal of Atmospheric and Oceanic Technology* 5: 640–649.
3. Kessler, E. 1969. On distribution and continuity of water substance in atmospheric circulations. *Meteorological Monographs* 10: 1–84.
4. Kollias, P., E.E. Clothiaux, M.A. Miller, B.A. Albrecht, G.I. Stephens, and T.P. Ackerman. 2007. Millimeter-wavelength radars. *Bulletin of the American Meteorological Society* 88: 1608–1624.

5. Liebe, H.J., G.A. Hufford, and M.G. Cotton. 1993. Propagation modeling of moist air and suspended water/ice particles at frequencies below 1000 GHz. In *AGARD Conference Proceedings*. Vol. 542, 3/1–10.
6. Maesaka, T., K. Iwanami, S.-I. Suzuki, Y. Shusse, and N. Sakurai. 2015. Cloud radar network in Tokyo metropolitan area for early detection of cumulonimbus Generation. Paper Presented at 37th Conference on Radar Meteorology. Norman: American Meteorological Society. https://ams.confex.com/ams/37RADAR/webprogram/Paper275910.html.

Part II
Spaceborne Remote Sensing

The Tropical Rainfall Measuring Mission

Mircea Grecu and David T. Bolvin

1 Introduction

Designed to provide accurate estimates of the precipitation falling in the Tropics and its diabatic contribution to the atmospheric energy budget, the Tropical Rainfall Measuring Mission's (TRMM) contribution to the science and technology of satellite remote sensing of precipitation significantly exceeded the mission's initial objectives. A major factor in TRMM's remarkable success is undoubtedly the quality and longevity of its instruments that provided an uninterrupted record of high quality infrared and microwave radiometer and precipitation radar observations that span over 17 years, i.e. from 1 December 1997 to 15 April 2015. This chapter's objectives are to describe the TRMM instruments and associated observations, the main products derived from TRMM observations and the use of TRMM products and observations in geophysical applications.

M. Grecu (✉)
Goddard Earth Sciences Technology and Research, Morgan State University, Baltimore, MD, USA

NASA Goddard Space Flight Center, Greenbelt, MD, USA
e-mail: mircea.grecu-1@nasa.gov

D. T. Bolvin
Science Systems and Applications, Inc., Lanham, MD, USA

NASA Goddard Space Flight Center, Greenbelt, MD, USA
e-mail: david.t.bolvin@nasa.gov

C. Andronache (ed.), *Remote Sensing of Clouds and Precipitation*, Springer Remote Sensing/Photogrammetry,
https://doi.org/10.1007/978-3-319-72583-3_6

2 Observations and Main Products

One remarkable fact about TRMM relative to previous satellite platforms was that it featured the first satellite Precipitation Radar (PR). Meant to provide the informational content required to make TRMM a "flying rain gauge" [60], i.e. a satellite platform facilitating the derivation of accurate precipitation estimates that may be used in the validation of other satellite precipitation estimates, TRMM PR proved itself a crucial instrument in the development and evaluation of novel improved methodologies to estimate precipitation from satellite observations. The main PR characteristics are summarized in Table 1 (adapted from [33]). While able to provide higher resolution and more specific information about precipitation, space borne radars have never been intended to replace satellite microwave radiometers as the instruments of choice to provide global information about precipitation. This is because, while significantly more expensive, space borne radars can provide significantly less coverage as their high resolution is intrinsically associated with significantly narrower swaths even for 3D radars such as the TRMM PR. It is worth mentioning that at monthly and longer temporal scales the PR sampling is no longer a limitation. To mitigate sampling limitations at shorter temporal scales, in addition to the PR, TRMM featured a microwave imager (TRMM Microwave imager or TMI) meant to provide information complementary to that provided by the PR within the common TMI/PR swath and facilitate the derivation of precipitation products consistent with those derived from PR observations but significantly less affected by temporal sampling errors. The main TMI characteristics are summarized in Table 2 (adapted from [33]). As apparent from Table 2, the TMI instrument was similar to previous microwave radiometers, i.e. the Special Sensor Microwave/Imager SSM/I) instruments aboard the U.S. Defense Meteorological Satellite Program (DMSP) satellites since 1987, but included an additional 10-GHz frequency channel. Moreover, because TRMM operated at much lower orbital height (350 km before August 2001 and 402.5 km afterwards) than that of the DMSP satellite, the TMI Effective Field of View (EOV) i.e. the field of view integrated over the sampling interval [29], was almost twice as small as that of a typical SSM/I EOV. A schematic of the PR's and TMI's scanning geometries is shown in [33]. From the scanning perspective, a major difference between the TMI and PR instruments is that while the TMI is a conically scanning instrument, i.e. characterized by a constant angle between the line of sight and the nadir (defined as the vertical

Table 1 PR main characteristics

Feature	Specification
Frequency	13.8 GHz
Sensitivity	17 dBZ (0.7 mm/h)
Vertical resolution	250 m
Horizontal resolution	4.3 km
Swath width	215 km
Number of rays	49

Table 2 TMI main characteristics

Channel	1	2	3	4	5	6	7	8	9
Frequency	10.65	10.65	19.35	19.35	21.3	37	37	85.5	85.5
Polarization	V	H	V	H	V	V	H	V	H
EFOV-CT (km)	9.1	9.1	9.1	9.1	9.1	9.1	9.1	4.6	4.6
EFOV-DT (km)	63.2	63.2	30.4	30.4	22.6	16	16	7.2	7.2
EFOV per scan	104	104	104	104	104	104	104	208	208

CT and DT stand for cross-track and down-track

from the instrument to the center of the Earth), the PR is a cross-track scanning instrument with an angle between the line of sight and nadir varying from $-17°$ to $17°$. Retrievals from observations collected by cross-tracking scanning instruments are usually more challenging than those from observations by conically scanning instruments, because the EOV resolution degrades as the instruments points away from nadir. However, the PR's EOV dependence on the pointing angle is minimal and, practically, the same retrieval algorithm can be applied irrespective of the pointing angle. Nevertheless, artifacts related to the EOV dependency on pointing angle are expected in the TRMM PR products because, among other factors, the extent of ground clutter are the strength of the bright band are functions of the pointing angle. In addition to the PR and the TMI, TRMM featured a Visible and Infrared Radiometer System (VIRS) and a Lightning Imaging Sensor (LIS). Although not as intensively used for satellite precipitation estimation as the PR and the TMI observations, the VIRS and LIS instruments were, nevertheless, widely used in a large variety of studies.

2.1 *Observations*

While strongly related to precipitation, the observations collected by the TRMM PR and TMI are not unambiguously invertible. That is, no unique, accurate mathematical relationship to relate the TRMM observations to the underlying precipitation variables exist. However, well developed theories and methodologies to derive optimal estimates from non-uniquely interpretable observations have been developed and applied to problems arising in a large variety of disciplines including the remote sensing of precipitation. But before dwelling into these theories, is it beneficial to qualitatively analyze the TRMM PR and TMI observations to understand their strengths and limitations.

Shown in the top panel of Fig. 1 is a cross section through three-dimensional (3D) reflectivity field provided the TRMM PR during an overflight of Hurricane Felix that occurred over the Atlantic Ocean, on 16 September 2001. As apparent in Fig. 1, the TRMM PR provides detailed and insightful information regarding the vertical structure of storms within its observing volume. The vertical extent

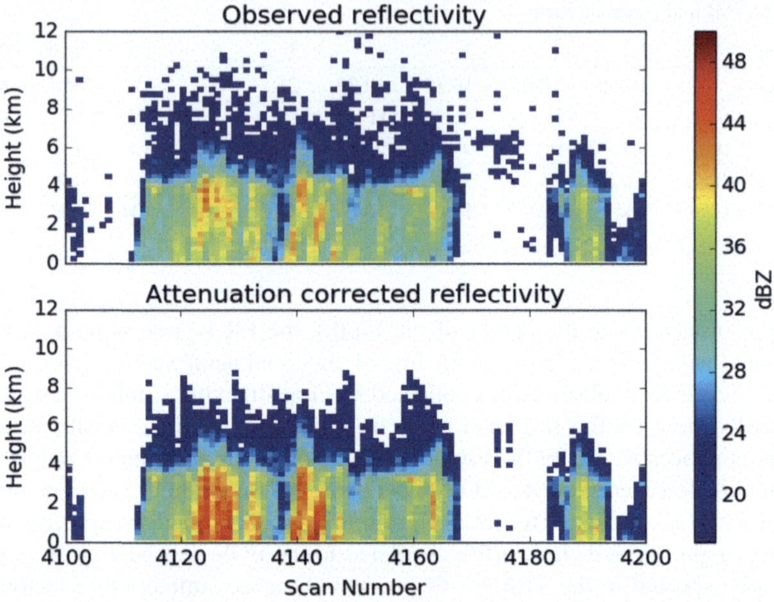

Fig. 1 Nadir cross section through the observed and attenuation corrected reflectivity fields for orbit 21899 on 16 September 2001. Version 7 of the TRMM PR products are used in this figure

and the intensity of precipitation above the freezing are readily noticeable in Fig. 1. Moreover, a bright band structure (reflectivity enhancement due to the melting of frozen hydrometeors; [2]) is apparent in the figure. Also apparent in some scans (around scan 4125) are the rather low reflectivity values at low levels (below 1 km) associated with quite large reflectivity values at heights above 3.0 km. While cloud physical processes such as evaporation and wind shear can explain reflectivity at lower levels significantly smaller than those above, attenuation due to rain and cloud water of the received rather signal may be also responsible for this phenomenon. This is because the TRMM PR operates at a frequency of 13.8 GHz at which the signal can be significantly attenuated by rain and cloud water. Consequently, the TRMM PR observations are used in quantitative analysis only after attenuation correction. Some information, such as the depth of observed precipitation structure, their type (e.g. convective vs stratiform; [22]) is derived before the attenuation correction procedure, but the derivation of precipitation estimates from the TRMM PR observations involves an attenuation correction step. This is described in the next section. Shown in the bottom panel of Fig. 1 are the attenuated corrected reflectivities. As apparent in the figure, there are significant differences between the observed and attenuated corrected reflectivities, which illustrates the necessity to account for attenuation in the interpretation of observed TRMM PR reflectivities.

Unlike the TRMM PR, the TRMM TMI provides observations that are significantly less localized. Specifically, the TRMM TMI measures radiation emitted by space, the Earth's surface and the atmosphere in large observing volumes (extending tens of kilometers horizontally and from the Earth's surface to the top of the atmosphere) and scattered or absorbed by the atmosphere, while the TRMM PR measures self-emitted radiation backscattered by precipitation particles within rather small observing volumes (extending a few kilometers in the horizontal direction and 250 m in the vertical direction). To facilitate the 3D reconstruction of the observed phenomena the TMI relies on multiple channels operating at multiple frequencies and in dual polarization. Early space borne microwave radiometers were primarily designed to provide observations useful in deriving estimates of geophysical variables such as the sea surface temperature, the sea surface wind speed, and the integrated atmospheric liquid water and water vapor over seas [12, 53], but it was quickly realized that space borne radiometer observations may be successfully used to detect heavy precipitation and even quantify its intensity [64]. As space-borne radiometers improved, it became obvious that multiple radiometer channels can provide complementary information about the gross vertical structure of precipitation systems over oceans [65]. Specifically, while low frequency channels (10- and 19-GHz) provide information mainly on the amount of precipitation below the freezing, the 85-GHz provides information about the amount of solid precipitation above the freezing level. The 36-GHz channel is sensitive to both emission by liquid precipitation below the freezing level and scattering of radiation by solid precipitation particle above the freezing level. The 22-GHz is also sensitive to the amount water vapor atmosphere and is useful in incorporating information about the amount of water vapor into the estimation process. The polarization information may be useful in minimizing the impact of the variability of surface's emissivity on the radiometer's signal interpretation.

Shown in Fig. 2 are the observed 19-GHz V and 85-GHz V brightness temperatures for a portion of TRMM orbit 21899 over Hurricane Felix on 16 September 2001. The vortex structure of the hurricane is apparent in both channels. However, as predicted by the theoretical calculations carried out in various studies [1, 63, 65], the 19-GHz V brightness temperature observations exhibit primarily emission, i.e. increase in the amount of radiation detected by the radiometer due to emission by raindrops below the freezing level, while the 85-GHz V brightness temperature observations exhibit mostly scattering, i.e. reduction in the amount of radiation detected by the radiometer due to scattering processes making the incoming radiation at the top of the atmosphere more relevant that emitted by hydrometeors and the surface of the Earth. The narrow rain-bands on the left side of the hurricane exhibit emission in both channels which is an indication of weak ice processes. Also, obvious in Fig. 2 is the much better resolution of the 85-GHz frequency channel relative to that of the 19-GHz channel.

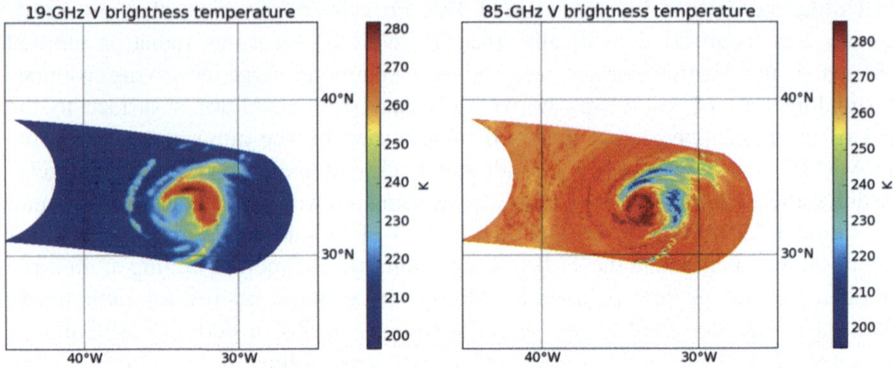

Fig. 2 Observed 19-GHz V and 85-GHz V brightness temperatures for a portion of TRMM orbit 21899 over Hurricane Felix on 16 September 2001

2.2 *Main Algorithms and Products*

2.2.1 The Radar Algorithm and Precipitation Products

As described above and apparent in Fig. 1, the PR reflectivity observations may be affected by significant attenuation. It this therefore imperative that any methodology meant to make quantitative use of the PR reflectivity observations incorporate an attenuation procedure correction. The attenuation correction is not a trivial process, although well-established methodologies exist [18], because the attenuation is a multivariate function of the Particle Size Distribution (PSD) within the radar observing volume, which makes the process uncertain. The procedure used in the official TRMM PR algorithm [27] makes use of a methodology commonly referred to as Hitschfeld and Bordan [18]. Specifically, starting from the relationship that relates the measured reflectivity at range r, $Z_m(r)$, to the true effective reflectivity at range r, $Z_e(r)$, i.e.

$$Z_m(r) = Z_e(r)exp(-0.2\ln(10)\int_0^r k(s)ds) \tag{1}$$

where $k(s)$ is the specific attenuation (in dB/km), and assuming that the specific attenuation and the true reflectivity are related through a power law of the type $k = \alpha Z^{\beta}$, a differential equation with an analytical solution can be derived. The user is referred to the study of [26] for the actual derivation. The analytical expression relating $Ze(r)$ to $Z_m(r)$ is

$$Z_e(r) = Z_m(r)/(1 - \epsilon q\beta \int_0^r \alpha(s)Z_m^{\beta}(s)ds)^{1/\beta} \tag{2}$$

where q=0.2ln(10) and ϵ is a factor to make the Path Integrated Attenuation (PIA), i.e. $\int_0^{r_s} k(s)ds$ where r_s is the surface range consistent with estimates of PIA based on the Surface Reference Technique (SRT). When ϵ=1, the original formulation of Hitschfeld and Bordan [18] is obtained. Interpolation techniques may be used to derive from observations of the surface return, i.e. $P_m(r_s)$, in rain free areas what the magnitude of the rain-affected surface return would be in the absence of rain $P_m(r_s;\text{no rain})$. Then the PIA can be estimated as

$$PIA_{SRT} = 10\log_{10}(P_m(rs;\text{no rain})/P_m(rs;\text{rain})) \tag{3}$$

[26, 48]. Discrepancies between the PIA estimate provided by Eq. (2) and that provide by Eq. (3) may be indicative of many potential deficiencies in the attenuation correction procedure, which are usually caused by inaccurate attenuation-reflectivity relationships. To relate the attenuation due to precipitation and the associated reflectivity within a radar observing volume (determined by the radar pulse width, the travel time of the pulse and the antenna beamwidth) knowledge regarding the distribution of precipitation particles as a function of their size within the observing volume is required [47]. Specifically, if the number of precipitation particles of size D per unit diameter is known, $N(D)$, one can calculate the specific attenuation and effective reflectivity as [47]

$$k = 0.434\int \sigma_t(D)N(D)dD \tag{4}$$

and

$$Z_e = \frac{\lambda^4}{\pi^5|K_w|^2}\int \sigma_b(D)N(D)dD \tag{5}$$

where σ_t and σ_b are the total extinction and backscattering cross-sections, λ is the radar wavelength and K_w is the water's dielectric constant. The total extinction and backscattering cross-sections may be derived using the Mie solution of Maxwell's electromagnetic equations [5] and, usually, are not a large source of uncertainties in attenuation reflectivity relationships in the rain phase. However, the PSD function, $N(D)$, is a source of considerable source of uncertainties. Two describe the PSD variability, functions of two or more variables are necessary. That is, two or more variables are required to calculate k and Z_e, while the single frequency TRMM PR systems provides only one measurement Z_m per observing volume. Therefore, function $N(D)$ cannot be uniquely determined from Z_m. Moreover, no unique coefficients exist to accurately describe the specific attenuation as a function of effective reflectivity using a $k = \alpha Z^\beta$ power law relationship. As a consequence the PIA derived by the analytical attenuation procedure is not necessarily consistent with the PIA estimate derived from the SRT. Iguchi and Meneghini [26] showed that adjusting the analytical attenuation correction based on the SRT PIA estimates results in improved rain estimation accuracy. It was shown [10] that the adjustment factor can be related to a $N(D)$ parameter, namely, the intercept in a is the normalized function [69]. Because the SRT PIA estimate is not uncertainty free, it is not optimal to make the analytical correction $[1-\epsilon q\beta \int_0^{r_s} \alpha(s)Z_m^\beta(s)ds]^{1/\beta}$ exactly match the SRT PIA. Instead a probabilistic procedure is applied to make optimal use of the "a priori" information regarding the retrieved precipitation and the "observed" SRT PIA [27]. That is, for heavy precipitation the SRT PIA estimates are expected to be reliable. In such situations, ϵ can be set to $\frac{1-10^{\beta PIA_{SRT}/10}}{\beta \int_0^{r_s} \alpha(s)Z_m^\beta(s)ds}$. However, for light precipitation SRT PIA estimates are expected to be uncertain, cases in which ϵ needs to be set 1.0. For situations between the two extremes, ϵ is set based on a probabilistic procedure to a value between 1.0 and $\frac{1-10^{\beta PIA_{SRT}/10}}{\beta \int_0^{r_s} \alpha(s)Z_m^\beta(s)ds}$.

Once the true effective reflectivity Z_e is derived, the precipitation rain rates are derived using power law relationships, i.e. $R = aZ_e^b$, consistent with the calculations in Eqs. (4) and (5). If the attenuation adjustment factor ϵ is not 1.0, then coefficients a and b are adjusted. The actual values of a and b and the adjustment procedure are described in the official TRMM PR algorithm manual. While this adjustment is procedurally different from the adjustment proposed by Ferreira et al. [10], it should be recognized that both types of adjustments are the outcome of the same postulate, namely, differences between the analytical PIA and the SRT estimate are indicative of particle size distributions different from those assumed in the default k-Z and R-Z relationships.

The official TRMM PR precipitation estimates are three-dimensional. The horizontal dimensions are the scan and ray number, while the vertical direction is the bin number. A radar bin is defined by the width of a received energy pulse and has a range resolution of 250 m. Because the sea-level location geographical location of every ray in the scan as well as the ray's inclination are specified in the output file, one can readily derive three-dimensional precipitation fields from the information saved in the TRMM PR algorithm output files. The phase of the precipitation fields can be inferred from a five-node structure specified for every ray in the output file. The five node is defined as follows [27]: (1) A is the highest-altitude bin with detectable echo at the Ku band, (2) B is the top of the mixed-phase layer, (3) C is the peak of the bright band for stratiform precipitation and the bin corresponding to estimated freezing height for convective precipitation, (4) D is the bottom of the mixed-phase layer, and (5) E is the lowest bin above surface clutter. The TRMM PR observation and product files are hosted by the NASA Precipitation Processing System (PPS) on a public server, i.e. ftp://arthurhou.pps.eosdis.nasa.gov. They are readily available to registered users. The registration process is trivial and grants instant access to the data to anyone with a valid email address. Additional variables of potential interest to the users, e.g. the precipitation type, the value of ϵ in Eq. (2), the final value of the PIA, are also saved in the output files of various PR products. A detailed description of these products and the associated file structure is provided by NASA PPS at https://pps.gsfc.nasa.gov/ppsdocuments.html.

2.2.2 The TMI Algorithm and Precipitation Products

The TMI observations are less specific and more difficult to quantitatively and unambiguously interpret than the radar observations. This is because the TMI observations are the outcome of more complex radiative processes that occur in larger observing volumes than those associated with the PR observations. While ad-hoc parameterizations and models may be used to uniquely related a set of TMI observations to precipitation structures, statistical approaches such as the Bayesian estimation theory [61] are preferable because they provide rigorous methodologies to quantify and mitigate the uncertainties associated with the estimation process. The evolution of radiometer precipitation algorithms from ad-hoc to Bayesian is not entirely clear, but by the mid 1990s at least one rigorous Bayesian retrieval formulation existed [9], shortly followed by formulations operationally applicable [32, 56]. A distinct feature of operational formulations relative to more general formulations is that they rely on "a priori" databases of physically coherent precipitation profiles and associated brightness temperatures. In the retrieval process, for each multispectral set of observed brightness temperatures in the "a priori" database, the estimation procedure assigns to each precipitation profile in the database the probability that it is the "observed" profile. The final estimate is a probabilistic combination of precipitation profiles in the database. Mathematically, this can be written as

$$Pr(\mathbf{R}|\mathbf{Tb}) \propto Pr(\mathbf{R}) \cdot Pr(\mathbf{Tb}|\mathbf{R}) \tag{6}$$

where $Pr(\mathbf{R}|\mathbf{Tb})$ is the conditional probability of precipitation profile $\mathbf{R}$ given the brightness temperature observations $\mathbf{Tb}$, $Pr(\mathbf{R})$ is the probability with which a certain profile $\mathbf{R}$ will be observed and $Pr(\mathbf{Tb}|\mathbf{R})$ is the probability of observing the brightness temperature vector, $\mathbf{Tb}$, given a particular rain profile $\mathbf{R}$ [34], and

$$\mathbf{R} = \sum_{i=1}^{n} \mathbf{R}_i Pr(\mathbf{Tb}|\mathbf{R}_i) / \sum_{i=1}^{n} Pr(\mathbf{Tb}|\mathbf{R}_i) \tag{7}$$

where $\mathbf{R}_i$ are the precipitation profiles in the "a priori" database and $Pr(\mathbf{T_b}|\mathbf{R}_i)$ are the conditional probabilities of $\mathbf{Tb}$ given $\mathbf{R}_i$. While precipitation profiles in the database are associated with unique deterministic values, derived through forward radiative transfer calculations (e.g. [34]) or directly observed, probabilities $Pr(\mathbf{Tb}|\mathbf{R}_i)$ are not delta-functions (i.e. 1 for $\mathbf{Tb} = \mathbf{Tb}(\mathbf{R}_i)$, where the $\mathbf{Tb}(\mathbf{R}_i)$ is the deterministic value uniquely associated to precipitation profile $\mathbf{R_i}$ in the "a priori" database and 0 otherwise), but Gaussian functions [34]. This is because neither the observations nor the radiative transfer model used in deriving the brightness temperatures associated with a given profile in the database are uncertainties free, and ignoring these uncertainties would result in large random errors (possibly even biases) in the estimates produced by Eq. (7). Early versions of the

TRMM TMI precipitation algorithm described above, commonly referred to as the Goddard profiling algorithm (GPROF), employed "a priori" brightness temperature precipitation databases derived from Cloud Resolving Models (CRM) simulations. However, as the mission progressed, the focus changed towards the derivation of observation based "a priori" database, as CRMs were deemed somewhat deficient in reproducing the entire distribution of precipitation profiles and the joint distribution of precipitation profiles and brightness temperatures at TMI characteristics [35]. Specifically, systematic comparisons of CRM simulated reflectivity data with observed radar reflectivity data using contoured frequency with altitude diagrams (CFADs) revealed that the resulting CRM snow contents tend to be too large [36]. This is consistent with the observations of [3] who found that simulated 85-GHz brightness temperatures are different from those observed by the TMI. Moreover, the CRMs are initialized using a relatively small set of large-scale environmental conditions that may not be statistically representative of the distribution of the large-scale environments in nature. Although the vertical distribution of CRM simulated hydrometeors is continuously improving (e.g. [37]), CRM derived "a priori" databases are still not as robust as those derived from observations, which motivated the use of observation-based "a priori" databases in the latest versions of the GPROF algorithm [35]. An example of TMI monthly precipitation product is shown in Fig. 3.

2.2.3 Combined Radar-Radiometer Algorithm and Precipitation Products

Although apparently different, the radar and radiometer algorithms are instances of the same general algorithm. That is, Eq. (6) can be more generally written as

$$Pr(\mathbf{R}|\,\mathbf{Y}) \propto Pr(\mathbf{R}) \cdot Pr(\mathbf{Y}|\mathbf{R}) \tag{8}$$

where $\mathbf{Y}$ is a set of observations that can be provided by any generic sensor or combinations or sensors. However, instead of evaluating the expected value of $\mathbf{R}$ by integrating over the conditional probability $Pr(\mathbf{R}|\,\mathbf{Y})$ one can estimate $\mathbf{R}$ by maximizing the conditional probability $Pr(\mathbf{R}|\,\mathbf{Y})$. It should be noted that the maximization of $Pr(\mathbf{R}|\,\mathbf{Y})$ is not the only procedure to derive generalized radar-radiometer algorithms, but for the sake of simplicity, only this procedure will be discussed.

If $\mathbf{Y}$ is a profile of observed space radar reflectivities, i.e. $\mathbf{Y} = (Z_1, Z_2, \ldots, Z_n)$ where the subscripts indicate the radar bins, it is convenient to approximate Eq. (8) as

$$\begin{aligned} Pr(\mathbf{R}|\,\mathbf{Y} \propto Pr(\mathbf{R}) \cdot Pr(\mathbf{Y}|\mathbf{R}) = Pr(\mathbf{R}) \cdot Pr(Z_1|R_1) \cdot \\ Pr(Z_2|R_1, R_2) \ldots \cdot Pr(Z_n|R_1, R_2, \ldots, R_n) \end{aligned} \tag{9}$$

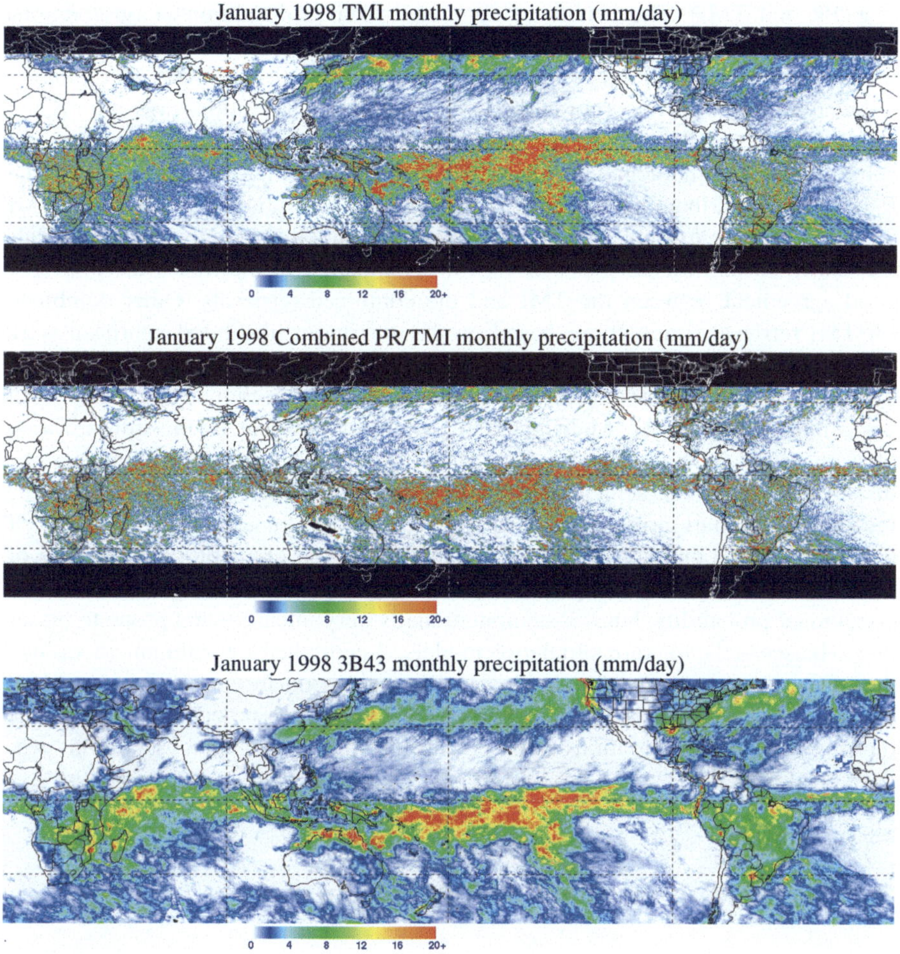

Fig. 3 TRMM TMI, combined and multisatellite analysis monthly precipitation estimates for January 1998

where the precipitation profile R consists of a set of precipitation values at locations defined by the radar bins. The likelihood function $Pr(\mathbf{R}|\ \mathbf{Y})$ can be assumed to be a separable function and optimized sequentially by optimizing $Pr(Z_1|R_1)$ as a function of R_1, $Pr(Z_2|R_1, R_2)$ as a function of R_2 and so on. This is equivalent to sequentially correcting for attenuation and estimate precipitation R_i from attenuated corrected reflectivity at bin i, which under the same assumptions regarding the relationships between attenuation and reflectivity and between reflectivity and precipitation is exactly the Hitschfeld and Bordan solution. But obviously, the formulation in Eqs. (8) and (9) are more general as they can handle uncertainties in the attenuation-reflectivity and reflectivity-precipitation relationships. Moreover they provide a rigorous framework to combine observations from multiple sources (e.g. PR and TMI). The benefit of combined radar and radiometer observations is that the uncertainties inherent in both radar and radiometer estimates can be reduced if the information is considered jointly. The first TRMM precipitation retrieval algorithm to incorporate both PR and TMI observations was the official day-1 TRMM combined algorithm [15]. Similarly to the radiometer algorithm, the combined algorithm integrates the possible precipitation values over the conditional $Pr(\mathbf{Z}, \mathbf{Tb}|\mathbf{R})$ distribution where **Z** are the PR radar observations and **Tb** are the TMI observations. An example of TRMM combined monthly precipitation retrievals are shown in the middle panel of Fig. 3. As apparent in the figure there is very good agreement between the TMI and the combined retrievals. Other combined PR/TMI retrieval formulations based on the maximum likelihood approach were formulated soon after launch (e.g. [13, 50]). These formulations are based on explicit optimization procedures that maximize the logarithm of the conditional probability $Pr(\mathbf{Z}, \mathbf{Tb}|\mathbf{R})$. Mathematically, this is equivalent to the minimization of a quadratic function that consists of two terms. The first term penalizes the disagreement between actual TRMM observations and their simulations from the retrieved precipitation profiles, while the second term penalizes the departure of the retrieved precipitation profiles from their expected climatology. This type of formulation has the advantage of not requiring an explicit integration over the conditional probability, but it is computationally very intensive and prone to biases due to biases in the forward simulation models. A combined algorithm methodology that does not require the simulation of observations on fly and relies instead on cloud-resolving model simulations and an iterative search procedure to determine the precalculated observations most consistent with the actual observations was developed in [45].

3 TRMM Multisatellite Precipitation Analysis

A major challenge regarding the use of precipitation estimates from Low Earth Orbit (LEO) satellites in applications is their limited spatial and temporal sampling. Furthermore, TRMM only covers the tropics and lower mid latitudes, which precludes applications requiring mid and high latitude precipitation estimates. To mitigate these limitations and in tune with other efforts [24, 28], the TRMM Multisatellite Precipitation Analysis (TMPA) was developed [25]. The TMPA incorporates estimates from other precipitation-related satellites of opportunity, with TRMM as the basis, to produce complete 3-hourly precipitation fields that cover the latitudes from 50°N to 50°S. Prior to their incorporation with TRMM, all other satellite precipitation estimates are calibrated to the combined PR/TMI product as it was shown to produce the most consistent estimates compared with in-situ observations. This intercalibration occurs in three steps: (1) all other satellite precipitation estimates are calibrated to TMI using fixed, season climatological calibrations based on matched histograms, (2) the TMI is calibrated to the combined

PR/TMI using monthly, dynamic calibrations based on histogram matching and, (3) the other satellites from (1) are calibrated to the combined PR/TMI using the calibration from (2). This multistep process is necessary as the other satellite estimates cannot be directly calibrated to the combined PR/TMI estimates due to sampling limitations. All satellite estimates are then merged to produce a single "high-quality" estimate of precipitation. Though this high-quality precipitation field can have many contributing LEO satellites, there are still gaps in the 3-h field. Half-hourly NOAA CPC 4-km infrared (IR) estimates are used to fill in these gaps in the high-quality field. IR brightness temperatures are histogram matched directly to the high-quality precipitation estimates to create a monthly, dynamic calibration. This calibration is then applied to the IR brightness temperatures and used to complete the TMPA precipitation field. Unlike the high-quality microwave estimates that provide "direct" observation of precipitation, the calibrated IR estimates are inferred from cold clouds and not as accurate and therefore have lower quality. For each calendar month, the TMPA 3-hourly estimates are then summed to the monthly scale and merged with a monthly Global Precipitation Climatology Center gauge analysis, based on inverse error weighting, to compute the TMPA monthly estimate. To maintain consistency between the 3-hourly and monthly products, the 3-hourly estimates are summed to match the monthly estimate so the 3-hourly estimates have gauge "influence". The TMPA suite includes both real time monitoring and post real time research products. The record begins in January 1998 and continues to the present. Over this period, the input satellites of opportunity have evolved but the intercalibration process seeks to preserve the homogeneity of the record. Furthermore, TMPA continues to be produced despite the loss of the TRMM satellite due to the transition to climatological calibrations.

4 Applications

TRMM observations and products have been used in a far larger variety of applications than envisioned before launch, however in this section the discussion will be mainly focused on atmospheric science applications. Even within this limited application domain, it is difficult to objectively and unambiguously stratify applications based on their areas of focus, because a large number of applications were driven by multiple objectives. However, several generic objectives were so frequently formulated in TRMM studies that they can be used by themselves or in combination with other for classification purposes. These include:

1. Statistical description of precipitation and associated variables derived from TRMM observations and products.
2. Use of statistical analyses to formulate, refine and validate conceptual models of various atmospheric processes.
3. Use of TRMM products to validate Numerical Weather Prediction (NWP) and climate models.

4. Development of new remote sensing techniques and retrieval methodologies applicable to TRMM or other mission/instruments.
5. Operational and quasi-operational applications.

The number of peer-reviewed journal articles that make significant use of TRMM observations is on the order of thousands, which makes it practically impossible to provide a comprehensive review of how TRMM observations are used in scientific applications. Nevertheless, it is worth mentioning that an analysis using the Publish or Perish (PoP) software [17] indicates that with the exception of several papers on the TRMM mission and its algorithms, studies preponderantly addressing objective (1) above are among the most cited TRMM studies. Example of such studies include [51, 52, 54] and [72]. These studies are mostly concerned with the temporal and spatial distribution of precipitation in the Tropics. This is not surprising, given that TRMM provide unprecedented information on the temporal and spatial structure of precipitation in the Tropics and a logical step in assessing our ability to understand and reproduce atmospheric processing through models consists naturally of establishing a reference. Studies addressing objective 2 usually address objective 1 as well to certain degree. Examples of such studies are [58] who studied the distribution of stratiform rain in the Tropics and related it to large scale atmospheric circulations, [55] who related the vertical distribution of precipitation in Amazonia to large scale regimes, and [23] who analyzed the three-dimensional structure of summer monsoon convection in the Himalayan region as depicted by TRMM in relationship to the underlying meteorology. Other examples include the use of TRMM observations to study the Madden-Julian Oscillation and formulate new possible mechanism of propagation [46] and the use of TRMM to investigate the effects of vertical wind shear and storm motion on tropical cyclone rainfall asymmetries [6].

Even before the launch of TRMM, several studies, e.g. [11, 66], revealed that significant discrepancies between the distribution of observed radar reflectivities and those derived from CRMs simulations may occur. However, a systematic evaluation of CRMs and existing cloud microphysical scheme was not easily achievable given that CRM performance is regime dependent and high-quality radar observations representative of multiple regimes were not readily available. With the launch of TRMM, the availability of radar observations was no longer a consideration and numerous studies (falling in category 3 above) have been carried out to evaluate performance of CRMs in various regimes using radar and radiometer observations. Examples of such studies include [16, 36, 39, 71] and [37]. As a result of these studies, several deficiencies in cloud microphysical schemes have been identified and addressed to large extent. But obviously the utility of TRMM observations and products is as means of evaluating models is not limited to CRMs only, but it extends to global circulation and climate models as well. Specifically, TRMM observations and products may and have been used to evaluate the ability of global circulation models to reproduce the spatial distributions of observed precipitation [7] and its diurnal cycle [8, 57] and the merit of new parametrizations relative to old ones (e.g. [67, 68]). Several studies (e.g. [30, 40]) have shown that global circulation models have difficulties in reproducing the MaddenJulian oscillation MJO [42]. From this perspective, the TMPA precipitation product described in Sect. 3 is an extremely important data set as its temporal and spatial resolution enables the rigorous evaluation of the ability global circulation models to reproduce the MJO phenomenon [31].

As apparent from the brief description of the radiometer and combined radar-radiometer algorithms in Sect. 2.1.1, retrievals methodology continuously evolved throughout the TRMM lifetime. While passive radiometer retrievals evolved from CRM derived databases to observations derived databases [35], combined retrievals developed in more than one way [13, 46, 50]. While the main objective of these efforts was to derived more accurate precipitation estimates from observations, their outcomes transcended the TRMM realm as the techniques developed in the processes were used in the development of retrieval methodologies applicable to other missions as well. For example, the Global Precipitation Measurement Mission (GPM) radar-radiometer algorithm [14], while deeply rooted in combined retrieval methodologies developed in the TRMM era, is a procedure specifically designed for a combination of instruments significantly different in many respects from the TRMM instruments [62] and is an example of TRMM applications falling in category 4 listed above. Other efforts include the development of variational retrieval methodologies applicable TRMM observations, but extendable to observations from other platforms (e.g. [49, 59]).

It should also be mentioned that, although not an operational mission, TRMM enabled the development of tools and methodologies operationally applicable. Data assimilation studies belong to applications in this category, i.e. category 5 in the list above. While assimilation of satellite observations in Numerical Weather Prediction (NWP) models has been a very active area of research even before TRMM, the high-resolution of TMI observations and their ready availability made them a popular choice in data assimilation studies. One such study [44] revealed that TRMM TMI observations enabled a better analysis than SSMI observations due to the additional low frequency channel and improved resolution of the TMI instrument that allowed a better estimate of surface parameters and of water vapour in tropical conditions. Other studies [20, 21, 43] showed that the assimilation of both TMI and SSM/I rain rates improved the forecasts and analyses of tropical cyclones as well as that of other tropical and mid-latitude precipitation systems. Similarly, it has been showed that TRMM PR observations although providing much narrower coverage than the TMI observation have a positive impact on the analysis of forecast of tropical cyclones [4] and that the assimilation of the TRMM TMPA precipitation products into a global model improves the 5-day forecast [41]. In addition to the advancement of knowledge and capabilities in data assimilation in NWP models, TRMM observations and precipitation products have been used for near-real time tropical cyclone monitoring [38], real-time time global flood estimation [70] and real-time landslide hazard assessment [19].

5 Summary

This chapter describes the Tropical Rainfall Measuring Mission. The longevity and the quality of observations provided by TRMM, its unique combination of instruments featuring the first space-borne precipitation radar, and the enthusiasm and the commitment of the its leadership and science community enabled outstanding progress in our ability to understand and quantify fundamental atmospheric processes related to precipitation from satellite observations. This chapter briefly describes some of this progress and elements that made it possible.

The TRMM radar and radiometer and their specifications are briefly described, followed by a presentation of the main algorithms used to derive precipitation from the actual satellite observations. The PR precipitation estimation algorithm and its challenges related to the multivariate variability of PSDs within radar observing volumes are presented first. Next, the official radiometer algorithm is presented. Unlike the radar estimation algorithm, the radiometer algorithm is significantly more statistical in nature (although incorporating complex physical models), as radiometer observations are less specific than radar observations. The motivation behind the official TRMM combined radar-radiometer algorithm and its abstract formulation are also presented. In addition to individual precipitation estimates from single or combined TRMM instrument observations, the TRMM mission featured a Multisatellite Precipitation Analysis product that consistently unified precipitation products derived from TRMM and non-TRMM instruments in an effort to mitigate temporal sampling errors in the individual precipitation estimates. This product and its significance is also described in the chapter.

The chapter includes a discussion on the use of TRMM observations and products in scientific and societal applications. TRMM generated so much science and peer-reviewed publication that are practically impossible to describe in their entirety even in a very concise manner. Instead the focus here is on the presentations of directions and examples of applications. It is hoped that such presentation, even though incomplete in many respects, contains sufficient information to embolden readers to explore on their own everything the TRMM mission has to offer.

Acknowledgements The authors thank Dr. Ramesh Kakar (NASA headquarters) and the TRMM project scientists Drs. Christian Kummerow (Colorado State University), Robert Adler (University of Maryland) and Scott Braun (NASA Goddard Space Flight Center) for their support of all TRMM related science projects. The authors would also like to thank Dr. Hirohiko Masunaga of Nagoya University for reviewing this book chapter and providing constructive comments.

References

1. Adler, R., H. Yeh, N. Prasad, W. Tao, and J. Simpson. 1991. Microwave simulations of a tropical rainfall system with a three-dimensional cloud model. *Journal of Applied Meteorology* 30: 924–953. https://doi.org/10.1175/1520-0450-30.7.924.

2. Battan, L.J. 1973. *Radar Observation of the Atmosphere*, 324 pp. Chicago: University of Chicago Press.
3. Bauer, P. 2001. Over-ocean rainfall retrieval from multisensor data of the Tropical Rainfall Measuring Mission. Part I: design and evaluation of inversion databases. *Journal of Atmospheric and Oceanic Technology* 18: 1315–1330.
4. Benedetti, A., P. Lopez, P. Bauer, and E. Moreau. 2005. Experimental use of TRMM precipitation radar observations in 1D+4D-Var assimilation. *Quarterly Journal of the Royal Meteorological Society* 131: 2473–2495.
5. Bohren, C.F., and D.R. Huffman. 1983. *Absorption and Scattering of Light by Small Particles*. New York: Wiley.
6. Chen, S.S., J.A. Knaff, and F.D. Marks. 2006. Effects of vertical wind shear and storm motion on tropical cyclone rainfall asymmetries deduced from TRMM. *Monthly Weather Review* 134: 3190–3208. https://doi.org/10.1175/MWR3245.1.
7. Deng, Y., K.P. Bowman, and C. Jackson. 2007. Differences in rain rate intensities between TRMM observations and community atmosphere model simulations. *Geophysical Research Letters* 34: L01808. https://doi.org/10.1029/2006GL027246.
8. Dirmeyer, P.A., B.A. Cash, J.L. Kinter, T. Jung, L. Marx, M. Satoh, C. Stan, H. Tomita, P. Towers, N. Wedi, and D. Achuthavarier. 2012. Simulating the diurnal cycle of rainfall in global climate models: resolution versus parameterization. *Climate Dynamics* 39 (1–2): 399–418.
9. Evans, K., J. Turk, T. Wong, and G. Stephens. 1995. A Bayesian approach to microwave precipitation profile retrieval. *Journal of Applied Meteorology* 34: 260–279. https://doi.org/10.1175/1520-0450-34.1.260.
10. Ferreira, F., et al. 2001. Study and tests of improved rain estimates from the TRMM precipitation radar. *Journal of Applied Meteorology* 40 (11): 1878–1899.
11. Ferrier, B.S., W. Tao, and J. Simpson. 1995. A double-moment multiple-phase four-class bulk ice scheme. Part II: simulations of convective storms in different large-scale environments and comparisons with other bulk parameterizations. *Journal of the Atmospheric Sciences* 52: 1001–1033. https://doi.org/10.1175/1520-0469(1995)052<1001:ADMMPF>2.0.CO;2$.
12. Gloersen, P., and F.T. Barath. 1977. A scanning multichannel microwave radiometer for Nimbus-G and SeaSat-A. *IEEE Journal of Oceanic Engineering* 2: 172–178.
13. Grecu, M., W.S. Olson, and E.N. Anagnostou. 2004. Retrieval of precipitation profiles from multiresolution, multifrequency active and passive microwave observations. *Journal of Applied Meteorology* 43: 562–575.
14. Grecu, M., W.S. Olson, S.J. Munchak, S. Ringerud, L. Liao, Z. Haddad, B.L. Kelley, and S.F. McLaughlin. 2016. The GPM combined algorithm. *Journal of Atmospheric and Oceanic Technology* 33: 2225–2245. https://doi.org/10.1175/JTECH-D-16-0019.1.
15. Haddad, Z.S., et al. 1997. The TRMM 'day-1' radar/radiometer combined rain-profiling algorithm. *Journal of the Meteorological Society of Japan. Series II* 75 (4): 799–809.
16. Han, M., S.A. Braun, W.S. Olson, P.O. Persson, and J. Bao. 2010. Application of TRMM PR and TMI measurements to assess cloud microphysical schemes in the MM5 for a winter storm. *Journal of Applied Meteorology and Climatology* 49: 1129–1148. https://doi.org/10.1175/2010JAMC2327.1.
17. Harzing, A.W. 2007. Publish or Perish. Available from http://www.harzing.com/pop.html.
18. Hitschfeld, W., and J. Bordan. 1954. Errors inherent in the radar measurement of rainfall at attenuating wavelengths. *Journal of Meteorology* 11: 58–67.
19. Hong, Y., R. Adler, and G. Huffman. 2006. Evaluation of the potential of NASA multi-satellite precipitation analysis in global landslide hazard assessment. *Geophysical Research Letters* 33 (22). http://dx.doi.org/10.1029/2006GL028010.
20. Hou, A.Y., S.Q. Zhang, A.M. da Silva, and W.S. Olson. 2000. Improving assimilated global datasets using TMI rainfall and columnar moisture observations. *Journal of Climate* 13: 4180–4195. https://doi.org/10.1175/1520-0442(2000)013<4180:IAGDUT>2.0.CO;2$.
21. Hou, A.Y., S.Q. Zhang, and O. Reale. 2004. Variational continuous assimilation of TMI and SSM/I rain rates: impact on GEOS-3 hurricane analyses and forecasts. *Monthly Weather Review* 132: 2094–2109. https://doi.org/10.1175/1520-0493(2004)132<2094:VCAOTA>2.0.CO;2.

22. Houze, R.A. Jr. 2014. *Cloud Dynamics*. Vol. 104. New York: Academic.
23. Houze, R.A., Wilton, D.C., and B.F. Smull. 2007. Monsoon convection in the Himalayan region as seen by the TRMM precipitation radar. *Quarterly Journal of the Royal Meteorological Society* 133 (627): 1389–1411.
24. Huffman, G.J., R.F. Adler, M. Morrissey, D.T. Bolvin, S. Curtis, R. Joyce, B. McGavock, and J. Susskind. 2001. Global precipitation at one-degree daily resolution from multisatellite observations. *Journal of Hydrometeorology* 2: 36–50.
25. Huffman, G.J., D.T. Bolvin, E.J. Nelkin, D.B. Wolff, R.F. Adler, G. Gu, Y. Hong, K.P. Bowman, and E.F. Stocker. 2007. The TRMM multisatellite precipitation analysis (TMPA): quasi-global, multiyear, combined-sensor precipitation estimates at fine scales. *Journal of Hydrometeorology* 8: 38–55. https://doi.org/10.1175/JHM560.1.
26. Iguchi, T., and R. Meneghini. 1994. Intercomparison of single-frequency methods for retrieving a vertical rain profile from airborne or spaceborne radar data. *Journal of Atmospheric and Oceanic Technology* 11: 1507–1516. https://doi.org/10.1175/1520-0426(1994)011.
27. Iguchi, T., T. Kozu, R. Meneghini, J. Awaka, and K. Okamoto. 2000. Rain-profiling algorithm for the TRMM precipitation radar. *Journal of Applied Meteorology* 39: 2038–2052.
28. Joyce, R.J., J.E. Janowiak, P.A. Arkin, and P. Xie. 2004. CMORPH: a method that produces global precipitation estimates from passive microwave and infrared data at high spatial and temporal resolution. *Journal of Hydrometeorology* 5: 487–503.
29. Kidder, S.Q., and T.H. von der Haar. 1995. *Satellite Meteorology - An Introduction*, 446. New York: Academic.
30. Kim, D., Sperber, K., Stern, W., Waliser, D., Kang, I.S., Maloney, E., Wang, W., Weickmann, K., Benedict, J., Khairoutdinov, M., Lee, M.I., Neale, R., Suarez, M., Thayer-Calder, K., and G. Zhang. 2009. Application of MJO simulation diagnostics to climate models. *Journal of Climate* 22: 6413–6436.
31. Klingaman, N.P., and S.J. Woolnough. 2014. Using a case-study approach to improve the MaddenJulian oscillation in the Hadley Centre model. *Quarterly Journal of the Royal Meteorological Society* 140: 2491–2505. https://doi.org/10.1002/qj.2314.
32. Kummerow, C., W.S. Olson, and L. Giglio. 1996. A simplified scheme for obtaining precipitation and vertical hydrometeor profiles from passive microwave sensors. *IEEE Transactions on Geoscience and Remote Sensing* 34 (5): 1213–1232.
33. Kummerow, C., W. Barnes, T. Kozu, J. Shiue, and J. Simpson. 1998. The tropical rainfall measuring mission (TRMM) sensor package. *Journal of Atmospheric and Oceanic Technology* 15: 809–817.
34. Kummerow, C., Y.Y. Hong, W.S. Olson, S.S. Yang, R.F. Adler, J.J. McCollum, R.R. Ferraro, G.G. Petty, D.B. Shin, and T.T. Wilheit. 2001. The evolution of the Goddard profiling algorithm (GPROF) for rainfall estimation from passive microwave sensors. *Journal of Applied Meteorology* 40: 1801–1820.
35. Kummerow, C.D., S. Ringerud, J. Crook, D. Randel, and W. Berg. 2011. An observationally generated a priori database for microwave rainfall retrievals. *Journal of Atmospheric and Oceanic Technology* 28: 113–130.
36. Lang, S.E., W.-K. Tao, R. Cifelli, W. Olson, J. Halverson, S. Rutledge, and J. Simpson. 2007. Improving simulations of convective systems from TRMM LBA: easterly and westerly regimes. *Journal of the Atmospheric Sciences* 64: 1141–1164.
37. Lang, S.E., W. Tao, J. Chern, D. Wu, and X. Li. 2014. Benefits of a fourth ice class in the simulated radar reflectivities of convective systems using a bulk microphysics scheme. *Journal of the Atmospheric Sciences* 71: 3583–3612. https://doi.org/10.1175/JAS-D-13-0330.1.
38. Lee, T.F., F.J. Turk, J. Hawkins, and K. Richardson. 2002. Interpretation of TRMM TMI images of tropical cyclones. *Earth Interactions* 6: 1–17. https://doi.org/10.1175/1087-3562(2002)006<0001:IOTTIO>2.0.CO;2.
39. Li, X., W.-K. Tao, T. Matsui, C. Liu, H. Masunaga. 2010. Improving a spectral bin microphysical scheme using TRMM satellite observations. *Quarterly Journal of the Royal Meteorological Society* 136: 382–399. https://doi.org/10.1002/qj.569.

40. Lin, J.L., G.N. Kiladis, B.E. Mapes, K.M. Weickmann, K.R. Sperber, W. Lin, M. Wheeler, S.D. Shubert, A. Del Genio, L.J. Donner, S. Emori, J.F. Gueremy, F. Hourdain, P.J. Rasch, E. Roeckner, and J.F. Scinocca. 2006. Tropical intraseasonal variability in 14 IPCC AR4 climate models. Part I: convective signals. *Journal of Climate* 19: 2665–2690.
41. Lien, G., T. Miyoshi, and E. Kalnay. 2016. Assimilation of TRMM Multisatellite precipitation analysis with a low-resolution NCEP global forecast system. *Monthly Weather Review* 144: 643–661. https://doi.org/10.1175/MWR-D-15-0149.1.
42. Madden, R.A., and P.R. Julian. 1971. Detection of a 40–50 day oscillation in the zonal wind in the tropical Pacific. *Journal of the Atmospheric Sciences* 28: 702–708.
43. Mahfouf, J.F., P. Bauer, and V. Marécal. 2005. The assimilation of SSM/I and TMI rainfall rates in the ECMWF 4DVar system. *Quarterly Journal of the Royal Meteorological Society* 131 (606): 437–458.
44. Marcal, V., É. Gérard, J.-F. Mahfouf, and P. Bauer. 2001. The comparative impact of the assimilation of SSM/I and TMI brightness temperatures in the ECMWF 4D-Var system. *Quarterly Journal of the Royal Meteorological Society* 127: 1123–1142. https://doi.org/10.1002/qj.49712757322.
45. Masunaga, H., and C.D. Kummerow. 2005. Combined radar and radiometer analysis of precipitation profiles for a parametric retrieval algorithm. *Journal of Atmospheric and Oceanic Technology* 22: 909–929.
46. Masunaga, H., T.S. L'Ecuyer, and C.D. Kummerow. 2006. The Madden-Julian oscillation recorded in early observations from the tropical rainfall measuring mission (TRMM). *Journal of the Atmospheric Sciences* 63: 2777–2794. https://doi.org/10.1175/JAS3783.1.
47. Meneghini, R., and T. Kozu. 1990. *Spaceborne Weather Radar*, 199 pp. London: Artech House.
48. Meneghini, R., T. Iguchi, T. Kozu, L. Liao, K. Okamoto, J.A. Jones, and J. Kwiatkowski. 2000. Use of the surface reference technique for path attenuation estimates from the TRMM Precipitation radar. *Journal of Applied Meteorology* 39: 2053–2070.
49. Moreau, E. 2003. Variational retrieval of rain profiles from spaceborne passive microwave radiance observations. *Journal of Geophysical Research* 108: D16.
50. Munchak, S.J., and C.D. Kummerow. 2011. A modular optimal estimation method for combined radar–radiometer precipitation profiling. *Journal of Applied Meteorology* 50: 433–448.
51. Nesbitt, S.W., and E.J. Zipser. 2003. The diurnal cycle of rainfall and convective intensity according to three years of TRMM measurements. *Journal of Climate* 16 (10): 1456–1475.
52. Nesbitt, S.W., E.J. Zipser, and D.J. Cecil. 2000. A census of precipitation features in the tropics using TRMM: radar, ice scattering, and lightning observations. *Journal of Climate* 13 (23): 4087–4106.
53. Njoku, E.G., J.M. Stacey, and F.T. Barath. 1980. The seasat scanning multichannel microwave radiometer (SMMR): instrument description and performance. *IEEE Journal of Oceanic Engineering* 5 (2): 100–115.
54. Petersen, W.A., and S.A. Rutledge. 2001. Regional variability in tropical convection: observations from TRMM. *Journal of Climate* 14: 3566–3586.
55. Petersen, W.A., S.W. Nesbitt, R.J. Blakeslee, R. Cifelli, P. Hein, and S.A. Rutledge. 2002. TRMM observations of intraseasonal variability in convective regimes over the amazon. *Journal of Climate* 15: 1278–1294. 10.1175/1520-0442(2002)015<1278:TOOIVI>2.0.CO;2
56. Pierdicca, N., F.S. Marzano, G. d'Auria, P. Basili, P. Ciotti, and A. Mugnai. 1996. Precipitation retrieval from spaceborne microwave radiometers using maximum a posteriori probability estimation. *IEEE Transactions on Geoscience and Remote Sensing* 34: 831–846.
57. Sato, T., H. Miura, M. Satoh, Y.N. Takayabu, and Y. Wang. 2009. Diurnal cycle of precipitation in the tropics simulated in a global cloud-resolving model. *Journal of Climate* 22: 4809–4826. https://doi.org/10.1175/2009JCLI2890.1.
58. Schumacher, C., and R.A. Houze. 2003. Stratiform rain in the tropics as seen by the TRMM precipitation radar. *Journal of Climate* 16: 1739–1756. https://doi.org/10.1175/1520-0442(2003)016<1739:SRITTA>2.0.CO;2.

59. Seo, E., S. Yang, M. Grecu, G. Ryu, G. Liu, S. Hristova-Veleva, Y. Noh, Z. Haddad, and J. Shin. 2016. Optimization of cloud-radiation databases for passive microwave precipitation retrievals over ocean. *Journal of Atmospheric and Oceanic Technology* 33: 1649–1671. https://doi.org/10.1175/JTECH-D-15-0198.1.
60. Simpson, J., R. Adler, and G. North. 1988. A proposed tropical rainfall measuring mission (TRMM) satellite. *Bulletin of the American Meteorological Society* 69: 278–295.
61. Sivia. D.S. 1996. *Data Analysis - A Bayesian Tutorial*. Oxford: Oxford University Press.
62. Skofronick-Jackson, G., W.A. Petersen, W. Berg, C. Kidd, E.F. Stocker, D.B. Kirschbaum, R. Kakar, S.A. Braun, G.J. Huffman, T. Iguchi, P.E. Kirstetter, C. Kummerow, R. Meneghini, R. Oki, W.S. Olson, Y.N. Takayabu, K. Furukawa, and T. Wilheit. 2017. The global precipitation measurement (GPM) mission for science and society. *Bulletin of the American Meteorological Society*. https://doi.org/10.1175/BAMS-D-15-00306.1.
63. Smith, E., H. Cooper, X. Xiang, A. Mugnai, and G. Tripoli. 1992. Foundations for statistical-physical precipitation retrieval from passive microwave satellite measurements. Part I: brightness-temperature properties of a time-dependent cloud-radiation model. *Journal of Applied Meteorology* 31: 506–531.
64. Spencer, R., W. Olson, W. Rongzhang, D. Martin, J. Weinman, and D. Santek. 1983. Heavy thunderstorms observed over land by the Nimbus 7 scanning multichannel microwave radiometer. *Journal of Climate and Applied Meteorology* 22: 1041–1046.
65. Spencer, R., H. Goodman, and R. Hood. 1989. Precipitation retrieval over land and ocean with the SSM/I: identification and characteristics of the scattering signal. *Journal of Atmospheric and Oceanic Technology* 6: 254–273.
66. Swann, H. 1998. Sensitivity to the representation of precipitating ice in CRM simulations of deep convection. *J. Atmos. Oceanic. Technol.* 47–48: 415–435.
67. Tao, W.-K., and M.W. Moncrieff. 2009. Multiscale cloud system modeling. *Reviews of Geophysics* 47: RG4002. https://doi.org/10.1029/2008RG000276.
68. Tao, W.-K., and J.-D. Chern. 2017. The impact of simulated mesoscale convective systems on global precipitation: a multiscale modeling study. *Journal of Advances in Modeling Earth Systems* 9. https://doi.org/10.1002/2016MS000836.
69. Testud, J., S. Oury, X. Dou, P. Amayenc, and R. Black. 2001. The concept of normalized distribution to describe raindrop spectra: a tool for cloud physics and cloud remote sensing. *Journal of Applied Meteorology* 40: 1118–1140.
70. Wu, H., R.F. Adler, Y. Tian, G.J. Huffman, H. Li, and J. Wang. 2014. Real-time global flood estimation using satellite-based precipitation and a coupled land surface and routing model. *Water Resources Research* 50: 2693.2717. https://doi.org/10.1002/2013WR014710.
71. Zhou, Y.P., W. Tao, A.Y. Hou, W.S. Olson, C. Shie, K. Lau, M. Chou, X. Lin, and M. Grecu. 2007. Use of High-resolution satellite observations to evaluate cloud and precipitation statistics from cloud-resolving model simulations. Part I: South China sea monsoon experiment. *Journal of the Atmospheric Sciences* 64: 4309–4329. https://doi.org/10.1175/2007JAS2281.1.
72. Zipser, E.J., C. Liu, D.J. Cecil, S.W. Nesbitt, and D.P. Yorty. 2006. Where are the most intense thunderstorms on Earth? *Bulletin of the American Meteorological Society* 87 (8): 1057–1071.

Global Precipitation Measurement (GPM): Unified Precipitation Estimation from Space

Gail Skofronick-Jackson, Wesley Berg, Chris Kidd, Dalia B. Kirschbaum, Walter A. Petersen, George J. Huffman, and Yukari N. Takayabu

1 Introduction

The Global Precipitation Measurement (GPM) mission is an international satellite mission initiated by the National Aeronautics and Space Administration (NASA) and the Japan Aerospace and Exploration Agency (JAXA) to unify and advance global precipitation measurements from space [13]. The mission centers on the deployment of a GPM Core Observatory (GPM-CO, launched February 2014) carrying a Ka/Ku-band Dual-frequency Precipitation Radar (DPR) and a multifrequency (10–183 GHz) microwave radiometer, the GPM Microwave Imager (GMI). The GPM-CO serves as a physics observatory of precipitating systems and a calibration reference for precipitation estimates from a constellation of microwave radiometers operated by a consortium of international partners [13]. The mission

G. Skofronick-Jackson (✉) · D. B. Kirschbaum · G. J. Huffman
NASA Goddard Space Flight Center, Greenbelt, MD, USA
e-mail: gail.s.jackson@nasa.gov; dalia.b.kirschbaum@nasa.gov; george.j.huffman@nasa.gov

W. Berg
Department of Atmospheric Science, Colorado State University, Fort Collins, CO, USA
e-mail: berg@atmos.colostate.edu

C. Kidd
University of Maryland, College Park, MD, USA

NASA Goddard Space Flight Center, Greenbelt, MD, USA
e-mail: chris.kidd@nasa.gov

W. A. Petersen
NASA Marshall Space Flight Center, Huntsville, AL, USA

Y. N. Takayabu
Center for Climate System Research, The University of Tokyo, Kashiwa, Chiba, Japan
e-mail: yukari@ccsr.u-tokyo.ac.jp

C. Andronache (ed.), *Remote Sensing of Clouds and Precipitation*,
Springer Remote Sensing/Photogrammetry,
https://doi.org/10.1007/978-3-319-72583-3_7

is designed to provide the next-generation, constellation-based global precipitation products for scientific research and societal applications [47].

As a science-discovery mission with integrated applications goals, GPM's mission objectives are: (1) establishing new reference standards for precipitation measurements from space using combined active and passive remote-sensing techniques, (2) improving knowledge of precipitation systems, water cycle variability, and freshwater availability with enhanced measurements of the space-time distribution of global precipitation, (3) improving climate modeling and prediction capabilities through better measurements of latent heating, precipitation microphysics, and surface water fluxes, (4) improving weather forecasting skills with more accurate estimates of instantaneous precipitation information and error characterizations, and (5) improving hydrological modeling and prediction of high-impact natural hazard events (e.g., flood, drought, landslide, and hurricanes) through improved temporal sampling and high-resolution model-downscaled precipitation products. By making data available in near-real-time (NRT) to operational agencies and stakeholders beyond the traditional science community, GPM enables the use of space-based precipitation observations in a variety of practical applications to directly benefit society [19].

Building upon the success of the U.S.-Japan Tropical Rainfall Measuring Mission (TRMM) [28], GPM is a partnership between NASA and JAXA. NASA and JAXA provided the GPM-CO, which was launched on 28 February 2014 from Tanegashima Island in Japan. The GPM-CO has completed its prime mission life of 3 years and is now in extended operations. Remaining fuel projections are for operations to last into the 2030s if the batteries and instruments remain functional.

This chapter presents the GPM mission description in Sect. 2, Core Observatory sensor capabilities in Sect. 3, constellation coverage and sampling in Sect. 4, inter-calibration of constellation radiometers in Sect. 5, precipitation retrieval methodologies in Sect. 6, ground validation in Sect. 7, applications in Sect. 8, and a summary in Sect. 9.

2 Mission Description

The GPM Core Observatory operates in a non-Sun-synchronous orbit at 65° inclination and an altitude of 407 km at the equator to permit diurnal observations of precipitation and coincident measurements with constellation radiometers to serve as a reference standard for the inter-calibration of both radiometric measurements and precipitation retrievals of the constellation members. The GPM-CO data are necessary for reducing the number of assumptions in the precipitation retrievals. This is because satellite rainfall estimates cannot be cast in terms of simple mathematical inversions. Precipitating clouds contain more free parameters (including their size, shape, and internal distribution of raining and non-raining water and ice particles) than can realistically be retrieved from a finite set of satellite measurements. Significant assumptions regarding the composition of clouds are therefore necessary [48]. GPM tackles this problem by first designing highly accurate instruments and second, by creating a common inventory of naturally occurring cloud states with details provided by the GPM-CO using the most advanced active and passive rainfall sensors. This inventory of clouds can then be used as the a priori state within a Bayesian or Optimal Estimation framework for constellation sensor retrievals (e.g., [29]). While the technique does not entirely eliminate the need for assumptions, it does employ the highly accurate GPM-CO data rather than less-capable radiometers or models to make the assumptions independently.

The 65° inclination for the GPM-CO was selected to offer broad latitudinal coverage without being locked into a Sun-synchronous Low Earth orbit (LEO), while maintaining a relatively short precession period to sample the diurnal variability of precipitation. TRMM, at 35° inclination, has shown that the additional sampling afforded by asynoptic observations in between observations by LEO satellites at fixed local times are important for near real-time monitoring and prediction of hurricanes. The GPM-CO provides, for the first time, asynoptic observations for tracking mid-latitude and extratropical storms. The non-Sun-synchronous orbit also provides multiple coincident overpasses with the mostly LEO constellation sensors to facilitate inter-sensor calibration over a wide range of latitudes.

The GPM mission is supported on the ground by (1) a NASA-provided mission operations system for the operation of the GPM Core, (2) a Ground Validation (GV) system consisting of dedicated and cooperative ground validation sites provided by NASA, JAXA, and an array of GPM ground validation partners, and (3) a NASA Precipitation Processing System (PPS) and a JAXA GPM Mission Operation System (GPM-MOS) in coordination with other GPM partner data processing sites to provide NRT and standard global precipitation products. For inter-satellite calibration, retrieval algorithm development, ground validation, and science applications, a NASA Precipitation Measurement Missions (PMM) Science Team and a JAXA PMM Science Team support GPM.

GPM is integral to international efforts to examine and understand the Earth's hydrological and energy cycles. The GPM mission is serving as the scientific cornerstone for a multi-national satellite constellation to monitor global precipitation under development by the Committee for Earth Observation Satellites (CEOS), and provides key measurements to advance the objectives of a host of international scientific programs and activities that include the Global Energy and Water Cycle Experiment (GEWEX), the Integrated Global Water Cycle Observations (IGWCO), and the International Precipitation Working Group (IPWG).

3 GPM Core Observatory Sensor Capabilities

TRMM focused primarily on moderate to heavy rain over tropical oceans. Since light rain ($<0.5\,\mathrm{mm\,h^{-1}}$) and falling snow account for significant fractions of the total precipitation events outside the tropics [37], improved measurements of these quantities are needed to gain a better understanding of global water fluxes and precipitation characteristics (e.g., frequency, intensity, distribution, etc.). The GPM-CO carries the first space-borne dual-frequency precipitation radar (DPR) at Ku and Ka bands capable of detailed microphysical measurements layer by layer within the cloud and a multi-spectral (10–183 GHz) GMI encompassing the range of frequencies typical of microwave imagers and humidity sounders. Specifications and on orbit performance is provided in Table 1. The active (DPR) and passive (GMI) sensors on board GPM-CO provide complementary information that are being used to improve the fidelity of physically-based precipitation retrievals. The increased sensitivity of the DPR relative to the TRMM Precipitation Radar (PR) and the high-frequency GMI channels enable GPM to measure light rain (down to 0.2 mm h^{-1}) and falling snow.

Table 1 GPM-CO GMI and DPR specifications and on-orbit performance

Channel (GHz)	Spatial resolution (km)	NEDT (K) (required)	NEDT (K) (on orbit)	Beamwidth (°) (on orbit)	Beam efficiency (%) (on orbit)
GMI 10.65 V-pol	19.4 × 32.1	0.96	0.77	1.72	91
GMI 10.65 H-pol	19.4 × 32.1	0.96	0.78	1.72	91
GMI 18.7 V-pol	10.9 × 18.1	0.84	0.63	0.98	91
GMI 18.7 H-pol	10.9 × 18.1	0.84	0.60	0.98	91
GMI 23.8 V-pol	9.7 × 16.0	1.05	0.51	0.85	93
GMI 36.64 V-pol	9.4 × 15.6	0.65	0.41	0.82	98
GMI 36.64 H-pol	9.4 × 15.6	0.65	0.42	0.82	98
GMI 89 V-pol	4.4 × 7.2	0.57	0.32	0.38	97
GMI 89 H-pol	4.4 × 7.2	0.57	0.31	0.38	97
GMI 166 V-pol	4.1 × 6.3	1.5	0.70	0.38	97
GMI 166 H-pol	4.1 × 6.3	1.5	0.66	0.37	97
GMI 183.31 ± 3 V-pol	3.8 × 5.8	1.5	0.56	0.37	95
GMI 183.31 ± 7 V-pol	3.8 × 5.8	1.5	0.47	0.37	95
DPR 13.6 (Ku)[a]	5 km horizontally				
	250 m vertically[b]				
DPR 35.5 (Ka)[c]	5 km horizontally				
	250 m vertically				
	500 m vertically[d]				

[a]Minimum detectable reflectivity ∼12–13 dBZ
[b]Oversampled to 125 km
[c]Minimum detectable reflectivity ∼17–18 dBZ (∼12–13 dBZ in high sensitivity mode)
[d]In high sensitivity mode (oversampled to 125 km and 250 km, respectively)

3.1 Dual-Frequency Precipitation Radar (DPR)

The DPR instrument, which provides measurements of 3-dimensional precipitation structure, consists of a Ka-band Precipitation Radar (KaPR) at 35.5 GHz and a Ku-band Precipitation Radar (KuPR) at 13.6 GHz being an updated version of the Ku-band unit flown on TRMM [27]. JAXA and the National Institute of Information and Communications Technology (NICT) of Japan built the DPR instrument. The KuPR and the KaPR are co-aligned such that their 5 km footprints coincide on the Earth surface, with cross-track swath widths of 245 and 125 km, respectively (see Fig. 1), and a vertical range resolution of 250 m, except when the KaPR is operating in a high-sensitivity, spatial over-sampling mode, in which case the range resolution is 500 m. Both Ka and Ku oversample vertically to reduce the range distance by half. The variable pulse repetition frequency technique used by the DPR increases the number of samples at each Instantaneous Field-Of-View (IFOV) to achieve a detection threshold of 12 dBZ, or 0.2 mm h^{-1}. The DPR has been calibrated both internally and externally through a series of Active Radar Calibration overpasses of calibrated ground sources provided by JAXA.

With information from the DPR, precipitation measurements are more accurate for retrieving two parameters of the bulk particle size distributions by using (1) the sensitivity of Ka to small particles, (2) the sensitivity of Ku to larger particles, and (3) a retrieval technique based on the differential attenuation between observations

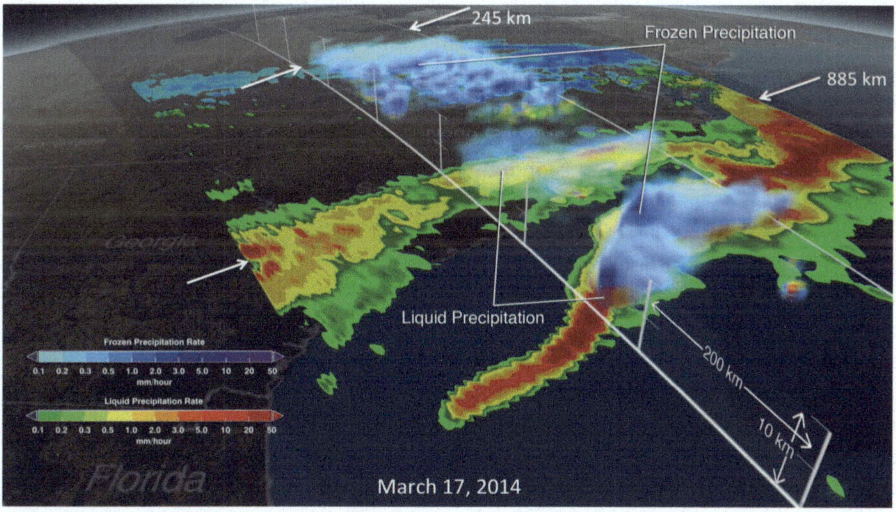

Fig. 1 GMI and DPR precipitation retrievals from the GPM Core Observatory observing a March 17, 2014 snow storm

at the two frequencies. The dual-frequency returns give us insight into microphysical processes (evaporation, collision/coalescence, aggregation) and also allow us to distinguish regions of liquid, frozen and mixed-phase precipitation, in addition to providing bulk properties of the precipitation such as water flux (rain rate) and water content in the measurement column. The improved accuracy and detailed microphysical information from the dual-wavelength radar is used to constrain precipitation retrievals from the GMI and to develop an a priori Bayesian cloud model database to be used for radiometer-based retrievals. This database is used by the constellation radiometers to unify precipitation retrievals globally.

3.2 GPM Microwave Imager (GMI)

The GMI instrument is a multi-channel, conical-scanning, microwave radiometer characterized by thirteen channels ranging in frequency from 10 to 183 GHz (Table 1), and a 1.2 m diameter antenna to provide better spatial resolution than TMI. The GMI has an off-nadir-viewing angle of 48.5°, which corresponds to an earth-incidence-angle of 52.8°, thus maintaining a similar geometry with the predecessor TRMM radiometer. The GMI main reflector rotates at 32 revolutions per minute to collect microwave radiometric brightness temperature (T_b) measurements over a 140° sector centered on the spacecraft ground track, giving a cross-track swath of 885 km on the Earth's surface from the GPM-CO 407 km altitude (see Fig. 1). The GMI was built by the Ball Aerospace and Technology Corporation under contract with NASA Goddard Space Flight Center. The central portion of the GMI swath overlaps the radar swaths with an approximately 67-s difference between the GMI and radar observation times (due to geometry and spacecraft motion). The measurements within the overlapped swaths of DPR and GMI are important for improving and cross-validating precipitation retrievals, especially the radiometer-based retrievals.

With the role of a reference radiometer in mind, the GMI is designed to achieve greater instrument accuracy and stability by employing several innovative engineering features. A high quality reflective coating was used on the main and cold calibration sub-reflectors. Noise diodes were included in the calibration scheme for the channels from 10.7 to 36.64 GHz enabling an explicit solution for non-linearity in the radiometer response functions [7]. The warm load has 14 temperature sensors imbedded in it to better characterize its physical temperature and any thermal gradients. The cold calibration sub-reflector is somewhat oversized to reduce spillover issues and the warm calibration load is shrouded to minimize solar intrusion [6].

After launch, a series of spacecraft attitude maneuvers were performed to provide checks on and refinements to the calibration. One of these maneuvers involved pitching up the GPM spacecraft so that both the main beam and the cold calibration mirror viewed cold space. The voltage from the radiometers was determined to be identical to within measurement uncertainty for both the main and sub-reflectors, indicating that either there are no significant emissivity issues for either reflector. A pitch down maneuver was also done to verify consistency between the vertical and horizontal polarizations at nadir for the dual-polarized channels and corrections were developed for calibration errors resulting from sensitivity of the radiometer to magnetic fields as well as updates to the pre-launch derived spillover corrections. This combination of the instrument design and detailed on-orbit calibration checks and refinements have resulted in an extremely stable and accurate calibration to within 0.5 K [50].

4 Coverage and Sampling

Given the current capabilities of microwave remote-sensing technology, the GPM mission takes a constellation-based approach by combining measurements from a constellation of microwave sensors in LEO to achieve global coverage and the high temporal sampling needed for a wide range of science and societal applications (Fig. 2). The GPM constellation was initially conceived to comprise conical-scanning radiometers, which have been the mainstay for rainfall estimation from space for many decades. Microwave sounding instruments, whose observations are less directly related to precipitation than conically scanning imagers, were not included in the original mission design. However, as retrieval algorithms for microwave sounders have continued to improve in recent years, the quality of

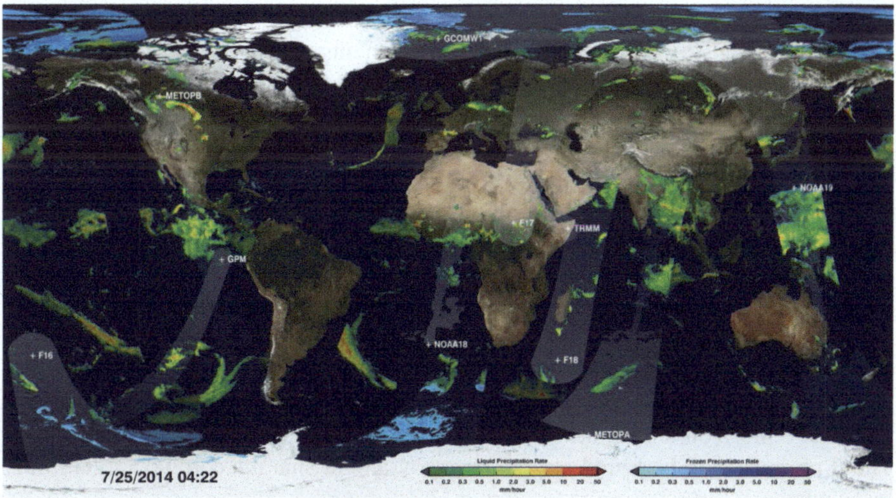

Fig. 2 The GPM constellation in 2017 includes the GPM Core Observatory and constellation radiometers on partner satellites (Megha-Tropiques, GCOM-W1, DMSP F18 and F19, NPP, NOAA-19, MetOp-B, and NPOESS C1)

current rainfall retrievals from microwave sounders with high-frequency channels between 150 and 183 GHz has been shown to be comparable to those from conical-scanning radiometers over land [33]. The baseline GPM constellation was reconfigured in 2006 to include microwave sounding instruments to augment the sampling by conical-scanning radiometers over land, where revisit time is between 1 and 2 h over the GPM mission life [13]. The high-frequency sounding channels on GMI will also provide a bridge in using GPM-CO sensor measurements to further improve sounder retrievals in the GPM constellation.

The build-up of the GPM constellation follows a rolling wave strategy with a flexible architecture to take advantage of satellites of opportunity to provide the best possible sampling and coverage with assets provided by cooperating partners. The baseline constellation partner configuration in 2017 has 10 sensors including these conical-scanning imagers: the Advanced Microwave Scanning Radiometer-2 (AMSR-2) on Japan's Global Change Observation Mission-Water 1 (GCOM-W1) [35], Special Sensor Microwave Imager/Sounder (SSMIS) instruments on three U.S. Defense Meteorological Satellite Program (DMSP) satellites [30]. Cross-track scanning temperature/humidity sounders in the constellation in 2017 include: the Sondeur Atmospherique du Profil d'Humidite Intertropicale par Radiometrie (SAPHIR) on the French-Indian Megha-Tropiques satellite [42], the Advanced Technology Microwave Sounder (ATMS) instrument on the National Polar-orbiting Operational Environmental Satellite System (NPOESS) Preparatory Project (NPP) [9], and four Microwave Humidity Sounder (MHS) instruments on NOAA and the European MetOp satellites [22]. Table 2 provides basic sensor and frequency details for the constellation sensors in the baseline configuration. Each constellation partner satellite radiometer has its own scientific or operational objectives but contributes microwave T_b observations to the GPM mission for unified constellation-based global precipitation products. The inter-satellite calibration (at both radiance and retrieval levels) using GPM-CO sensor measurements as reference standards within a consistent framework is a key objective of the GPM mission.

With such a constellation of varying precipitation sensors, inter-satellite calibration is an important aspect of the GPM mission. Indeed, GPM's multi-satellite Integrated Multi-satellitE Retrievals for GPM (IMERG; [15]) provide precipitation estimates nearly globally every 30 min by using infrared (IR) observations to morph the precipitation estimates from the constellation sensors. The IMERG product is available at a surface resolution of $0.1° \times 0.1°$ (or about 10 km × 10 km) in NRT about 5 h after the precipitation event. All GPM-CO products are available in NRT for application users and also later as research products that include all ancillary data for better performance and lower uncertainty. The Japanese counterpart of IMERG is the Global Satellite Mapping of Precipitation (GSMaP), provided also with $0.1° \times 0.1° \times 0.1°$ horizontal resolution and with various latencies; nowcast, 4 h, etc. [25]. These products can be found at gpm.nasa.gov and at sharaku.eorc.jaxa.jp/GSMaP, respectively.

Table 2 GPM-CO and constellation partner passive microwave radiometer sensor capabilities in 2017

	Channel[a]								
	6–7 GHz	10 GHz	19 GHz	23 GHz	31–37 GHz	50–60 GHz	89–91 GHz	150–167 GHz	183–190 GHz
GMI		10.65	18.70	23.80	36.50		89.0	165.6	183.31±3 V
		V/H	V/H	V	V/H		V/H	V/H	183.31±8 V
Resolution[b]		26 km	15 km	12 km	11 km		6 km	5 km	5 km
AMSR 2	6.925/7.3	10.65	18.70	23.80	36.50		89.0		
	V/H	V/H	V/H	V/H	V/H		V/H		
Resolution[b]	62/58 km	42 km	22 km	26 km	12 km		5 km		
SSMIS			19.35	22.235	37.0	50.3-63.28	91.65	150	183.31±1 H
			V/H	V	V/H	V/H	V/H	H	183.31±3 H
Resolution[b]			59 km	59 km	36 km	22 km	14 km	14 km	14 km
SAPHIR									183.31±0.2 QH
									183.31±1.1 QH
									183.31±2.8 QH
									183.31±4.2 QH
									183.31±6.8 QH
									183.31±11 QH
Resolution[b]									10 km
MHS							89 QV	157 QV	183.31±1 QH
									183.31±3 QH
									190.311 QV
Resolution[b]							17 km	17 km	17 km

(continued)

Table 2 (continued)

	Channel[a]								
	6–7 GHz	10 GHz	19 GHz	23 GHz	31–37 GHz	50–60 GHz	89–91 GHz	150–167 GHz	183–190 GHz
ATMS				23.8	31.4	50.3-57.29	87-91	165.5 H	183.31±1 QH
				QV	QV	QH	QV	QH	183.31±1.8 QH
									183.31±3 QH
									183.31±4.5 QH
									183.31±7 QH
Resolution[b]				74 km	74 km	32 km	16 km	16 km	16 km

[a]Channel Center Frequency (GHz): V—Vertical Polarization, H—Horizontal Polarization, QV—Quasi-Vertical, QH—Quasi-Horizontal (QV and QH indicates mixed polarization for cross-track scanners)
[b]Mean spatial resolution (km)

5 Radiance Intercalibration of Constellation Sensors

Climate quality data sets need to be accurate and consistent across long time periods spanning multiple observing platforms. Any T_b biases among the different GPM microwave constellation instruments might impede consistent global precipitation estimates. Despite similarities among these constellation sensors, differences exist in terms of center frequencies, viewing geometries, and spatial resolutions (Table 2), which must be reconciled in order to produce uniform global precipitation estimates. An important by-product of the cross calibration effort is the ability to monitor each instrument in the GPM constellation for abrupt and/or gradual calibration changes as well as other degradations.

Achieving consistency through intersatellite calibration requires several steps. First, each instrument provider is responsible for providing well-calibrated T_b products adhering to established standards. Based on these standard T_b products provided by partners, GPM examines the data to find and remove pixel-to-pixel biases that can be caused by pitch and roll errors, spacecraft interference and more subtle causes. For conical scanners, detecting biases is conceptually straightforward; the average T_b of each beam position should be the same. Biases due to latitude and small incidence angle anomalies are readily removed [10]. For cross-track scanners detecting and correcting for biases to achieve self-consistency is much more difficult because the dependence on incidence angle must be very accurately modeled.

Once an instrument is shown to be self-consistent across a scan, it must be self-consistent through the orbit. For this purpose, environmental data such as Global Data Assimilation System (GDAS) is used to compute T_b for non-raining oceanic pixels throughout the orbit. While these calculations may not be adequate for absolute calibration, they are suitable for identifying T_b anomalies. When a consistent pattern of anomalies is detected, an approach is developed to correct for them.

Once individual instrument T_b are self-consistent, the multiple radiometers in the constellation are inter-calibrated to produce a uniform global T_b product within a consistent framework. To this end, GMI observations are spatially and temporally matched to the other constellation sensors. For the matched data sets, algorithm development precedes along two approaches (1) a system wherein co-located T_bs are translated to a common basis and compared [1, 12], and (2) deriving limiting values for each radiometer independently and comparing them vicariously through a model (e.g., [3, 43]). Corrections indicated by these matched comparisons are subsequently applied to constellation T_bs in the intercalibrated data files that are used as input for the radiometer retrieval algorithm [1].

6 Precipitation Retrievals and Products

Unified precipitation estimation from a heterogeneous constellation of radiometric sensors requires not only robust intercalibration of T_b but also a physical framework for deriving rainfall with respect to a common standard. The GPM concept achieves this by using the GPM-CO to generate the best possible observed a priori statistics of precipitation profiles and their corresponding T_b at all channels observed by the diverse constellation radiometers. The GPM constellation radiometers then reference their own observations to this common a priori database to be physically consistent as envisioned in the mission concept. The GPM mission produces three types of precipitation products based on (1) DPR retrievals, (2) combined DPR/GMI retrievals, and (3) radiometer retrievals based upon a priori databases generated by the combined DPR/GMI retrievals. The DPR retrieval has been shown to provide the best retrieval when directly compared to surface radars [18] and associated microphysical data that has historically been obtained from multi-parameter surface radars. The combined DPR/GMI retrieval is designed to begin with the DPR solution but modify this solution as necessary to ensure consistency not only with DPR reflectivities, but GMI radiances as well. This approach improves the overall results—particularly in scenes where the DPR is unable to fully describe the drop size distributions, the cloud liquid water content, or the rainfall falling below the detection threshold of the Ka band radar.

The radar algorithms take advantage of the sensitivity of the Ka and Ku bands to the underlying precipitation particles. The DPR's reflectivities provide two degrees of freedom at each range gate for determining the drop size distribution in the vertically sampled profile [32]. This is a significant step forward, since TRMM and surface based radars rely on a single frequency and therefore require assumptions about the drop size distribution. The radar retrievals estimate profiles of the rain rate, liquid water content and parameters of the rain and snow particle size distributions as well as their associated error statistics. Special emphasis is focused on retrieval methods for very light rainfall cases, very heavy rainfall cases, and for falling snow cases (which were not retrieved with TRMM's orbit). With measurements at two frequencies, the DPR retrieval also benefits from the use of differential attenuation correction techniques instead of the more restrictive surface reference technique [36].

The combined radar-radiometer retrieval conceptually provides the best estimates of precipitation structure from a spaceborne platform through the combination of DPR and GMI information [11]. The retrieval naturally serves as the a priori cloud/precipitation database for precipitation retrievals from constellation radiometers. While this approach was used already with TRMM over tropical oceans, one of the challenges in GPM combined algorithm development was the creation of updated physical parameterizations for use at higher latitudes and for the enhanced channel sets of GMI and the DPR. Since freezing levels are lower at these latitudes, there is a need for the DPR to better identify the altitudes where the transitions from ice-phase to mixed-phase to liquid occur, in both convective and stratiform precipitation regions. In addition, improved descriptions of the extinction and scattering of ice and mixed-phase precipitation that are applicable to GMI's 10–183 GHz frequency range have been developed (e.g., [31, 39]). For GPM algorithm development improved physical parameterizations were supported by a series of ground validation field campaigns (see Sect. 7).

The DPR, GMI, and Combined algorithms all estimate falling snow utilizing the higher frequency channels (166, 183 on GMI and Ka on DPR) to sense the characteristics of the frozen particles. These higher frequency channels were added to the design in 2006 after a series of papers showed that they were useful for detecting and estimating falling snow (e.g., [5, 23, 34, 45]). Since launch, evaluations of GPM's falling snow retrieval performance are reported in [4] and [49].

Radiometer algorithms for uniform global precipitation estimates must be physically consistent across multiple frequencies and all constellation members. Bayesian or Optimal Estimation algorithms are well suited to retrieve precipitation and its vertical structure by comparing the information content of any multi-frequency radiometer to that supplied by a common a priori database of vertical profiles of precipitating clouds together with simulated T_b for each profile and sensor. By associating the combined algorithm's retrieved precipitation with DPR reflectivity profiles and GMI T_b, a robust global a priori database constrained by both DPR and GMI measurements is constructed, eliminating the need for a priori databases that rely solely on cloud resolving models. When the DPR does not detect precipitation but a precipitation signal is present in the 166 and 183 GHz channels of the GMI, CloudSat data is incorporated in the a priori database. This approach, designated as "radar-enhanced" radiometer algorithms, improves the accuracy and physical consistency among radiometer retrievals. While imagers and sounders typically have different channel sets and scanning procedures (conical versus cross-track), algorithm development is conceptually the same for all the passive sensors, with the Bayesian database reflecting the differences in instrumentation. Database T_b are computed from the geophysical parameters derived by the combined DPR/GMI retrievals so that differences in sensor channels, incidence angles and spatial resolutions are fully accounted for. Because of the variable surfaces features and impacts on precipitation, the Bayesian database is split into about 14 land surface classifications helping to improve the retrievals by minimizing the surface emissivity impacts. Since the same methodology can be used for previous as well as future sensors it is possible to extend GPM rainfall climatologies back to 1987 using SSM/I observations and potentially forward beyond the era of the GPM-CO satellite itself. A longer time series product will be released to the public in 2018.

While frequent coverage afforded by multiple constellation microwave radiometers is important, there are as always, applications such as hydrology that require even greater temporal and spatial resolution. To this end, a number of multi-satellite techniques have been developed (see [8]). One approach to ensuring uniform, frequent coverage is to use the passive microwave rainfall estimates as enhanced by geostationary-based IR observations such as IMERG and GSMaP (e.g., [24, 25]). GPM ensures that the microwave products being used for bias correction in these merged multi-satellite microwave/IR methods are consistent with one another. GPM also plans to explore ways to develop very-high spatial and temporal resolution (1–2 km and 5–10 min) precipitation estimates through statistical downscaling or data assimilation using cloud-resolving models for certain hydrological applications.

In order to meet the objective (3) of the GPM mission, we also produce three-dimensional latent heating data. There are two representative latent heating algorithms, namely, Convective-Stratiform Heating (CSH) algorithm, which basically uses GMI data and utilizes convective/stratiform ratio with regionally produced latent heating tables, and Spectral Latent Heating (SLH) algorithm, which uses GPM DPR products with spectrally expressed tables for convective and stratiform precipitation separately. Latent heating data can be obtained as GPM standard products.

7 Ground Validation

The NASA GPM program established a vigorous ground validation (GV) plan to use ground-based observations (including aircraft measurements) to support pre-launch satellite algorithm development and post-launch precipitation product evaluation. Three key components of the GPM GV infrastructure include numerous field campaigns [14, 16, 38, 46] and field site measurements, exploitation of NOAA's Multi-Radar Multi-Sensor (MRMS) products [51], and development of a Validation Network (VN) architecture described below. The field campaigns fully sampled precipitation at the surface, precipitation physics in the clouds, and conducted active/passive microwave remote sensing from above the clouds using ground, in situ and remote sensing aircraft, and satellite measurements. These field campaign data were instrumental both in developing the satellite retrieval algorithms by reducing the assumptions and in measuring integrated quantities for hydrometeorological applications. MRMS data are rain gauge bias-adjusted radar estimates of precipitation rate and type (rain/snow) over the U.S. and neighboring territory (130–60°W, 20–55°N). GV MRMS datasets are derived from the NOAA MRMS products but are further processed (e.g., [20, 21]) to produce gauge-corrected reference precipitation rate datasets augmented with information on precipitation types and data quality at the native 2-min/0.01° $\times$ 0.01° resolutions. The Validation Network (VN) radar data consists of $\sim$80 U.S. network, oceanic, and/or other national/international research site-specific dual polarimetric (dual-pol) radars and their derived precipitation rate, DSD and hydrometeor type estimates that are footprint and geometrically volume-matched to GPM-CO overpasses [2, 40, 44].

GPM GV precipitation measurement equipment inventory is vast [47]. Most equipment is designed to be portable for moving to field campaign locations as GPM continues to participate in precipitation related field campaigns worldwide. When not deployed for GPM-related field campaigns, these instruments operate within the NASA Wallops Flight Facility (WFF) network (or partner sites), with special emphasis placed on targeted and user-adapted data collections during GPM-CO overpasses to meet the needs and requests of the GPM algorithm and GV teams. All GV data is freely available from a link on the GPM webpage (gpm.nasa.gov).

8 Applications

Through improved measurements of rain and snow the GPM suite of products contributes to a wide range of societal applications such as: tropical and extratropical cyclone location and rainfall monitoring, famine early warning, drought monitoring, water resource management, agricultural forecasting, numerical weather prediction, land surface modeling, global climate modeling, disease tracking, economic studies, and animal migration; many of which were initially developed with TRMM data. Several of these applications require NRT data as well as longer-term, well-calibrated merged-satellite precipitation information [41]; the GPM mission supports both of these product latencies. IMERG is being used as an input for forecasts (e.g., floods, landslides, agriculture/famine, precipitation-induced disease outbreaks) in other regions of the world, especially areas lacking adequate ground-based information [17]. Selected applications are reported in [19] and [26]. GPM's remotely-sensed precipitation data enable a diverse range of applications across agencies, research institutions and the global community.

9 Summary

GPM is an international satellite mission to unify and advance precipitation measurements from a constellation of dedicated and operational microwave sensors to provide the next-generation unified global precipitation products for research and applications. GPM is a partnership between NASA and JAXA, with additional domestic and international partnerships. In February 2014, NASA and JAXA launched the GPM-CO in a 65°-inclined orbit carrying the first dual-frequency radar in space together with an advanced microwave radiometer serving as a precipitation physics observatory and calibration reference for all constellation radiometric measurements and retrievals.

GPM is a science mission with integrated applications goals to advance the knowledge of the global water/energy cycle variability and freshwater availability and improve weather, climate, and hydrological prediction capabilities. Within the framework of GEO and GEOSS, GPM serves as a cornerstone for the development of the CEOS Precipitation Constellation (PC). During its mission life, GPM is a mature realization of the CEOS-PC to provide state-of-the-art global precipitation products for scientific research and societal applications.

Acknowledgements Contributions of the GPM Inter-Satellite Calibration Working Group, the Ground Validation Working Group, the GPM algorithm teams, and the PMM Science Team are gratefully acknowledged. Imagery assistance from the NASA Science Visualization Studio is greatly appreciated. Funding for these activities is provided by NASA Headquarters, primarily Ramesh Kakar.

References

1. Berg, W., S. Bilanow, R. Chen, S. Datta, D. Draper, H. Ebrahimi, S. Farrar, W.L. Jones, R. Kroodsma, D. McKague, V. Payne, J. Wang, T. Wilheit, and J.X. Yang. 2016. Intercalibration of the GPM Radiometer constellation. *Journal of Atmospheric and Oceanic Technology* 33: 2639–2654. https://doi.org/10.1175/JTECH-D-16-0100.
2. Bolen, S., and V. Chandrasekar. 2003. Methodology for aligning and comparing spaceborne radar and ground-based radar observations. *Journal of Atmospheric and Oceanic Technology* 20: 647–659. https://doi.org/10.1175/1520-0426(2003)20<647:MFAACS>2.0.CO;2$.
3. Brown, S.T., and C.S. Ruf. 2005. Determination of a hot blackbody reference target over the amazon rainforest for the on-orbit calibration of microwave radiometers. *AMS Journal of Oceanic and Atmospheric Technology* 22(9): 1340–1352.
4. Casella, D., G. Panegrossi, P. Sanò, A.C. Marra, S. Dietrich, B.T. Johnson, and M.S. Kulie. 2017. Evaluation of the GPM-DPR snowfall detection capability: comparison with CloudSat-CPR. *Atmospheric Research*. https://doi.org/10.1016/j.atmosres.2017.06.018.
5. Chen, F.W., and D.H. Staelin. 2003. AIRS/AMSU/HSB precipitation estimates. *IEEE Transactions on Geoscience and Remote Sensing* 41 (2): 410–417. https://doi.org/10.1109/TGRS.2002.808322.
6. Draper, D.W., D.A. Newell, D.S. McKague, and J.R. Piepmeier. 2015. Assessing calibration stability using the global precipitation measurement (GPM) microwave imager (GMI) noise diodes. *IEEE Journal of Selected Topics in Applied Earth Observations and Remote Sensing* 8: 4239–4247. https://doi.org/10.1109/JSTARS.2015.2406661.
7. Draper, D.W., D.A. Newell, F.J. Wentz, S. Krimchansky, and G. Skofronick-Jackson. 2015. The global precipitation measurement (GPM) microwave imager (GMI): instrument overview and early on-orbit performance. *IEEE Journal of Selected Topics in Applied Earth Observations and Remote Sensing* 8: 3452–3462. https://doi.org/10.1109/JSTARS.2015.2403303.
8. Ebert, E., J. Janowiak, and C. Kidd. 2007. Comparison of near real time precipitation estimates from satellite observations and numerical models. *Atmospheric Research* 88: 47–64.
9. Goldberg, M.D., H. Kilcoyne, H. Cikanek, and A. Mehta. 2013. Joint polar satellite system: the United States next generation civilian polar-orbiting environmental satellite system. *Journal of Geophysical Research—Atmospheres* 118: 13463–13475. https://doi.org/10.1002/2013JD020389.
10. Gopalan, K., W.L. Jones, S. Biswas, S. Bilanow, T. Wilheit, and T. Kasparis. 2009. A time-varying radiometric bias correction for the TRMM microwave imager. *IEEE Transactions on Geoscience and Remote Sensing* 47 (11): 3722–3730.
11. Grecu, M., W.S. Olson, S.J. Munchak, S. Ringerud, L. Liao, Z. Haddad, B.L. Kelley, and S.F. McLaughlin. 2016. The GPM combined algorithm. *Journal of Atmospheric and Oceanic Technology* 33: 2225–2245. https://doi.org/10.1175/JTECH-D-16-0019.1.
12. Hong, L., W.L. Jones, T.T. Wilheit, and T. Kasparis. 2009. Two approaches for inter-satellite radiometer calibrations between TMI and WindSat. *Journal of the Meteorological Society of Japan. Series II* 87: 223–235.
13. Hou, A.Y., R.K. Kakar, S. Neeck, A.A. Azarbarzin, C.D. Kummerow, M. Kojima, R. Oki, K. Nakamura, and T. Iguchi. 2014. The global precipitation measurements mission. *Bulletin of the American Meteorological Society* 95: 701–722. https://doi.org/10.1175/BAMS-D-13-00164.1.

14. Houze, R.A., L.A. McMurdie, W.A. Petersen, M.R. Schwaller, W. Baccus, J. Lundquist, C. Mass, B. Nijssen, S.A. Rutledge, D. Hudak, S. Tanelli, G.G. Mace, M. Poellot, D. Lettenmaier, J. Zagrodnik, A. Rowe, J. DeHart, L. Madaus, and H. Barnes. 2017. The olympic mountains experiment (OLYMPEX). *Bulletin of the American Meteorological Society*. https://doi.org/10.1175/BAMS-D-16-0182.1.
15. Huffman, G.J., D.T. Bolvin, D. Braithwaite, K. Hsu, R. Joyce, C. Kidd, E.J. Nelkin, S. Sorooshian, J. Tan, and P. Xie. 2017. Algorithm theoretical basis document (ATBD) version 4.6 for the NASA global precipitation measurement (GPM) Integrated Multi-satellitE Retrievals for GPM (I-MERG), 32 pp. GPM Project: Greenbelt, MD. Available online at https://pmm.nasa.gov/sites/default/files/document_files/IMERG_ATBD_V4.6.pdf.
16. Jensen, M.P., W.A. Petersen, A. Bansemer, N. Bharadwaj, L.D. Carey, D.J. Cecil, S.M. Collis, A.D. Del Genio, B. Dolan, J. Gerlach, S.E. Giangrande, A. Heymsfield, G. Heymsfield, P. Kollias, T.J. Lang, S.W. Nesbitt, A. Neumann, M. Poellot, S.A. Rutledge, M. Schwaller, A. Tokay, C.R. Williams, D.B. Wolff, S. Xie, and E.J. Zipser. 2016. The midlatitude continental convective clouds experiment (MC3E). *Bulletin of the American Meteorological Society* 97: 1667–1686. https://doi.org/10.1175/BAMS-D-14-00228.1.
17. Kidd, C., G.J. Huffman, A. Becker, G. Skofronick-Jackson, D. Kirschbaum, P. Joe, and C. Muller. 2017. So, how much of the Earth's surface is covered by rain gauges? *Bulletin of the American Meteorological Society* 98: 69–78. https://doi.org/10.1175/BAMS-D-14-00283.1.
18. Kidd, C., J. Tan, P. Kirstetter, and W.A. Petersen. 2017. Validation of the version 05 level 2 precipitation products from the GPM core observatory and constellation satellite sensors. *Quarterly Journal of the Royal Meteorological Society.* First Published: 4 December 2017. http://dx.doi.org/10.1002/qj.3175.
19. Kirschbaum, D.B., G.J. Huffman, R.F. Adler, S. Braun, K. Garrett, E. Jones, A. McNally, G. Skofronick-Jackson, E. Stocker, H. Wu, and B.F. Zaitchik. 2017. NASA's remotely-sensed precipitation: a reservoir for applications users. *Bulletin of the American Meteorological Society* 98: 1169–1184. https://doi.org/10.1175/BAMS-D-15-00296.1.
20. Kirstetter, P., Y. Hong, J.J. Gourley, S. Chen, Z.L. Flamig, J. Zhang, M. Schwaller, W. Petersen, and E. Amitai. 2012. Toward a framework for systematic error modeling of spaceborne precipitation radar with NOAA/NSSL ground radar-based National Mosaic QPE. *Journal of Hydrometeorology* 13: 1285–1300. https://doi.org/10.1175/JHM-D-11-0139.1.
21. Kirstetter, P.E., Y. Hong, J.J. Gourley M. Schwaller, W. Petersen, and Q. Cao. 2015. Impact of sub-pixel rainfall variability on spaceborne precipitation estimation: evaluating the TRMM 2A25 product. *Quarterly Journal of the Royal Meteorological Society* 141: 953–966. https://doi.org/10.1002/qj.2416.
22. Klaes, K.D., M. Cohen, Y. Buhler, P. Schlüssel, R. Munro, A. Engeln, E. Clérigh, H. Bonekamp, J. Ackermann, J. Schmetz, and J. Luntama. 2007. An introduction to the EUMETSAT polar system. *Bulletin of the American Meteorological Society* 88: 1085–1096. https://doi.org/10.1175/BAMS-88-7-1085.
23. Kongoli, C., P. Pellegrino, R.R. Ferraro, N.C. Grody, and H. Meng. 2003. A new snowfall detection algorithm over land using measurements from the advanced microwave sounding unit (AMSU). *Geophysical Research Letters* 30: 1756. https://doi.org/10.1029/2003GL017177,14.
24. Kubota, T., S. Shige, H. Hashizume, K. Aonashi, N. Takahashi, S. Seto, M. Hirose, Y.N. Takayabu, K. Nakagawa, K. Iwanami, T. Ushio, M. Kachi, and K. Okamoto. 2007. Global precipitation map using satelliteborne microwave radiometers by the GSMaP project: production and validation. *IEEE Transactions on Geoscience and Remote Sensing* 45: 2259–2275.
25. Kubota, T., K. Aonashi, T. Ushio, S. Shige, Y.N. Takayabu, Y. Arai, T. Tashima, M. Kachi, R. Oki. 2017. Recent progress in global satellite mapping of precipitation (GSMaP) product. In *Proceedings of IGARSS2017*, Fort Worth.
26. Kucera, P.A., E.E. Ebert, F.J. Turk, V. Levizzani, D. Kirschbaum, F.J. Tapiador, A. Loew, M. Borsche. 2013. Precipitation from space: advancing earth system science. *Bulletin of the American Meteorological Society* 94: 365–375. https://doi.org/10.1175/BAMS-D-11-00171.1.

27. Kummerow, C., W. Barnes, T. Kozu, J. Shiue, and J. Simpson. 1998. The tropical rainfall measuring mission (TRMM) sensor package. *Journal of Atmospheric and Oceanic Technology* 15: 809–817.
28. Kummerow, C., J. Simpson, O. Thiele, W. Barnes, A.T.C. Chang, E. Stocker, R.F. Adler, A. Hou, R. Kakar, F. Wentz, P. Ashcroft, T. Kozu, Y. Hong, K. Okamoto, T. Iguchi, K. Kuriowa, E. Im, Z. Haddad, G. Huffman, B. Ferrier, W.S. Olson, E. Zipser, E.A. Smith, T.T. Wilheit, G. North, T. Krishnamurti, and K. Nakamura. 2000. The status of the tropical rainfall measuring mission (TRMM) after two years in orbit. *Journal of Applied Meteorology* 39 (12, Part 1): 1965–1982.
29. Kummerow, C.D., D.L. Randel, M. Kulie, N.Y. Wang, R. Ferraro, S.J., Munchak, and V. Petkovic. 2015. The evolution of the Goddard profiling algorithm to a fully parametric scheme. *Journal of Atmospheric and Oceanic Technology* 32: 2265–2280. https://doi.org/10.1175/JTECH-D-15-0039.1.
30. Kunkee, D.B., G. Poe, D. Boucher, S. Swadley, Y. Hong, J. Wessl, and E. Uliana. 2008. Design and evaluation of the first special sensor microwave imager/sounder. *IEEE Transactions on Geoscience and Remote Sensing* 46: 863–883.
31. Kuo, K., W.S. Olson, B.T. Johnson, M. Grecu, L. Tian, T.L. Clune, B.H. van Aartsen, A.J. Heymsfield, L. Liao, and R. Meneghini. 2016. The microwave radiative properties of falling snow derived from nonspherical ice particle models. Part I: an extensive database of simulated pristine crystals and aggregate particles, and their scattering properties. *Journal of Applied Meteorology and Climatology* 55: 691–708. https://doi.org/10.1175/JAMC-D-15-0130.1.
32. Liao, L., and R. Meneghini. 2005. A study of air/space-borne dual-wavelength radar for estimation of rain profiles. *Advances in Atmospheric Sciences* 22 (6): 841–851.
33. Lin, X., and A.Y. Hou. 2008. Evaluation of coincident passive microwave rainfall estimates using TRMM PR and ground-based measurements as references. *Journal of Applied Meteorology and Climatology* 47: 3170–3187.
34. Liu, G. 2004. Approximation of single scattering properties of ice and snow particles for high microwave frequencies. *Journal of the Atmospheric Sciences* 61: 2441–2456. https://doi.org/10.1175/1520-0469(2004)061<2441:AOSSPO>2.0.CO;2.
35. Maeda, T., Y. Taniguchi, and K. Imaoka. 2016. GCOM-W1 AMSR2 Level 1R product: dataset of brightness temperature modified using the antenna pattern matching technique. *IEEE Transactions on Geoscience and Remote Sensing* 54 (2): 770–782. https://doi.org/10.1109/TGRS.2015.2465170.
36. Meneghini, R., T. Iguchi, T. Kozu, L. Liao, K. Okamoto, J.A. Jones, and J. Kwiatkowski. 2000. Use of the surface reference technique for path attenuation estimates from the TRMM precipitation radar. *Journal of Applied Meteorology* 39: 2053–2070.
37. Mugnai, A., et al. 2007. Snowfall measurements by the proposed European GPM Mission. In *Measuring Precipitation from Space: EURAINSAT and the Future*, ed. by V. Levizzani, P. Bauer, and J. Turk, 750 pp. Berlin: Springer.
38. Nayak, M.A., G. Villarini, and A.A. Bradle. 2016. Atmospheric rivers and rainfall during NASA's iowa flood studies (IFloodS) campaign. *Journal of Hydrometeorology* 17: 257–271. https://doi.org/10.1175/JHM-D-14-0185.1.
39. Olson, W.S., L. Tian, M. Grecu, K.-S. Kuo, B.T. Johnson, A.J. Heymsfield, A. Bansemer, G.M. Heymsfield, J.R. Wang, and R. Meneghini. 2017. The microwave radiative properties of falling snow derived from nonspherical ice particle models. Part II: initial testing using radar, radiometer and in situ observations. *Journal of Applied Meteorology and Climatology* 55: 709–722. https://doi.org/10.1175/JAMC-D-15-0131.1.
40. Pippitt, J., D.B. Wolff, W.A. Petersen, and D. Marks. 2015. Data and operational processing for NASA's GPM ground validation program. In *37th Conference on Radar Meteorology*, Norman, OK. Boston: American Meteorological Society.
41. Reed, P.M., N.W. Chaney, J.D. Herman, M.P. Ferringer, and E.F. Wood. 2015. Internationally coordinated multi-mission planning is now critical to sustain the space-based rainfall observations needed for managing floods globally. *Environmental Research Letters* 10: 024010. https://doi.org/10.1088/1748-9326/10/2/024010.

42. Roca, R., H. Brogniez, P. Chambon, O. Chomette, S. Cloché, M.E. Gosset, J.-F. Mahfouf, P. Raberanto, and N. Viltard. 2015. The Megha-tropiques mission: a review after three years in orbit. *Frontiers in Earth Science* 3: https://doi.org/10.3389/feart.2015.00017.
43. Ruf, C.S., Y. Hu, and S.T. Brown. 2006. Calibration of windsat polarimetric channels with a vicarious cold reference. *IEEE Transactions on Geoscience and Remote Sensing* 44 (3): 470–475.
44. Schwaller, M.R., and K.R. Morris. 2011. A ground validation network for the global precipitation measurement mission. *Journal of Atmospheric and Oceanic Technology* 28: 301–319. https://doi.org/10.1175/2010JTECHA1403.1.
45. Skofronick-Jackson, G.M., M.J. Kim, J.A. Weinman, and D.E. Chang. 2004. A physical model to determine snowfall over land by microwave radiometry. *IEEE Transactions on Geoscience and Remote Sensing* 42 (5): 1047–1058.
46. Skofronick-Jackson, G., D. Hudak, W. Petersen, S.W. Nesbitt, V. Chandrasekar, S. Durden, K.J. Gleicher, G. Huang, P. Joe, P. Kollias, K.A. Reed, M.R. Schwaller, R. Stewart, S. Tanelli, A. Tokay, J.R. Wang, and M. Wolde. 2015. Global precipitation measurement cold season precipitation experiment (GCPEX): for measurement's sake, let it snow. *Bulletin of the American Meteorological Society* 96: 1719–1741. https://doi.org/10.1175/BAMS-D-13-00262.1.
47. Skofronick-Jackson, G., W.A. Petersen, W. Berg, C. Kidd, E.F. Stocker, D.B. Kirschbaum, R. Kakar, S.A. Braun, G.J. Huffman, T. Iguchi, P.E. Kirstetter, C. Kummerow, R. Meneghini, R. Oki, W.S. Olson, Y.N. Takayabu, K. Furukawa, and T. Wilheit. 2017. The global precipitation measurement (GPM) mission for science and society. *Bulletin of the American Meteorological Society*. https://doi.org/10.1175/BAMS-D-15-00306.1.
48. Stephens, G.L., and C.D. Kummerow. 2007. The remote sensing of clouds and precipitation from space: a review. *Journal of the Atmospheric Sciences* 64: 3742–3765.
49. Tang, G., Y. Wen, J. Gao, D. Long, Y. Ma, W. Wan, and Y. Hong. 2017. Similarities and differences between three coexisting spaceborne radars in global rainfall and snowfall estimation. *Water Resources Research* 53. https://doi.org/10.1002/2016WR019961.
50. Wentz, F., and D. Draper. 2016. On-orbit absolute calibration of the global precipitation mission microwave imager. *Journal of Atmospheric and Oceanic Technology* 33: 1393–1412. https://doi.org/10.1175/JTECH-D-15-0212.1.
51. Zhang, J., K. Howard, C. Langston, B. Kaney, Y. Qi, L. Tang, H. Grams, Y. Wang, S. Cocks, S. Martinaitis, A. Arthur, K. Cooper, and J. Brogden. 2016. Multi-radar multi-sensor (MRMS) quantitative precipitation estimation: initial operating capabilities. *Bulletin of the American Meteorological Society* 97: 621–638. https://doi.org/10.1175/BAMS-D-14-00174.1.

Cloud Remote Sensing by Active Sensors: New Perspectives from CloudSat, CALIPSO and EarthCARE

Hajime Okamoto and Kaori Sato

1 Introduction

Clouds are one of the most important contributors to the spread of climate change predictions by coupled atmosphere-ocean models [8]. Disparities in the representation of clouds among the general circulation models are large. Differences in simulated liquid water paths were more than ten times among 20 models [18] and those for simulated ice water path are even larger (20 times) [45]. These can be attributed to the poor understanding of cloud generation and dissipation mechanism.

Concerning the satellite remote sensing of clouds, there are in general two types of sensors: passive and active. The former sensors detect natural sources from the clouds such as reflected sunlight or thermal emission and observe integrated energy described above along the line of sight. Thus it is in principle difficult to obtain high resolution of the vertical structure of clouds. Passive sensors can provide large horizontal coverage. The passive remote sensing of clouds has a long history. International Satellite Cloud Climatology Project (ISCCP) is an international satellite project using several passive sensors on weather satellites have provided the archive of cloud properties since 1 July 1983 (e.g., [32]).

Active sensors imply cloud profiling radar and lidar, which emit electromagnetic wave and detect returned signal from the target. Active sensor technique is superior, being able to identify the cloud top and bottom altitudes and multi-layer structures of clouds compared with passive remote sensing. A drawback of active instruments is the small footprint size, i.e. they have narrow swath compared with passive instruments. There are differences between the scattering properties of the same target such as cloud at the cloud radar and lidar wavelengths. Owing to these

H. Okamoto (✉) · K. Sato
Research Institute for Applied Mechanics, Kyushu University, Kasuga, Fukuoka, Japan
e-mail: okamoto@riam.kyushu-u.ac.jp; sato@riam.kyushu-u.ac.jp

C. Andronache (ed.), *Remote Sensing of Clouds and Precipitation*,
Springer Remote Sensing/Photogrammetry,
https://doi.org/10.1007/978-3-319-72583-3_8

particularities, the sensitivity to particles makes the distinction between the cloud radar and lidar, leading to the differences in the detectability of clouds.

Cloud profiling radar at 94 GHz (a W-band radar) has been first developed by Lhermitte [17]. The frequency is selected that atmospheric gasses absorption is small at the wavelength. The wavelength of 3.16 mm is much shorter than the precipitation radar, e.g. 13.8 GHz (2.17 cm). When we consider small particles compared with the wavelength, the backscattered intensity is proportional to λ^{-4}. That is, shorter wavelength such as 3.16 mm has higher sensitivity than precipitation radars when other factors are the same. The 94 GHz cloud radar enables to achieve information of multi-layer structure of hydrometeors since attenuation is generally small except for heavy precipitation.

Ground-based polarization lidar for cloud studies has a longer history. It started in the decades of the 1970s. Scotland et al. [40] showed the capability to obtain depolarization ratio is inevitable for the discrimination of cloud phase. Since then extensive observations using polarization lidar have been conducted for cloud studies (e.g. [34]). Lidar, in general, has more sensitivity to small effective size than to large ones when total mass of ensemble of particles are same. Lidar can detect aerosols and clouds when attenuation is small between the target and the lidar.

CloudSat and CALIPSO satellites were launched on 28 April 2006 and have been operational since June 2006. Though the original lifetime was 22 months for CloudSat and 3 years in orbit for CALIPSO, respectively, they are still providing data after 11 years. They are flying NASA afternoon constellation of satellites, called A-Train so that CloudSat radar and CALIPSO lidar are able to see the same place at the same time with other satellites [41]. Their orbits are Sun-synchronous and altitude of about 705 km at the equator with the equator-crossing time of 1:30 PM and a 16 day repeat cycle. CloudSat carries the first 94 GHz radar in space, and it measures the radar reflectivity factor (Z_e) from clouds and precipitation [42]. Cloud-Aerosol Lidar and Infrared Pathfinder Satellite Observation (CALIPSO) carries a backscatter-type lidar, the Cloud-Aerosol Lidar with Orthogonal Polarization (CALIOP), as the first polarization lidar in space. CALIOP can measure the attenuated backscattering coefficient (β_{att}) at 532 and 1064 nm and the depolarization ratio at 532 nm [46]. CALIPSO also carries the infrared imager radiometer (IIR) and wide-field-camera (WFC). CloudSat and CALIPSO offer data on the internal structure of clouds and precipitation from space.

To evaluate the role of clouds on climate system through interactions of cloud radiative effects and convection, we need to know (1) three-dimensional structure of cloud occurrence and multi-layer structure, (2) cloud particle phase and (3) cloud microphysics, together with (4) atmospheric conditions such as temperature, pressure and water vapor amount. For this purpose, we introduce these algorithms to provide above mentioned physical quantities. In Sect. 2, we describe observational data by CloudSat and CALIPSO and auxiliary data for cloud analysis. In Sect. 3, the suite of algorithms for cloud studies are introduced. Cloud mask, cloud particle type and cloud microphysics are treated in Sects. 3.1, 3.2 and 3.3, respectively. Two scattering theories, the discrete dipole approximation (DDA) and physical optics are introduced. These are relevant for the interpretation of observables of CloudSat radar and CALIPSO lidar for the retrieval of cloud microphysics. In Sect. 4, analysis of cloud mask, cloud types and ice microphysics are shown. A summary, remaining issues in the analysis and possible future directions are also given in the discussion in Sect. 5.

2 Input Data for Cloud Products

We used CloudSat data from 2B GEOPROF product (release R04) and CALIPSO lidar level 1B (version 3) from CALIPSO with atmospheric profile data from European Center for Medium-range Weather Forecasting (ECMWF).

CloudSat has been set to 0.16° off-nadir since 15 August 2006. Vertical resolution is 480 m, and the measured return backscatter signal is oversampled, i.e. sampled at 240 m. Returned pulses are averaged about 0.16 s along the nadir track, yielding 1.4 km × 1.7 km [21]. CPR profiles of Z_e of clouds and precipitation are reported at every 1.1 km along the track. The minimum detectable signal of CloudSat is about −30 dBZ.

CALIOP uses a solid-state neodymium-doped yttrium aluminum garnet (Nd: YAG) laser and emits pulses at 532 and 1064 nm. Resultant beam diameter is about 70 m at the earth's surface, and the receiver footprint diameter is 90 m at the ground [47]. The fundamental horizontal resolutions are 1/3 km below 8.2 km and 1.0 km above 8.2 km, respectively for 532 and 1064 nm. And the vertical resolution is 30 and 60 m below and above 8.2 km, respectively for 532 nm. The vertical resolution for 1064 nm is 60 m below 20.2 km.

We created the merged data sets in which the observables from CloudSat CPR and CALIPSO lidar have the same horizontal and vertical resolutions. Here the CALIPSO data were sampled when the along-track and cross-track distances between the footprints of CloudSat and CALIPSO were within 0.55 and 0.7 km, respectively [9]. Then attenuated lidar backscattering coefficients for parallel and perpendicular channels at 532 nm and total backscattering coefficient were averaged to have the same horizontal and vertical resolutions as CloudSat. These data are called KU-merged dataset where KU denotes Kyushu University.

3 Algorithms to Study Cloud Macroscale and Microphysical Properties

3.1 Cloud Mask Schemes for CloudSat and CALIPSO

There are four cloud mask schemes. The schemes are for the clouds detected by CloudSat (C1), those detected by CALIPSO lidar (C2), those detected by both of CloudSat and CALIPSO and those detected by either CloudSat or CALIPSO. They were originally developed by ship-based observations in the western Pacific Ocean near Japan and the tropical western Pacific Ocean [27, 28]. By definition, fractions of C3-cloud and C4-cloud are minimum and maximum, respectively.

We first describe cloud detection scheme for C1 considers the signal to noise ratio (SNR) and special continuity test. We follow the CloudSat standard algorithm for cloud mask. The initial detection relies on the mean power of noise (P_N) and its standard deviation (σ_N). The values P_N and σ_N at each record are estimated by averaging the echo powers in the stratosphere [21]. When the echo power at a range bin of interest exceeds the sum of P_N and σ_N, the target bin is assigned as a candidate of a cloudy bin. Then a special continuity test is applied to reduce the uncertainties in cloud detection. By the detection algorithm, we can infer the hydrometer occurrence, i.e. clouds and precipitation. To further discriminate clouds and precipitation, it is necessary to introduce the empirical threshold such as Z_e of clouds should be smaller than -15 dBZe. It is generally difficult to determine cloud bottom boundary from radar signal alone because of the contamination of precipitation. Another difficulty is the detection of low-level clouds in C1 due to the effect of surface clutter below about 720 m.

CALIPSO C2 scheme [9] is different from the standard mask that was provided as in the CALIPSO lidar level 2 Vertical Feature Mask (VFM) [44]. The C2 mask requires two criteria. The first criterion is the backscatter power of clouds at should exceed the threshold value in Eq. (1)

$$P_{obs}(R, r) > P_{th}(R) \tag{1}$$

where $P_{th}(R)$ is estimated using mean noise signals and standard deviation of it, for the original CALIPSO lidar resolution and as expressed by Eq. (2)

$$\begin{aligned} P_{th}(R, r) = & \frac{1}{2}\left(P_{aerosol}(r) + (P_m(R, r) + P_n + \sigma_n)\right) \\ & -\frac{1}{2}\left(P_{aerosol}(r) - (P_m(R, r) + P_n + \sigma_n)\right)\tanh(R - 5) \end{aligned} \tag{2}$$

where R (km) denotes the altitude of the layer from ground. P_m and β_{mol} denote molecule signals and attenuated backscattering coefficient due to molecules estimated by using ECMWF data. Here P_m is estimated by Eq. (3)

$$P_m(R, r) = \beta_{mol}(R)/r^2 \tag{3}$$

The threshold value for aerosols is determined by the ship-based lidar data (Eq. (4)),

$$P_{aerosol}(r) = \beta_{aerosol}/r^2 \quad \text{and} \quad \beta_{aerosol} = 10^{-5.25}\left[\frac{1}{m \cdot sr}\right] \tag{4}$$

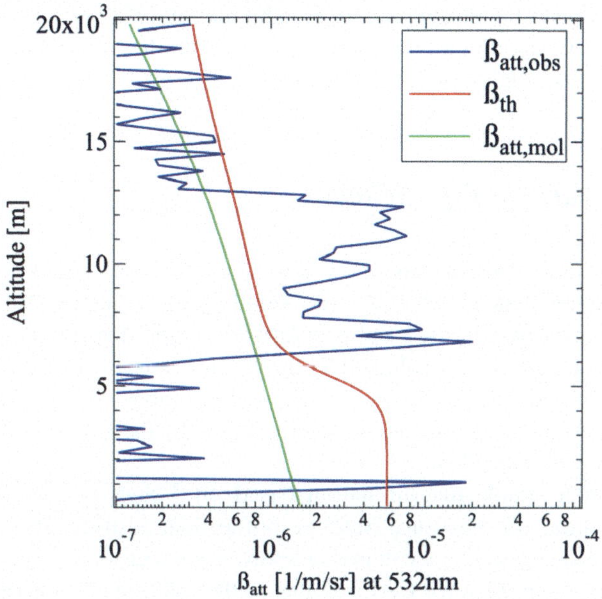

Fig. 1 Profile of attenuated backscattering coefficient β_{att} at 532 nm from CALIPSO (blue), the threshold profile used in the CALIPSO lidar mask (red) and the profile of attenuated backscattering coefficient due to molecules (green)

The backscattering coefficient corresponding to P_{th} is estimated as $\beta_{th}(R, r) = P_{th}(R) \cdot r^2$. The coefficient β_{th} is close to $\beta_{aerosol}$ for R lower than about 4 km and $\beta_{mol}(R) + (P_n + \sigma_n) \cdot r^2$ for R higher than about 8 km, respectively. Example of a vertical profiles of β_{th}, β_{mol} and $\beta_{att,obs}$ are displayed in Fig. 1. There are high clouds at a 7–12 km and a geometrically thin low-level cloud layer at 1 km. There some spurious signals at about 17 km in Fig. 1. In order to remove the spurious signals due to noise, we introduce a second criterion, the special continuity test. We consider five times 5 dimensional pixels in horizontal and vertical directions for a target pixel of interest. If more than 13 pixels fulfill the first criterion of cloud mask, the pixel is considered to be cloudy. This turned out to be effective to remove the spurious signals.

Hagihara et al. [9] identified the CALIPSO standard cloud mask product, Vertical Feature Mask (VFM), has an issue to produce too many low-level clouds for CALIPSO version2 data was used. Rossow and Zhang [33] compared cloud layer occurrence from the International Satellite Cloud Climatology Project (ISCCP), the results from two merged data sets detected by CloudSat and CALIPSO, i.e., one from 2B-Geoprof-lidar product [19] and one from KU-mask. They found that the disagreement in low-level clouds between ISCCP and 2B-Geoprof-lidar product, which is a merged data set of CloudSat 2B-Geoprof product and VFM product and interpreted the most of the disagreement was caused by the CALIPSO overestimation of low cloud amounts in the 2B-Geoprof-lidar product. KU-mask significantly reduced the disagreement. Hagihara et al. [10] also conducted similar comparisons for CALIPSO level1 version 3 data. And they still identified the same tendency of overestimation of low-level clouds.

3.2 Cloud Particle Type Scheme

Ground-based lidar observations show the depolarization ratio of water clouds are mostly smaller than about 10%. The small value indicated that the multiple scattering contribution is limited and depolarization ratio does not deviate from the value of spherical particle as explained by the single scattering based on Mie theory. In contrast, ice clouds produce large depolarization ratio (~40%). In ground-based lidar observations, cloud phase can be discriminated by the δ based on observations and theoretical estimates [34].

However, such simple discrimination based on δ does not work in the case of spaceborne lidar observations. Satellite-borne lidar instruments observe much larger footprints on Earth surface than does conventional ground-based lidar. Former footprint size is about 90 m for CALIOP due to the satellite altitude of 700 km, and also the field of view of 0.13 mrad and the latter one is about 1 m at an attitude of 1 km. Because of the CALIOP's large footprint size, the contribution of multiple scattering to the lidar signals becomes much larger than in ground-based ones for the same extinction profile. Due to the effect of multiple scattering, the δ of water clouds measured by CALIOP becomes comparable to that of ice clouds as optical thickness increases. It is concluded that there is a fundamental difficulty in discriminating ice and water clouds based solely on depolarization ratio.

Yoshida et al. [49] introduced two-dimensional diagram to discriminate cloud particle type. They used δ and the ratio of attenuated backscattering coefficients (β_{att}) for two vertically consecutive layers. Here we use common logarithmic form $\chi(R_i)$ as follows

$$\chi(R_i) = \log_{10}\left[\frac{\beta_{att}(R_i)}{\beta_{att}(R_{i+1})}\right] \tag{5}$$

where R_i denotes the center altitude of layer i. When cloud microphysics are vertically homogeneous, χ is proportional to layer extinction.

For the temperature > 0 °C, water clouds are expected to be dominant and clouds in the temperature ranges show large attenuation and moderate depolarization ratio (~30%) by CALIPSO observations. When the temperature is below −40 °C, ice particles are dominant. The δ of clouds for these cold temperature is larger than 20% with small attenuation. The ice clouds are categorized as 3D-ice type, i.e. randomly oriented ice particles from the feature. When the temperature is between −5 and −20 °C, there are some clouds with very low δ (<3%) with small attenuation in addition to the feature of 3D-ice revealed by CALIPSO observations. The clouds are categorized as 2D-plate, i.e., horizontally oriented ice plates [35]. In the same

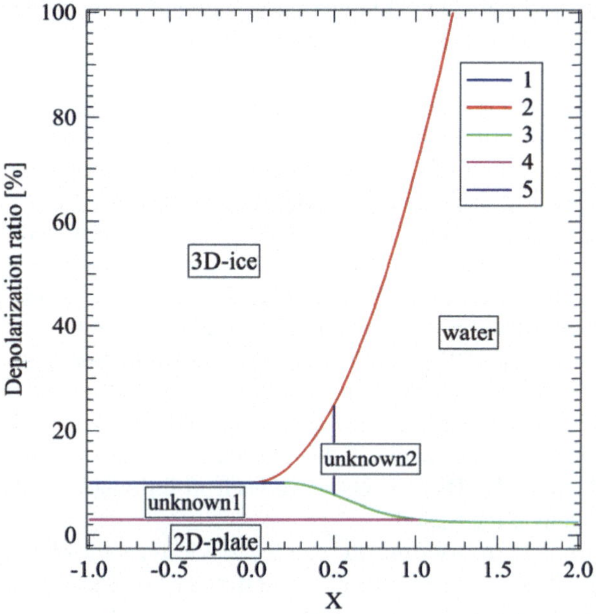

Fig. 2 Depolarization ratio versus χ

temperature range, water and ice particles can exist and the similar feature found in $T > 0°$ for water can be also found [12, 49].

Exact boundaries of cloud particle categories are determined by using three lines and two curves as function of χ and δ [%] (Fig. 2). (1) $\delta = 10$ for $\chi \leq 0.2$ (blue line in Fig. 2), (2) $\delta = 60\chi^2 + 10$ for $0 \leq \chi \leq \sqrt{1.5}$ (red curve), (3) $\delta = 7.5exp(-4(\chi - 0.2)^2) + 2.5$ for $0.2 \leq \chi \leq 2$ (green curve), (4) $\delta = 3$ for $\chi \leq 1.0228$ (magenta line), (5) $7.7326 \leq \delta \leq 25$ for $\chi = 0.5$ (purple line).

Area of 3D-ice type is determined by blue line 1 and red curve 2. The values of δ of 3D-ice type is larger than 10% and separation between 3D-ice and water particle is done by the red curve 2. The separation assures the large depolarization with small attenuation of 3D-ice.

Water particle category is determined by red curve 2, green curve 3 and purple line 5. Red and green curves are determined by the Monte Carlo simulations for water clouds with CALIOP geometry [49]. Contrary to the 3D-ice, the separation of red curve reflects the relatively large δ with large attenuation of water clouds in case of spaceborne lidar observations as written above. Once water categories are determined by the two-dimensional diagram, water clouds are further categorized to warm water or super-cooled one by the temperature.

2D-plate category exists below magenta line 4 for $\chi \leq 1.0228$ and green line 3 for $\chi \geq 1.0228$. The value 1.0228 corresponds to χ for a crossing point of magenta line 4 and green curve 3, i.e., given by $3 = 7.5exp(-4(\chi - 0.2)^2) + 2.5$. The conditions assure small δ of the 2D-plate. Although there is practically no condition for χ, the actual distribution of the 2D-plate category observed from CALIOP shows relatively small χ.

There are some ambiguities to determine ice or water categories. There are two categories of unknown (unknown 1 and 2) near the boundaries of water, 3D-ice and 2D-plate categories where robust discrimination of cloud particle type seems to be difficult. Unknown 1 is introduced as the type likely containing 3D-ice and 2D-plate. By definition, it exists for $10\% > \delta > 3\%$. The type is determined by blue line 1, green curve 3 and magenta line 4. Unknown 2 is considered to contain liquid water or 3D-ice. Its area is determined by red curve 2, green curve 3 and purple line 5. Actual populations of unknown type 1 and 2 turn out to be small.

It is noted that CALIOP pointed in a near-nadir direction (~0.3°) before 27 November 2007. Since then the laser has pointed 3° off-nadir. The occurrence frequencies of the 3D-ice and water are not significantly different in the two periods, but the frequency of occurrence of the 2D-plate category drastically decreases during off-nadir periods. There is certainly need to introduce another criterion to categorize 2D-plate in the period.

There is also need to discriminate cloud particle type for cloud portions only detected by CloudSat. It seems effective to introduce empirically determined two-dimensional diagram for Z_e and temperature to infer cloud particle type. The development of the diagram for CloudSat has been done by comparing the results of cloud particle type from CALIOP for clouds detected by both of CloudSat and CALIOP [16].

3.3 Retrievals of Cloud Microphysics

Once cloud detection and discrimination of cloud particle phase are done by cloud mask and cloud particle type algorithms, respectively, retrieval of cloud microphysics can be performed. In this subsection, we describe methods to retrieve ice cloud microphysics by using CloudSat and CALIPSO. In order to obtain microphysics information, we first need to establish relationships between observables for 94 GHz and 0.532 μm or 1.064 μm and cloud microphysics. Single scattering properties of homogeneous spherical particles can be obtained by Mie theory. Input parameters of the Mie theory are wavelength of interest, radius, and complex refractive indices at the wavelength of interest. When particles are non-spherical, Mie theory may not be adequate and scattering theories that can treat non-spherical particles are required. Number of parameters to characterize the scattering calculations increase for arbitrary shaped particles and the following difficulties appear. Since analytical solutions have been known only for homogeneous sphere, infinite cylinder, spheroids and aggregates of monomers and there is no analytical solutions for arbitrary shapes, we must introduce certain degree of approximation into scattering calculations. This is also true that we must approximate particle shapes since it is not possible to treat infinite number of shape variations. Consequently, two different scattering theories suitable for cloud radar and lidar wavelengths are chosen.

It is also noted that single scattering approximation is not always sufficient to interpret signals from those active sensors. When the optical thickness is large, multiple scattering occurs. In such cases, radar or lidar equations should be treated as time-dependent radiative transfer equation. Monte-Carlo method or some semi-analytical methods can be applied. Former is, in general, accurate as long as sufficient number of photons are used and it can also treat 3D-radiative transfer equation. Its computational cost is expensive. Contrary, latter method has some limitation in applicability but in general much faster than the former approach.

3.3.1 Discrete Dipole Approximation (DDA) for Cloud Radar

When the particles are small or comparable to the wavelength, the discrete dipole approximation (DDA) [31], T-matrix method [22] and Finite Difference Time Domain Method (FDTD) [48] are known as numerical approaches. Among them, we introduce the DDA here. DDA has been initially developed for scattering by interstellar dust particles. Later [5] extended the method. The particle is divided by many sub-volume elements whose sizes are taken to be smaller than the wavelength. Then sub-volume element is replaced by a point electric dipole. The particle are approximated by an ensemble of N dipoles. Inter-dipole spacing is set to be d. Chemical composition of the target is expressed by the polarizability α of the dipole. α is determined by Clausius-Mossotti relation or Lattice Dispersion Relation (LDR) and the choice of the methods is related to the accuracy of the estimated values [7]. Dipole moment induced at J-th dipole P_j due to incident electric field is

$$\mathbf{P}_j = \alpha_j \left(\mathbf{E}_{inc,j} - \sum_{k \neq j}^{N} \mathbf{A}_{jk} \mathbf{P}_k \right) \tag{6}$$

where the second term in the right hand side denotes the electric field at j-th dipole produced by N-1 dipoles except for j-th dipole. The term can be further obtained by Eq. (7). It is noted that the expression is exact since the retardation effect of electric field at each dipole is fully taken into account.

$$\mathbf{A}_{jk}\mathbf{P}_k = \frac{\exp(ikr_{jk})}{r_{jk}^3} \left\{ k^2 \mathbf{r}_{jk} \times (\mathbf{r}_{jk} \times \mathbf{P}_k) + \frac{1 - ikr_{jk}}{r_{jk}^2} \times \left[r_{jk}^2 \mathbf{P}_k - 3\mathbf{r}_{jk}(\mathbf{r}_{jk} \cdot \mathbf{P}_k) \right] \right\} \tag{7}$$

here k denotes wavenumber, r_{jk} is distance between j-th dipole and k-th dipole and is estimated by $r_j - r_k$. Equation (6) contains 3N unknown parameters. Since incident electric fields and dipole polarizability are known, we can construct 3N couples of linear equations and it is possible to solve the equations. Once all sum of scattered fields from N dipoles are obtained, scattering and absorption characteristic of the whole particle, including extinction and backscattering cross section, can be directly determined. This is a theoretical background of the DDA.

Errors of the solution of the DDA are from (1) to approximate the target boundary by the finite number of dipoles and (2) neglect of higher order than the electric dipoles [24]. Accuracy of the DDA have been evaluated for spherical particles, two-touching spheres where the analytical solution can be obtained [6] and for hexagonal columns and plates by using the converging solutions [23, 24]. Validity criterion is derived using m, λ and d (Eq. (8)),

$$2\pi \frac{d}{\lambda}|m| < 0.5 \tag{8}$$

It is apparent from Eq. (8), N should be taken to be sufficiently large when particle is larger compared with the wavelength in order to obtain accurate results. Equation (8) also implies that the DDA is suitable for calculation of the dielectric materials (m is relatively small) and the accuracy suffers for absorbing materials with imaginary part of m to be very large. It is also noted that the calculation should be performed for each orientation. That is, DDA calculations for randomly oriented particle become expensive. The DDA solutions for Z_e can be obtained within errors of a few percent when the criterion is satisfied and random orientations are taken into account [23]. Applicability of the DDA is not restricted to homogeneous particles but can be applicable to inhomogeneous particle such as cloud particle containing soot particles as long as Eq. (8) is fulfilled.

Size parameter X defined as Eq. (9) is often used to measure the range of applicability of a scattering theory. r_{eq} is melted equivalent radius.

$$X \equiv \frac{2\pi r_{eq}}{\lambda} \tag{9}$$

Tractable X of the DDA exceeds 100 and it is increasing due to the increase of available computer memory and speed (e.g., [50]). For the cloud particle sizes of 2000 μm, X for 94 GHz is about 4, which is much below the limitation of the DDA. Here Z_e can be written by λ and K value and radar backscattering coefficient σ_{bk}.

$$Z_e = \frac{\lambda^4}{\pi^5} \frac{1}{|K|^2} \sigma_{bk} \tag{10}$$

Z_e is in unit $\left[\frac{mm^6}{m^3}\right]$ and $K = \left|\frac{m^2-1}{m^2+2}\right|$ and m corresponds to the complex refractive index at 94 GHz for water droplet at 10 °C. Resultant $|K|^2$ is 0.75 [43]. σ_{bk} is estimated by using size distribution function $dn(r_{eq})/dr_{eq}$ and backscattering cross section $C_{bk}(r_{eq})$ (e.g., [23]).

$$\sigma_{bk} = \int \frac{dn(r_{eq})}{dr_{eq}} C_{bk}(r_{eq}) dr_{eq} \tag{11}$$

$$C_{bk} \equiv 4\pi \left. \frac{C_{sca}(\theta)}{d\Omega} \right|_{\vartheta=180^\circ} \tag{12}$$

where $dC_{sca}(\theta)/d\Omega$ denotes the differential scattering cross section and θ and Ω are scattering angle and solid angle, respectively. Modified-Gamma distribution function and log-normal size distribution are used for ice and water clouds.

The size distribution function is often parameterized by modified Gamma distribution function or log-normal size distribution function. The effective radius r_{eff} is defined as

$$r_{eff} = \frac{\int r_{eq}^3 \frac{dn(r_{eq})}{dr_{eq}} dr_{eq}}{\int r_{eq}^2 \frac{dn(r_{eq})}{dr_{eq}} dr_{eq}} \tag{13}$$

Instead of Z_e, logarithmic form of Z_e is usually used; $dBZe = log_{10}(Z_e)$. Z_e of ice clouds is generally dependent on r_{eff}, ice water content (IWC) and particle shape. We now consider hexagonal ice column oriented in horizontal plane (2D-column), hexagonal ice plate oriented in horizontal plane (2D-plate) and randomly oriented bullet-rosette and spherical ice to estimate Z_e (Fig. 3) [36]. Here IWC is set to be $1\,\mathrm{g\,m^{-3}}$. There are small differences in Z_e among different shapes when particle

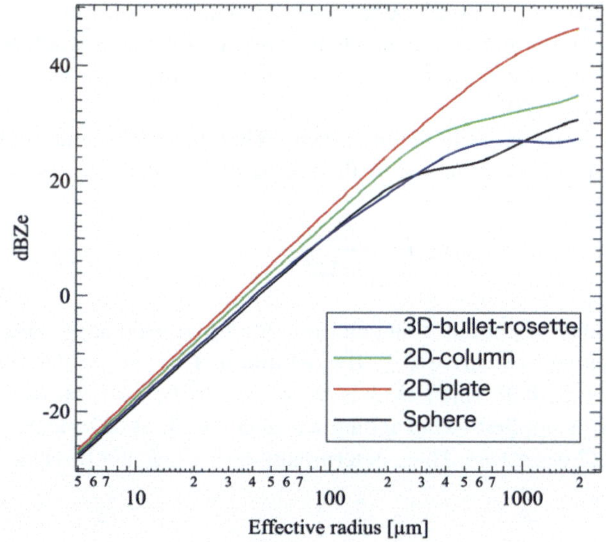

Fig. 3 Dependence of Z_e on r_{eff}. Comparisons of Z_e among four different ice particle models are also shown. IWC is assumed to be $1\,\mathrm{g\,m^{-3}}$

size is small. 2D-plate produces the largest Z_e and spheres and bullet rosette are similar.

As size increases, Z_e in general increases and differences in Z_e among different shapes increase. Those differences become significant for $r_{eff} > 100\,\mu m$, e.g., 6 and 16 dB differences at $r_{eff} = 200\,\mu m$ and $r_{eff} = 1$ mm, respectively. In conclusion, size is a control parameter to determine Z_e for $r_{eff} < 100\,\mu m$ and non-sphericity plays a major role for $r_{eff} > 100\,\mu m$.

3.3.2 Physical Optics for Lidar Analysis

It is difficult to apply the DDA for the analysis of lidar signals from ice clouds because X ranges from 68 for $r_{eff} = 5\,\mu m$ to 11810 for $r_{eff} = 1$ mm for the wavelength of 0.532 μm. Since large $X \gg 100$ cannot be calculated by the DDA, scattering theories based on geometrical optics are needed to simulate lidar signals for ice particles. Geometrical optics is derived when particle shape is much larger than the wavelength. Scattering electric fields created by incident fields at points inside a particle are assumed to behave as independent ones where interactions between the scattered waves are neglected in the geometrical optics. Incoming waves are reflected and refracted and such processes are repeated many times inside the particle. Reflectance and transmittance are determined according to Fresnel's law. These quantities determine a portion of the wave that remains in a particle and other portion that are reflected outside the particle.

To trace processes described above, ray tracing method based on Monte Carlo approach has been widely used [20]. The effect of diffraction is modelled by Fraunhofer diffraction theory. It depends on the particle size, wavelength of interest and scattering angle and is independent from m. Resultant scattered waves are obtained as a sum of electric fields due to reflection and refraction and fields due to the diffraction [3, 20].

One critical issue exists in the approach. When the backscattering cross section expressed in Eq. (12) is estimated by the ray-tracing method, C_{bk} is estimated by

$$C_{bk} \cong 4\pi \left. \frac{\Delta C_{sca}(\theta)}{\Delta\Omega} \right|_{\Delta\Omega \to 0, \theta \cong 180^\circ} \tag{14}$$

where, the number of photons that enter into the finite solid angle $\Delta\Omega$ at backward direction is counted to derive C_{bk}. By definition, C_{bk} can be determined only if $\Delta C_{sca}(\theta)$ becomes half when $\Delta\Omega$ is decreased to be half. However, when ray tracing method is applied to hexagonal ice plates such as 2D-plate, estimated C_{bk} increases as $\Delta\Omega$ decreases. Thus, determination of C_{bk} is not possible [15].

Physical optics is introduced in order to overcome the issue. Iwasaki and Okamoto [15] showed C_{bk} for thin rectangles and direction of incident laser beam is parallel to the normal of the surface of the rectangle. In such case, analytical solution for the shape can be obtained by Kirchhoff's diffraction theory [1].

$$C_{bk} = 4\pi \left| \frac{R_{fre} S}{\lambda} \right|^2 \tag{15}$$

where R_{fre} is the reflectance of the rectangle calculated by the Fresnel formula and S is the area of the rectangle. The approach was further extended to treat more realistic ice particle shapes by Borovoi et al. [2]. In their approach, a particle only consists of flat surfaces and electromagnetic fields inside the particle are assumed to be represented by geometrical approximation and once electric field on a surface is estimated, scattering waves radiated outward direction from the particle can be calculated by using analytical expressions for diffraction for general polyhedron particles [11]. Unlike a simple ray tracing method, physical optics method can offer converged solutions for C_{bk}. When $X < 100$, the errors in physical optics method might become large.

Intrieri et al. [14] has conducted the first analysis of ice particle sizing by combining 35 GHz radar and CO_2-lidar with a wavelength of 10 μm. They did not take into account attenuation correction for lidar signals. Okamoto et al. [25] proposed a forward algorithm for use of 94 GHz and lidar with 0.532 μm and comprehensive description, error analysis and analysis of ground-based data are conducted in [26]. Attenuation for 94 GHz and lidar signals are corrected in a consistent manner. In the first version, depolarization ratio is not included as input parameter. Then Okamoto et al. [29] introduced depolarization ratio into their algorithms to analyze CloudSat and CALIPSO signals. The retrieval algorithm is briefly described as follows.

We consider the mixture of two type of ice particles as ice model, i.e., 3D-ice and 2D-plate. 3D-ice consists of 50% of 2D-column and 50% of bullet-rosette. Mass mixing ratio of 2D-plate to the total IWC is allowed to vary. We first create look up tables for Z_e and extinction coefficient at 94 GHz for 3D-ice and 2D-plate by the DDA and those for β, extinction coefficient and depolarization ratio at 0.532 μm by the physical optics method for various r_{eff} but for a fixed IWC.

Since r_{eff} -dependencies of radar and lidar signals for a constant IWC are very different, it is possible to derive cloud microphysics by combining radar and lidar signals. r_{eff}, IWC and mass-mixing ratio of 2D-plate are retrieved by using Z_e, β and depolarization ratio for a grid box of interest. Extinction at 94 GHz and 0.532 μm are simultaneous retrieved. When only one of CloudSat or CALIPSO lidar information is available due to the strong attenuation for lidar or particle size is too small to be detected by CloudSat, we must introduce some additional processes. For example, when the lidar signals are fully attenuated and cloud radar signals are still observed, lidar signals are estimated by using the signals from upper layers and lidar-radar retrievals for the whole cloud layers are carried out by Levenberg-Marquardt method [37].

4 Analysis of CloudSat and CALIPSO: Case Study

We show a case study of cloud mask , cloud particle type and ice cloud microphysics in this section. The data used here is taken in 3 July 2007 at latitude between 31 °N–38 °N. Raw data for Z_e from CloudSat is shown in Fig. 4a and result after cloud mask described in previous section (Fig. 4b). There are high clouds at about 8–12 km and also low level clouds are found below 2 km. Similarly, CALIPSO detected clouds are shown in Fig. 5a and β for cloud mask results are displayed in Fig. 5b. CALIPSO signals due to aerosols, molecules and noise are removed by cloud mask. Cloud areas detected by CloudSat generally agree with those by CALIPSO lidar, except that some thin clouds detected by CALIPSO lidar are not detected by CloudSat at

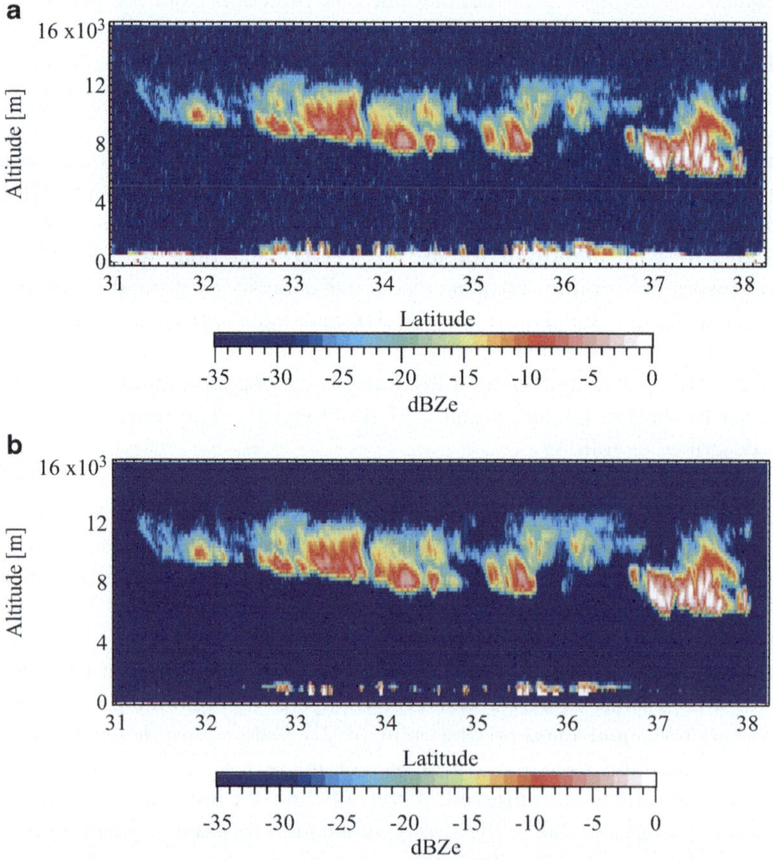

Fig. 4 (**a**) Altitude-latitude cross section of radar reflectivity factor Z_e observed by CloudSat, (**b**) Same as (**a**) but results after the application of cloud mask for CloudSat

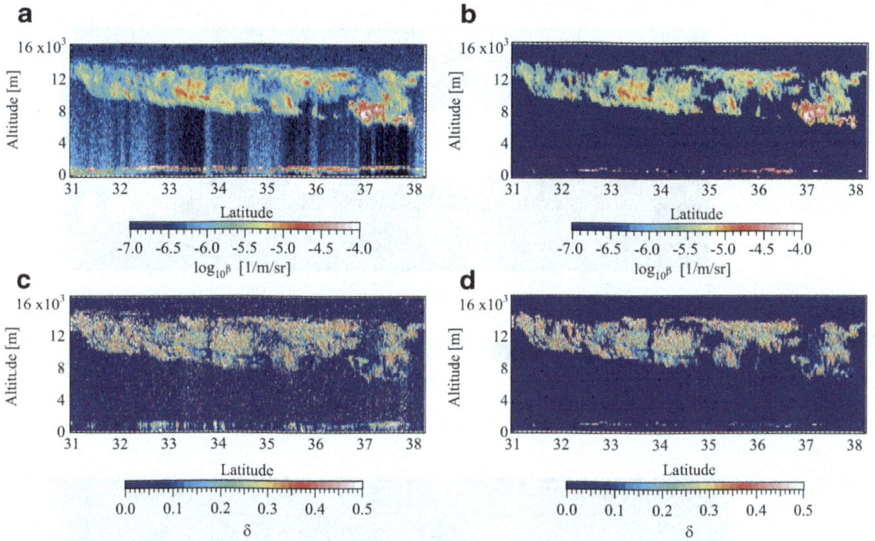

Fig. 5 (**a**) Same as Fig. 4a but for attenuated backscattering coefficient β_{att} by CALIPSO, (**b**) CALIPSO cloud mask, (**c**) depolarization ratio detected by CALIPSO. (**d**) same as (**c**) but after CALIPSO mask

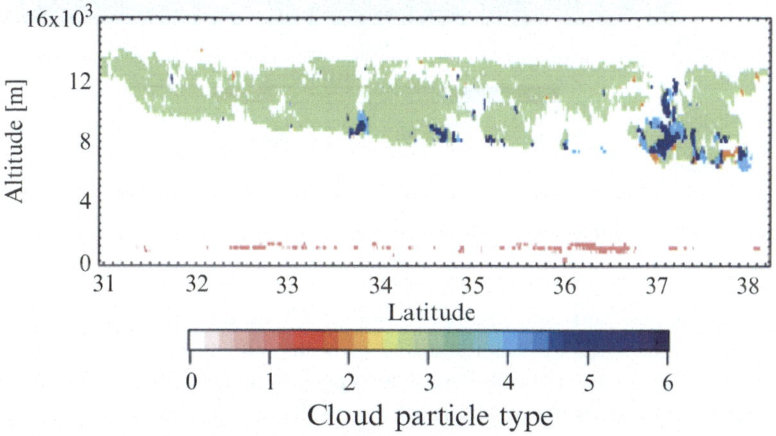

Fig. 6 Altitude-latitude plot of cloud particle type. 3D-ice, 2D-plate, warm water clouds are indicated by green, blue and red, respectively

around 13 km. Some of low-level clouds detected by CALIPSO lidar are also not detected by CloudSat at around 1 km.

Cloud particle type is derived for the same clouds described above (Fig. 5). Ice clouds are identified at 8–12 km. Cloud particle types are mostly 3D-ice(green) with small fraction of 2D-plate (blue) and unknown types. Warm water clouds ($T > 0\,^{\circ}$C) are found at 1 km (Fig. 6).

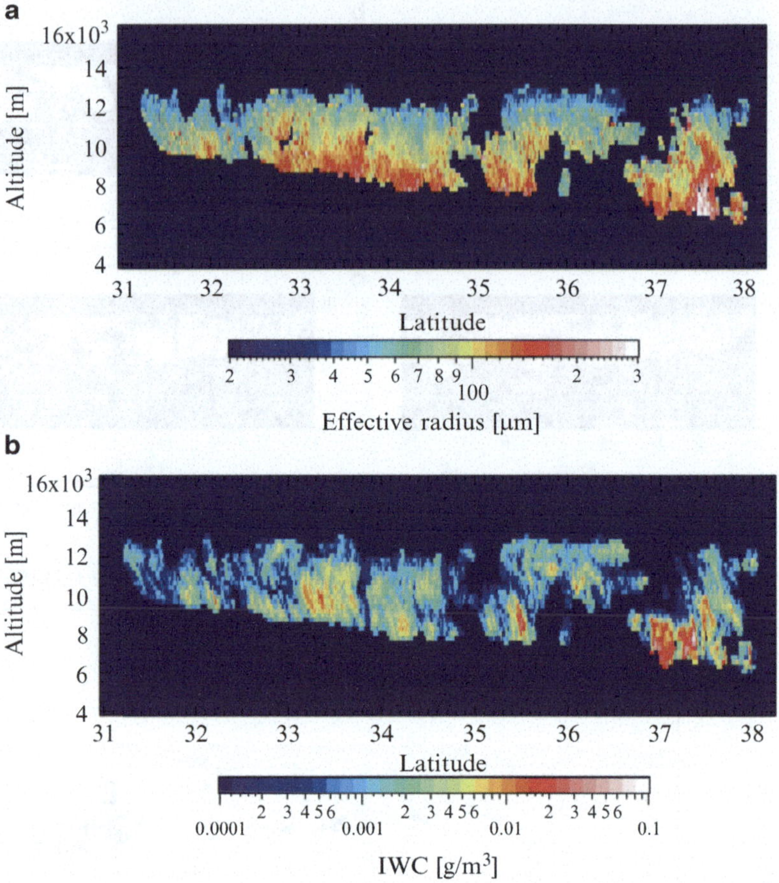

Fig. 7 (**a**) Altitude—latitude plot of effective radius r_{eff} and (**b**) that of IWC

We choose cloud portions detected by both of CloudSat and CALIPSO indicated as C3 mask data. Then clouds identified as water and unknowns are subtracted according to the results of cloud particle type scheme. Finally ice microphysics are retrieved. Figure 7a shows the retrieval results of r_{eff}. The values of r_{eff} tend to increase as the altitude decreases due to the growth and water vapor availability. Cloud top part consists of 50–70 μm at 11 km and above and lower layers at 8–9 km are about 150 μm. IWC patterns are somewhat different from those of r_{eff}, i.e., locations of its maxima are somewhat higher than those for r_{eff} as revealed by the ground-based radar-lidar analysis [26].

Cesana et al. [4] compared three global products of cloud ice and water fractions by CALIPSO and showed large differences exist among the three. Results for KU type were the smallest ice and water clouds fraction compared with NASA-standard product and GOCCP product. The differences in the three products are due to the different treatment of (1) noise, (2) horizontal and vertical resolutions (3) fully attenuated pixels, (4) cloud-aerosol discrimination. These differences originate from poor understanding of multiple scattering in space-borne lidar signals. It is also noted that there is no global product for water cloud microphysics from CALIPSO.

5 Summary and Discussions

The algorithms introduced in this article have been originally developed and tested by using ground-based cloud radar and lidar. As mentioned in previous sections, there exit essential differences between ground-based and spaceborne instruments. For example, area of about radius of 1 m at altitude of 1 km is observed by ground-based lidar and it is much smaller compared with that of CALIPSO lidar (90 m). These differences lead to our poor understanding of multiple scattering observed in space. To overcome the limitations of conventional ground-based lidar, Okamoto et al. [30] developed a multiple-field-of-view multiple-scattering polarization lidar (MFMSPL) system. The system was designed to detect lidar signals from optically thicker part of clouds that a conventional lidar could not penetrate and it is also designed to study similar lidar signals that are heavily influenced by multiple-scattering detected by space-borne lidar as CALIPSO. Actual observations have been conducted and observed depolarization ratios from water clouds were comparable to those by CALIPSO. The MFMSPL is currently used to evaluate algorithms for CALIPSO lidar data.

In 2019, JAXA-ESA cooperation mission EarthCARE will be launched. Following four sensors will be onboard; 94 GHz Doppler cloud radar (CPR), high spectral resolution lidar at 0.355 μm (ATLID), 7-channel multi-spectral imager (MSI) and broad band radiometer (BBR) [13]. There are several advantages in the EarthCARE mission compared with the CloudSat and CALIPSO. EarthCARE-CPR provides first Doppler information from space. Its minimum sensitivity is about 7 dB lower (−37 dBZ) than that of the CloudSat radar (−30 dBZ). ATLID provides extinction information that were not directly observed quantity in CALIPSO lidar. It is thus expected that more fraction of high thin cirrus and water clouds might be detected by both of EarthCARE-CPR and ATLID than CloudSat and CALIPSO. In addition, retrieval of vertical air-motion and terminal velocity of cloud particles are planned in a similar manner as demonstrated by using ground-based/ship-borne Doppler cloud radar and lidar [38, 39]. These new capabilities will greatly accelerate our knowledge of clouds and relationship between cloud microphysics and vertical air motion so that the ability to predict climate change will be improved by evaluating cloud parameterization with the new data sets.

Acknowledgements This work was supported by JSPS KAKENHI Grant Numbers JP17H06139 and JP15K17762. The CloudSat and CloudSat-collocated ECMWF data are downloaded from the CloudSat Data Processing Center (http://www.cloudsat.cira.colostate.edu) and CALIOP data are downloaded from the Atmospheric Science Data Center (https://eosweb.larc.nasa.gov).

References

1. Born, M., and E. Wolf. 1975. *Principles of Optics*. Oxford: Pergamon.
2. Borovoi, A., A. Konoshonkin, N. Kustova, and H. Okamoto. 2012. Backscattering Mueller matrix for quasi-horizontally oriented ice plates of cirrus clouds: application to CALIPSO signals. *Optics Express* 20: 28222–28233. https://doi.org/10.1364/OE.20.028222.
3. Cai, Q., and K. Liou. 1982. Polarized light scattering by hexagonal ice crystals: theory. *Applied Optics* 21 (19): 3569–3580.
4. Cesana, G., et al. 2016. Using in situ airborne measurements to evaluate three cloud phase products derived from CALIPSO. *Journal of Geophysical Research-Atmospheres*. 121. https:// 10.1002/2015JD024334.
5. Draine, B.T. 1988. The discrete-dipole approximation and its application to interstellar graphite grains. *Astrophysics Journal* 333: 848–872.
6. Draine, B.T., and P. J. Flatau. 1994. Discrete-dipole approximation for scattering calculations. *Journal of the Optical Society of America A* 11: 1491–1499.
7. Draine, B.T., and J. Goodman. 1993. Beyond Clausius-Mossotti: wave propagation on a polarizable point lattice and the discrete dipole approximation. *Astrophysics Journal* 405: 685–697.
8. Dufresne, J-L., and S. Bony. 2006. An assessment of the primary sources of spread of global warming estimates from coupled atmosphere-ocean models. *Journal of Climate*. https://10.1175/2008JCLI2239.1.
9. Hagihara, Y., H. Okamoto, and R. Yoshida. 2010. Development of a combined CloudSat-CALIPSO cloud mask to show global cloud distribution. *Journal of Geophysical Research* 115: D00H33. https://doi:10.1029/2009JD012344.
10. Hagihara, Y., H. Okamoto, and Z.J. Luo. 2014. Joint analysis of cloud top heights from CloudSat and CALIPSO: new insights into cloud top microphysics. *Journal of Geophysical Research-Atmospheres* 119: 4087–4106. https://doi:10.1002/2013JD020919.
11. Heffels, C., D. Heitzmann, E.D. Hirlemann, and B. Scarlett. 1995. Forward light scattering for arbitrary sharp-edged convex crystals in Fraunhofer and anomalous diffraction approximations. *Applied Optics* 34 (28): 6552–6560.
12. Hirakata, M., H. Okamoto, Y. Hagihara, T. Hayasaka, and R. Oki. 2014. Comparison of global and seasonal characteristics of cloud phase and horizontal ice plates derived from CALIPSO with MODIS and ECMWF. *Journal of Atmospheric and Oceanic Technology* 31: 2114–2130. https://10.1175/JTECH-D-13-00245.1.
13. Illingworth, A.J., H.W. Barker, A. Beljaars, M. Ceccaldi, H. Chepfer, N. Clerbaux, J. Cole, J. Delanoë, C. Domenech, D.P. Donovan, S. Fukuda, M. Hirakata, R.J. Hogan, A. Huenerbein, P. Kollias, T. Kubota, T. Nakajima, T.Y. Nakajima, T. Nishizawa, Y. Ohno, H. Okamoto, R. Oki, K. Sato, M. Satoh, M.W. Shephard, A. Velázquez-Blázquez, U. Wandinger, T. Wehr, and G. van Zadelhoff. 2015. The EarthCARE satellite: the next step forward in global measurements of clouds, aerosols, precipitation, and radiation. *Bulletin of the American Meteorological Society* 96: 1311–1332. https://doi.org/10.1175/BAMS-D-12-00227.1.
14. Intrieri, J.M., G.L. Stephens, W.L. Eberhard, and T. Uttal. 1993. A method for determining cirrus cloud particle sizes using lidar and radar backscatter technique. *Journal of Applied Meteorology* 32: 1074–1082.
15. Iwasaki, S., and H. Okamoto. 2001. Analysis of the enhancement of backscattering by nonspherical particles with flat surfaces. *Applied Optics* 40 (33): 6121–6129.
16. Kikuchi, M., H. Okamoto, K. Sato, K. Suzuki, G. Cesana, Y. Hagihara, N. Takahasi, T. Hayasaka, and R. Oki. 2017. Development of algorithm for discriminating hydrometeor particle types with a synergistic Use of CloudSat and CALIPSO. *Journal of Geophysical Research: Atmospheres* 122. https://dx.doi.org/10.1002/2017JD027113.
17. Lhermitte, R. 1987. A 94-GHz Doppler radar for cloud observations. *Journal of Atmospheric and Oceanic Technology* 4: 36–48.

18. Li, J.-L.F., D. Waliser, C. Woods, J. Teixeira, J. Bacmeister, J. Chern, B.-W. Shen, A. Tompkins, W.-K. Tao, and M. Köhler. 2008. Comparisons of satellites liquid water estimates with ECMWF and GMAO analyses, 20th century IPCC AR4 climate simulations, and GCM simulations. *Geophysical Research Letters* 35: L19710. https://doi:10.1029/2008GL035427.
19. Mace, G.G., Q. Zhang, M. Vaughan, R. Marchand, G. Stephens, C. Trepte, and D. Winker. 2009. A description of hydrometeor layer occurrence statistics derived from the first year of merged CloudSat and CALIPSO data. *Journal of Geophysical Research* 114: D00A26. https://doi:10.1029/2007JD009755.
20. Macke, A., J. Mueller, and E. Raschke. 1996. Single scattering properties of atmospheric ice crystals. *Journal of the Atmospheric Sciences* 53: 2815–2825.
21. Marchand, R., G.G. Mace, T. Ackerman, and G. Stephens. 2008. Hydrometeor detection using Cloudsat - an Earth-orbiting 94-GHz cloud radar. *Journal of Atmospheric and Oceanic Technology* 25: 519–533. https://doi:10.1175/2007JTECHA1006.1.
22. Mishchenko, M.I. 1993. Light scattering by size-shape distributions of randomly oriented axially symmetric particles of a size comparable to a wavelength. *Applied Optics* 32: 4652–4666.
23. Okamoto, H. 2002. Information content of the 95-GHz cloud radar signals: theoretical assessment of nonsphericity and error evaluation of the discrete dipole approximation. *Journal of Geophysical Research* 107 (D22): 4628. https://doi:10.1029/2001JD001386.
24. Okamoto, H., A. Macke, M. Quante, and E. Raschke. 1995. Modeling of backscattering by non-spherical ice particles for the interpretation of cloud radar signals at 94GHz. An error analysis. *Contributions to Atmospheric Physics* 68 (4): 319–334.
25. Okamoto, H., S. Iwasaki, M. Yasui, H. Horie, H. Kuroiwa, and H. Kumagai. 2000. 95-GHz cloud radar and lidar systems: preliminary results of cloud microphysics. *Proceedings of SPIE* 4152: 355–363.
26. Okamoto, H., S. Iwasaki, M. Yasui, H. Horie, H. Kuroiwa, and H. Kumagai. 2003. An algorithm for retrieval of cloud microphysics using 95-GHz cloud radar and lidar. *Journal of Geophysical Research* 108 (D7): 4226. https://doi:10.1029/2001JD001225.
27. Okamoto, H., T. Nishizawa, T. Takemura, H. Kumagai, H. Kuroiwa, N. Sugimoto, I. Matsui, A. Shimizu, S. EMori, A. Kamei, and T. Nakajima. 2007. Vertical cloud structure observed from shipborne radar and lidar: midlatitude case study during the MR01/K02 cruise of the research vessel Mirai. *Journal of Geophysical Research* 112: D08216. https://doi:10.1029/2006JD007628.
28. Okamoto, H., T. Nishizawa, T. Takemura, K. Sato, H. Kumagai, Y. Ohno, N. Sugimoto, A. Shimizu, I. Matsui, and T. Nakajima. 2008. Vertical cloud properties in the tropical western Pacific Ocean: validation of the CCSR/NIES/FRCGC GCM by shipborne radar and lidar. *Journal of Geophysical Research* 113: D24213. https://doi:10.1029/2008JD009812.
29. Okamoto, H., K. Sato, and Y. Hagihara. 2010. Global analysis of ice microphysics from CloudSat and CALIPSO: incorporation of specular reflection in lidar signals. *Journal of Geophysical Research* 115: D22209. https://doi:10.1029/2009JD013383.
30. Okamoto H., K. Sato, T. Makino, T. Nishizawa, N. Sugimoto, Y. Jin, and A. Shimizu. 2016. Depolarization ratio of clouds measured by multiple-field of view multiple scattering polarization lidar. *EPJ Web of Conferences* 119: 11007. https://doi.org/10.1051/epjconf/201611911007.
31. Purcell, E.M., and C.R. Pennypacker. 1973. Scattering and absorption of light by nonspherical dielectric grains. *Astrophysics Journal* 186: 705–714.
32. Rossow, W.B., and R.A. Schiffer. 1991. ISCCP cloud data products. *Bulletin of the American Meteorological Society* 72 (1): 2–20. https://doi.org/10.1175/1520-0477(1991)072<0002:ICDP>2.0.CO;2.
33. Rossow, W.B., and Y. Zhang. 2010. Evaluation of a statistical model of cloud vertical structure using combined CloudSat and CALIPSO cloud layer profiles. *Journal of Climate* 23: 6641–6653. https://doi:10.1175/2010JCLI3734.1.
34. Sassen, K. 1991. The polarization lidar technique for cloud research: a review and current assessment. *Bulletin of the American Meteorological Society* 72 (12): 1848–1866.

35. Sassen, K., and S. Benson. 2001. A midlatitude cirrus cloud climatology from the Facility for Atmospheric Remote Sensing. Part II: Microphysical properties derived from lidar depolarization. *Journal of the Atmospheric Sciences* 58: 2103–2112.
36. Sato, K., and H. Okamoto. 2006. Characterization of Ze and LDR of nonspherical and inhomogeneous ice particles for 95-GHz cloud radar: its implication to microphysical retrievals. *Journal of Geophysical Research* 111: D22213. https://doi:10.1029/2005JD006959.
37. Sato, K., and H. Okamoto. 2011. Refinement of global ice microphysics using spaceborne active sensors. *Journal of Geophysical Research* 116: D20202. https://doi:10.1029/2011JD015885.
38. Sato, K., H. Okamoto, M.K. Yamamoto, S. Fukao, H. Kumagai, Y. Ohno, H. Horie, and M. Abo. 2009. 95-GHz Doppler radar and lidar synergy for simultaneous ice microphysics and in-cloud vertical air motion retrieval. *Journal of Geophysical Research* 114: D03203. https://doi:10.1029/2008JD010222.
39. Sato, K., H. Okamoto, T. Takemura, H. Kumagai, and N. Sugimoto. 2010. Characterization of ice cloud properties obtained by shipborne radar/lidar over the tropical western Pacific Ocean for evaluation of an atmospheric general circulation model. *Journal of Geophysical Research* 115: D15203. https://doi:10.1029/2009JD012944.
40. Scotland, R.M., K. Sassen, and R.J. Stone. 1971. Observations by lidar of linear depolarization ratios by hydrometeors. *Journal of Applied Meteorology* 10: 1011–1017.
41. Stephens, G.L., et al. 2002. The CloudSat mission and the A-TRAIN: a new dimension to space-based observations of clouds and precipitation. *Bulletin of the American Meteorological Society* 83: 1771–1790. https://doi:10.1175/BAMS-83-12-1771.
42. Stephens, G.L., et al. 2008. CloudSat mission: performance and early science after the first year of operation. *Journal of Geophysical Research* 113: D00A18. https://doi:10.1029/2008JD009982.
43. Tanelli, S., et al. 2008. CloudSat's cloud profiling radar after two years in orbit: performance, calibration and processing. *IEEE Transactions on Geoscience and Remote Sensing* 46 (11): 3560–3573. https://10.1109/TGRS.2008.2002030.
44. Vaughan, M.A., D.M. Winker, and K.A. Powell. 2005. CALIOP Algorithm Theoretical Basis Document: Part 2; Feature detection and layer properties algorithms. Release 1.01, NASA Langley Research Center Doc. PC-SCI-202 Part 2, 87 pp. [Available online at http://calipsovalidation.hamptonu.edu/PC-SCI-202Part2rev1x01.pdf].
45. Waliser, D., et al. 2009. Cloud ice: a climate model challenge with signs and expectations of progress. *Journal of Geophysical Research* 114: D00A21. https://doi:10.1029/2008JD010015.
46. Winker, D.M., W.H. Hunt, and M.J. McGill. 2007. Initial performance assessment of CALIOP. *Geophysical Research Letters* 34: L19803. https://doi:10.1029/2007GL030135.
47. Winker, D.M., et al. 2009. Overview of the CALIPSO mission and CALIOP data processing algorithms. *Journal of Atmospheric and Oceanic Technology* 26: 2310–2323. https://10.1175/2009JTECHA1281.1.
48. Yang, P., and K. Liou. 1995. Finite-difference time domain method for light scattering by small ice crystals in three-dimensional space. *Journal of the Optical Society of America A* 13: 2072–2085.
49. Yoshida, R., H. Okamoto, Y. Hagihara, and H. Ishimoto. 2010. Global analysis of cloud phase and ice crystal orientation from Cloud-Aerosol Lidar and Infrared Pathfinder Satellite Observation (CALIPSO) data using attenuated backscattering and depolarization ratio. *Journal of Geophysical Research*. 115: D00H32. https://doi:10.1029/2009JD012334.
50. Yurkin, M.A., and A.G. Hoekstra. 2007. The discrete dipole approximation: an overview and recent developments. *Journal of Quantitative Spectroscopy and Radiative Transfer* 106: 558–589. https://doi:10.1016/j.jqsrt.2007.01.034.

Atmospheric Water Vapour Profiling over Ocean/Land and for Clear/Cloudy Situations Using Microwave Observations

Filipe Aires

1 Introduction

Water vapour is a critical component of the Earth water cycle: It represents about 99% of the atmospheric moisture and it constrains the amount of precipitation. As a major source of the atmospheric energy, water vapour strongly impacts weather (at short time scales) and climate (at longer time scales). It is the major greenhouse gas, trapping more heat than CO_2. The role of water vapour for clouds, circulation and climate sensitivity is of high interest, in particular in the Tropics [11]. Water vapour circulation and its associated latent heat represents about 50% of the heat transport from the tropics to the poles. Furthermore, water vapour is one essential variable of the Numerical Weather Prediction (NWP) models and therefore its monitoring strongly impacts the quality of weather forecasts.

The estimation of water vapour in the atmosphere can be done using NWP models constrained by *in situ* observations. However, it was shown that with its global coverage, and high frequency of observations (especially when combining several instruments), the use of satellite data has a very important impact on the NWP analyses quality. This is particularly true in regions with a low meteorological stations density, such as in the southern hemisphere. Furthermore, the construction of long time series based uniquely on satellite observations is an essential tool for the climate community [53, 56], to analyse long-term trends and validate the global climate models.

F. Aires (✉)
LERMA, Observatoire de Paris, Paris, France

Estellus, Paris, France
e-mail: filipe.aires@obspm.fr; filipe.aires@estellus.fr

C. Andronache (ed.), *Remote Sensing of Clouds and Precipitation*,
Springer Remote Sensing/Photogrammetry,
https://doi.org/10.1007/978-3-319-72583-3_9

Infrared satellite observations have been widely used for the retrieval of the atmospheric water vapour, with instruments such as the Infrared atmospheric Sounding Interferometer (IASI) [4, 13, 46] or the Atmospheric Infrared Sounder (AIRS) [49]. However, these measurements cannot penetrate clouds and therefore are limited to clear-sky scenes. Satellite passive microwave observations around the 183.311 GHz water vapour line can provide accurate monitoring of water vapour profiles with good temporal and spatial sampling for operational. They represent a good complement to the infrared sounding measurements when clouds are present. Furthermore, the synergy of both observation types, infrared and microwave, can be exploited to better retrieve the atmospheric water vapour profiles [1, 9, 43, 49], or near-infrared reflection [34].

Over ocean, the passive microwave measurements are now routinely assimilated in NWP systems and they provide interesting atmospheric profiling capabilities. Over land however, they are not fully exploited. First, land surface microwave emissivities are usually much higher than ocean emissivities, with the consequence that the surface contribution to the measured signal is much larger. Second, the land surface emissivities have a higher spatial heterogeneity than their oceanic counter part. In addition, they are much more difficult to model, being dependent upon a large number of variables (e.g., soil moisture, vegetation and snow properties). However, efforts are underway to use, invert and assimilate microwave radiances over land (e.g., [3, 30]), thanks essentially to the development of land surface emissivity databases, directly calculated from satellite observations (e.g., [6, 37, 38, 40, 54]).

Various retrieval algorithms have been developed for the retrieval of atmospheric water vapour from microwave instruments. SSM/I (Special Sensor Microwave Instrument) observations have been exploited for example in [10, 27, 50] mostly over ocean. In [17], an algorithm is built for ocean and land surfaces. Grody et al. [24] and Jethva and Srinivasan [28] use AMSU (Advanced Microwave Sounding Unit) observations over ocean to retrieve water vapour. TRMM (Tropical Rainfall Measuring Mission) Microwave Imager (TMI) observations are used in [23, 52] over the tropical belt and again mostly over ocean surfaces.

An usual retrieval technique is the 1D-Var method. This is an iterative process that uses a radiative transfer equation and its Jacobians to minimise a quality criteria (often the root mean square errors of the simulations versus the observations). Such an approach is described for instance in [12]. In this study, a Neural Network (NN) approach is used instead for several reasons: It is fast in terms of computations; easy to implement and test; its flexibility allows to test several instrumental configurations; it is able to exploit a first guess information (as the 1D-Var) [3]; it has been shown to work for many atmospheric and surface products from infrared or microwave observations or the fusion of both. The processing chain developed here operates over the ocean and land, for clear or cloudy/non-precipitating situations. This NN inversion is trained on a large collection of radiative transfer simulations using RTTOV code [45] and ECMWF analyses. It uses *a priori* information on the land surface emissivities to help solve the inversion problem over land [6]. It also relies on an innovative methodology to calibrate the observations, prior to the inversion [5]. This is necessary because there are systematic differences between radiative transfer simulations and satellite observations (both should be coherent in the retrieval process). A separate NN is used for four different configurations: land/ocean and clear/cloudy situations. A precipitation and cloud flag is required. The ECMWF temperature profile is also used as *a priori* information.

A processing chain for satellite microwave observations is developed in this study for the retrieval of the atmospheric humidity. This algorithm operates over both ocean and land, for clear and cloudy/non-precipitating situations, for three platforms that include water vapour microwave sounders, including the Megha-Tropiques mission. The theoretical assessment of the microwave retrievals are presented. The algorithm will then be tested on two platforms: the AQUA mission (AMSR-E and HSB instruments), and the MetOp (METeorological OPerational) platform with different companion instruments (AMSU-A and MHS), on which a near real-time scheme is tested.

Section 2 introduces the microwave instruments from the three satellite platforms considered in this study, and provides a preliminary assessment of their sounding capabilities based on their main characteristics. Section 3 presents the various datasets used in this work. The retrieval scheme is described in Sect. 4, and the theoretical results are discussed in Sect. 5, as compared to the results from the AQUA and MetOp platforms. The Water Vapour (WV) estimates are evaluated in Sect. 6. The validation in the space of the satellite observations is developed in Sect. 7. Finally, conclusions are drawn in Sect. 8.

2 The Microwave Instruments

The water vapour sounders from the three satellite missions are described, along with the other microwave radiometers available on the same platforms. A preliminary analysis of the sounding capabilities of the instruments is conducted, based on the calculation of the weighting function and the Jacobians, and on considerations on the instrument noise. Table 1 summarises the main characteristics of the instruments.

2.1 AQUA

The passive microwave measurements from the AQUA platform are composed of collocated observations from two radiometers: the Humidity Sounder for Brazil (HSB) (only operated from May 2002 to February 2003) and the Advanced Microwave Scanning Radiometer—Earth observing system (AMSR-E). The HSB cross-track sounder is a nearly identical copy of AMSU-B with four moisture-sounding channels (instead of five for AMSU-B). Three out of four are located around the strong water vapour absorption line at 183.31 GHz (±1.0, ±3.0 and ±7.0) and the fourth channel is a window channel at 150 GHz [32]. During one scan, HSB samples 90 scenes of 1.1° between ±49.5°, with a footprint size of 13.5 km at nadir. AMSR-E is a dual-polarised radiometer operating at frequencies

Table 1 Instrument characteristics for the three satellite missions considered in this study: Megha-Tropiques, AQUA, and MetOp

Platform	Megha-Tropiques			AQUA			MetOp		
Instrument	**Saphir**			**HSB**			**MHS**		
Scanning	Cross-track			Cross-track			Cross-track		
Spatial resolution	10 km at nadir			13.5 km at nadir			16 km at nadir		
Frequencies							89.0	2800	0.22
in GHz (left),				150.0	4000	1.0	157.0	2800	0.34
bandwidth in	183±0.2	2×200	2.03						
MHz (center)	183±1.1	2×350	1.53	183.311±1	2×500	1.0	183.311±1	2×500	0.51
& noise in K	183±2.8	2×500	1.37	183.311±3	2×1000	1.0	183.311±3	2×1000	0.40
(right)	183±4.2	2×700	1.25						
	183±6.8	2×1200	1.06	183.311±7	2×2000	1.2	190.311	2200	0.46
	183±11	2×2000	0.99						
Instrument	**Madras**			**AMSR-E**			**AMSU-A**		
Scanning	Conical			Conical			Cross-track		
Spatial resolution	40×67 km (lower freq) to 6×10 km (higher freq)			60 to 5 km			48 km at nadir		
Frequencies	18.7H		0.46	6.9V		0.3	23.8		0.30
in GHz (left)	18.7V		0.53	6.9H		0.3	31.4		0.30
& noise in K	23.8V		0.48	10.8V		0.6	50.3		0.40
(right)	36.5H		0.44	10.8H		0.6	52.8		0.25
	36.5V		0.49	18.7V		0.6	53.596±0.115		0.25
	89.0 H		0.63	18.7H		0.6	54.4		0.25
	89.0 V		0.58	23.8V		0.6	54.94		0.25
	157.0 H		1.75	23.8H		0.6	55.5		0.25
	157.0 V		1.65	36.5V		0.6	57.290344 (= f_0)		0.25
				36.5H		0.6	f_0±0.217		0.40
				89.0V		1.1	f_0±0.3222±0.048		0.40
				89.0H		1.1	f_0±0.3222±0.022		0.60
							f_0±0.3222±0.010		0.80
							f_0±0.3222±0.0045		1.20
							89.0		0.50

Notation: $f_0 \pm y \pm z$; f_0 is the center frequency (in GHz). If y appears (in GHz), the center frequency is not sensed but two bands, one on either side of the center frequency, are sensed; y is the distance of f_0 from the center of the two pass bands. If z appears, it is the width of the two pass bands (in GHz)

of 6.9, 10.7, 18.7, 23.8, 36.5, and 89 GHz designed for the retrieval of surface and atmospheric variables including water vapour, cloud liquid water, precipitation, sea surface temperature, near surface wind speed, and soil moisture. The instrument has a conically scanning antenna that provides multichannel observations at a constant incidence angle of 55° across a 1445 km swath. The spatial resolution of AMSR-E varies from approximately 60 km at 6.9 GHz to 5 km at 89 GHz [31].

2.2 MetOp

Launched on October 19, 2006, MetOp is the first European polar-orbiting satellite dedicated to operational meteorology. MetOp carries two passive microwave sounders, the Microwave Humidity Sounder (MHS), and AMSU-A (1 and 2) for temperature sounding.

Similar to the historical instruments on board DMSP (Defense Meteorological Satellite Program) and NOAA (National Oceanic and Atmospheric Agency) orbitors, MHS provides measurements in the 183.31 GHz water vapour absorption line, at ±1, ±3 and ±7 GHz and at 190 GHz, plus at two window channels at 89 and 157 GHz that enable deeper penetration through the atmosphere down to the Earth's surface. Each swath is made up of 90 contiguous individual pixels and scanned every 2.67 s. MHS pixels at nadir have a diameter of approximately 16 km.

AMSU-A is dedicated to the retrieval of atmospheric temperature profiles, with twelve sounding channels between 50 and 60 GHz O_2 band, and three other channels at 23.8, 31.4, and 89 GHz. It is a cross-track scanning radiometer, with ±48.3° from nadir and a total of 30 Earth fields-of-view of 3.3° per scan line, providing a nominal spatial resolution of 48 km at nadir. The swath is approximately 2000 km and the instrument realises one scan in 8 s. The MHS and AMSU scans are synchronised. Each AMSU-A pixels is covered by 3×3 MHS pixels, this facilitates their synergetic use [8].

2.3 Megha-Tropiques

Saphir is a passive microwave sounder with six channels around the 183.31 GHz absorbing line at ±0.2, ±1.1, ±2.8, ±4.2, ±6.8, and ±11 GHz (Table 1). Compared to the currently flying water vapour microwave instruments, it has more sounding channels, for a theoretically better vertical description of the water vapour profile. Note that it does not include window channels around 90 and 150 GHz such as the Advanced Microwave Sounding Unit-B (AMSU-B), but these frequencies are available on the Madras companion instrument described below. Saphir has a cross-track viewing geometry, with 130 pixels per scan line from nadir to ±42.96°. This means that the pixels size of 10 km at nadir increases with the scanning angle. It can be noted that Saphir has larger instrument noise than MHS, but it is expected that the additional channels in Saphir will allow for better description of the vertical distribution of water vapour.

Madras is a passive microwave imager that measures the radiation at nine frequency bands: 18.7, 23.8, 36.5, 89 and 157 GHz at vertical and horizontal polarisations, except for the 23.8 that is vertical-only. It has a conical viewing geometry: the incidence angle of the measure has is constant at 53.5° (i.e. about 45° on-board angle), which makes it easier to exploit the polarisation information. The size of the pixels is constant but the difference in the scan geometry complexifies the merging of the data with the cross-track measurements from Saphir. The scan coverage is ±65° and the swath is about 1700 km wide.

2.4 Preliminary Evaluation of the Sounding Capabilities

The weighting function of an instrument channel is defined as:

$$W(\nu) = \int \frac{\partial t}{\partial ln(P)} \tag{1}$$

where ν is the frequency, t is the transmission to the Top Of the Atmosphere (TOA), and P is the pressure. The integration is performed on the vertical, from the bottom to the upper part of the atmosphere. Analysing this quantity is a standard way to monitor the sounding characteristics of an instrument. Figure 1 represents the water vapour weighting functions at various frequencies and for different instruments, i.e., Saphir (left), HSB (middle), and MHS (right) for a clear sky scene (those figures would strongly change in the presence of clouds). These results are calculated with a standard tropical atmosphere. The water vapour weighting function of a particular channel, at a given pressure/altitude, indicates the change of the measured brightness temperature at the TOA for a perturbation of the

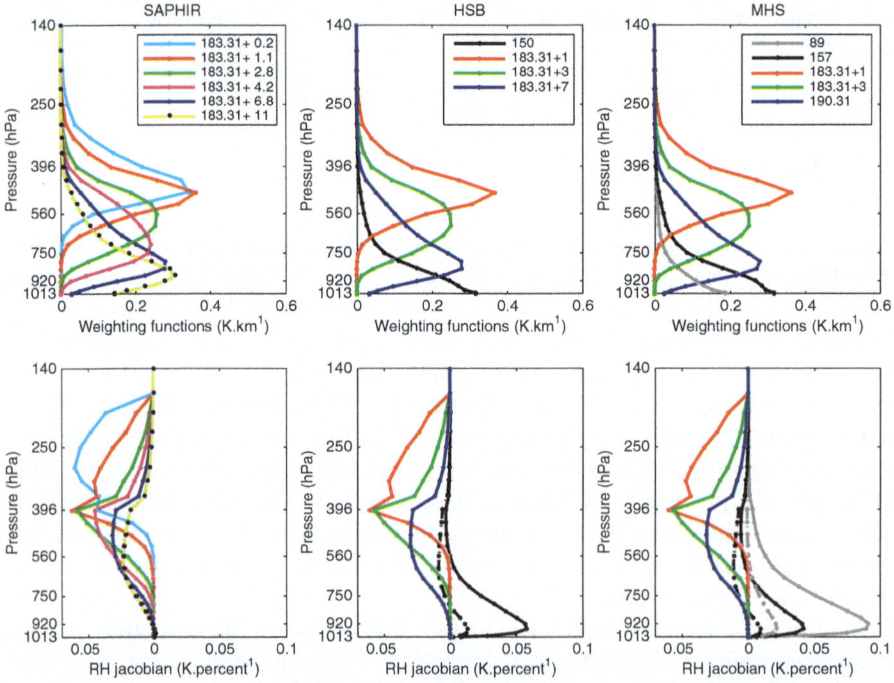

Fig. 1 Weighting functions (upper part) and Jacobians (bottom) for a standard tropical clear-sky situation, for Saphir (left), HSB (middle) and MHS (right column). The Jacobians are provided for an oceanic situation (continuous line) and a land surface (dashed line). The Jacobians are expressed in $K \cdot \%^{-1}$ because the water vapour is expressed in relative humidity (%)

water vapour at that pressure/altitude. They are here expressed in change of relative humidity and normalised to one km of atmosphere (K/km per unit change of relative humidity). The further the channel from the line center, the lower the atmospheric layer it sounds. Saphir has more channels than the other instruments close to the 183.311 GHz line, and the bandwidths of these channels are significantly narrower than for the other instruments. The channel at ± 11 GHz from the line center sees closer to the surface, but still with a limited contribution from the surface itself. As a consequence, it is expected that the lower layer of the atmosphere will not be well described by such an instrument. The bandwidths of the channels have two implications. On one hand, the narrower the bandwidth, the narrower the weighting function (up to a certain extent), and, the higher the vertical resolution of the profile. On the other hand, the instrument noise is inversely proportional to the square root of the channel bandwidth, with direct consequences on the retrieval accuracy. For drier atmospheres, the weighting function would peak at lower altitudes (less opacity in the atmosphere). Note that there is large overlap of the weighting functions, meaning that the informations from the different channels will not be independent. This can be illustrated by computing the number of principal components in each channel observations from the three sounders considered in this study: HSB, MHS and Saphir. When performing such principal component analysis, it can be seen that three (resp. four) principal components represent more than 99% (resp. 99.5%) of the total variability of the satellite observation datasets. These two numbers mean that there are four independent pieces of information out of the six Saphir measurements (results are identical for HSB and MHS instruments).

An alternative way to analyse sounding capacities of an instrument is given by the Jacobians, i.e., the first derivative of the Brightness Temperature (T_B) measured on each channel with respect to the geophysical variable of interest (the Relative Humidity, RH, in this case): $\frac{\partial T_B}{\partial RH}$. In practice, since the analytical Jacobians are not available for the water vapour when expressed in relative humidity, we estimate them with the "input perturbation" method where the RH is perturbed by 10%, independently for each atmospheric layer, and the resulting perturbations in TB are calculated. Jacobians are a direct measure of the information content of the observations, as used in the atmospheric inversion. The Jacobians are related to the weighting function, but integrated on the atmospheric layers on which the inversion is performed. This means that the Jacobian is dependent on the discretisation of the atmosphere that is chosen, not the weighting function. Another difference is that the Jacobian can be computed for each atmospheric variable, where the weighting function is traditionally computed for the total transmission that integrates the contribution of all the geophysical variables at a time (this represents of course less information). Figure 1 (bottom) represents such Jacobians computed using the RTTOV radiative transfer model [45]. This code will be used in this study to build the synthetic satellite observation data sets. These Jacobians can be interpreted in terms of profile retrieval, as compared to the instrument noise of the radiometers: When a first guess in relative humidity RH is close enough to the real profile, the perturbation of the T_Bs for each channel is given by the vector $\Delta T_B = \frac{\partial T_B}{\partial RH} \cdot \Delta RH$. This is to be compared to the instrument noise of the radiometer. These computations have been performed using a standard tropical situation over land (with an emissivity equal to 0.9 in dashed line), and over ocean (with an emissivity equal to 0.6 in continuous line). If no continuous line can be seen, it means that it is identical to the ocean case. The weighting functions are not sensitive to the surface type (i.e. to the emissivity), contrarily to the Jacobians of the channels sensitive to the surface: It can be seen that for the 150 GHz of HSB and the 89 and 157 GHz of MHS, the Jacobians over land surfaces are reduced compared to oceanic surfaces. Since Saphir does not possess a channel truly sensitive on the surface, there is no impact (but an impact would be observed on the surface-sensitive channels of Madras, not shown). Examination of the Jacobians shows also that the Saphir instrument would likely provide more information on the upper part of the atmosphere, due to its channel located closer to the line centre. However, the large radiometric uncertainty on this channel might limit its potential. Limited information will be derived from Saphir alone about the lower atmospheric layers. As a consequence, Saphir observations will have to be combined to Madras measurements to estimate water vapour content in the lower part of the atmosphere. This will be confirmed by the theoretical retrieval statistics in Sect. 5.

3 Datasets and Radiative Transfer Code

3.1 ECMWF Operational Analyses

A dataset of surface and atmospheric situations is developed in order to allow RT simulations of the satellite observations. We use the atmospheric profiles and surface properties from the 6-hourly operational global analyses [51] from the Integrated Forecasting System of the European Centre for Medium Range Forecasting (ECMWF) provided on a regular 1.125° grid (hereafter referred to as ANA database). Both clear-sky satellite observations and radiosondes are assimilated in this ANA dataset. In order to run accurate RT simulations, the following information are kept: the temperature, water vapour and ozone profiles on 21 pressure levels ranging from 1000 to 1 hPa, the cloud profiles (cloud cover, liquid and ice water), and surface properties (10 m horizontal wind, 2 m pressure and temperature, surface skin temperature, convective and large-scale precipitation, and total cloud cover).

In the following, only non-precipitating situations are considered to limit possible biases in the simulation due to the scattering of the up-welling radiation by droplets: a threshold of 0.1 mm in the precipitation fields is used to detect precipitation. Scenes above 1000 m are avoided to minimise the topographic effects, and pixels closer than 100 km from the coast are discarded, to reduce inconsistencies between the different frequency channels with different field-of-views.

In order to put the ECMWF data and the real satellite observations in coincidence, the analysis fields are interpolated using a bilinear spatial interpolation scheme. For each satellite observation, only the neighbours in the ECMWF grid with the same surface type (ocean or land) are considered to calculate the bilinear interpolation. The coincidence is accepted only if the time difference is less than 1 h and a half.

3.2 Global Land Surface Microwave Emissivities

For atmospheric profiling, surface-sensitive microwave observations are so far essentially used over ocean. Over land, the surface emissivity is difficult to estimate: it is usually high, limiting the contrast with the atmospheric contribution, very variable in space, and complex to model.

A parameterisation of the land surface microwave emissivities has been recently developed [39]. For each location and time of the year, it provides realistic first guess estimates of the land surface microwave emissivities from 19 to 100 GHz, for all scanning conditions, incidence angles, and polarisations. It is anchored to climatological monthly-mean maps of the emissivities at 19, 37, and 85 GHz, calculated from SSM/I [37, 38]. It is originally designed for frequencies between 19 and 85 GHz but tests proved that it is beneficial down to 5 GHz and up to 190 GHz [6]. The nominal spatial resolution of the emissivity estimates is $0.25° \times 0.25°$. The results have been thoroughly evaluated and the Root Mean Square (RMS) errors are usually within 0.02, with the noticeable exception of snow-covered regions where the high spatial and temporal variability of the emissivity signatures are difficult to capture. A tool based on this parameterisation has been developed for the EUMETSAT NWP Satellite Application Facility (SAF) [6]. This tool is interfaced with the RTTOV radiative transfer code.

3.3 Satellite Observation Datasets

The evaluation of the retrieval method will be based on existing observations from AMSR-E/HSB and AMSU-A/MHS. In addition, to perform accurate retrievals, the observations have to be calibrated and we adopt here an innovative calibration procedure based on a learning database. Datasets of collocated AMSR-E/HSB and AMSU-A/MHS are created and they are large enough to be used for both calibration and evaluation, the learning dataset for calibration accounting for less than 5% of the total database.

The collocation procedure between AMSR-E and HSB footprints is described in [5]. It assumes that for both sensors, the footprints are co-registered. Observations for AMSR-E are extracted from the level 2A data that provide a resampling of the nominal resolution observations into coarser resolution fields-of-view. In the present study, the chosen resolution is 21 km. The level 1B dataset provides the calibrated and geolocated HSB observations, along with scene information. The collocation method averages all the HSB scenes that fall into each AMSR-E elliptical footprint, accepting a maximum time difference of 70 s and assuming that the information for each HSB scene is concentrated at its centre. The final AQUA satellite database is composed of 2 full months of observations (September 2002 and January 2003) containing the sixteen collocated brightness temperatures, and the scene information (land fraction) from the MHS dataset.

The collocation procedure between AMSU-A and MHS is much simpler thanks to the similarity in the scanning mechanisms of both instruments (Sect. 2.2). The final MetOp satellite database is composed of 2 full months of observations (July 2007 and January 2008) containing the 20 collocated brightness temperatures.

3.4 The RTTOV Radiative Transfer

Our retrieval scheme is based on a radiative transfer model. The model that is chosen is part of the retrieval scheme and is used directly in the retrieval algorithm within an iterative inversion. Furthermore, it is used to create the learning datasets (for inversion and calibration) in statistical algorithms. The RTTOV-9.3 radiative transfer model can simulate the microwave instrument on board Megha-Tropiques, AQUA, and MetOp platforms. This model, originally developed at ECMWF [21] and now supported by the EUMETSAT NWP-SAF [35, 45], allows for rapid simulations of radiances for satellite infrared and microwave radiometers for a given atmospheric state vector.

In order to compare real satellite observations and RT simulations, RTTOV is run using all the analysis information (Sect. 3.1) together with the surface emissivity database over land (Sect. 3.2), while over ocean the emissivities are computed by the FASTEM-3 [14] surface emissivity model. To keep the spatial variability of the land surface emissivities, the 1.125° grid point ANA database is interpolated for each pixel of the 0.25° emissivity database. Based on these inputs, RTTOV is used to perform simulations of the T_B for each microwave instrument. For comparison purposes and to build up the calibration learning database, the RT is calculated for each real observation in coincidence with the ANA database and the surface emissivity dataset.

4 Retrieval Scheme

The retrieval scheme presented in this study is composed of a classification procedure to identify the type of scene being treated and two processing steps. The first processing step uses a statistical Neural Network (NN) model to calibrate the satellite observations. The second step is based on another NN that performs the actual retrieval, using the calibrated datasets (Fig. 2). The choice for this approach is described in the following section.

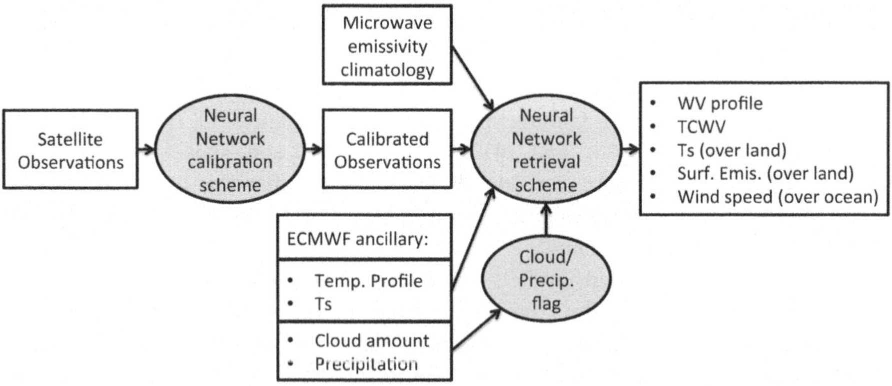

Fig. 2 Schematic organisation of the processing chain

4.1 General Strategy

In order to train the NN for remote sensing purpose, two strategies can be used: First, the training can be done using a learning database composed of real satellite observations and collocated WV profiles from, for example, the ECMWF operational analysis. This type of scheme is said to be an "empirical" inversion because no RT model is used to solve the inverse problem. Second, the training can also be done using a learning database composed of the same WV profiles from ECMWF operational analysis with its simulated satellite observation, provided by a RT model, instead of real observations. This type of inversion is said to be a "physical" inversion because a RT model is being used. The first approach involves only one transformation of the real observations: It mixes the calibration and the retrieval in a unique procedure. The second approach involves explicitly two transformations: The calibration of the data and the actual retrieval. The calibration of the real satellite observations is necessary because systematic differences exist between RT simulations (used in the training of the retrieval scheme) and the real observations. The second one is preferred here because the first one assumes that the WV profiles (from ECMWF) are true, which is not the case. We prefer here to rely on the RT model, as in [5]. In this approach, there is no introduction of coincidence or resolution errors contrarily to the empirical approach that needs to put in coincidence the ECMWF profiles and the satellite observations.

The choice of the NN for the retrieval of water vapour has been made for various reasons. One of them is that weighting functions tend to go higher in the atmosphere when the water vapour content increases. This means that the dependencies between the satellite observations and the water vapour content at a given altitude varies with the actual atmospheric situation. The NN is a non-linear method meaning that its outputs/inputs relationships are state-dependent. The training will teach the NN to adapt its retrieval to the water vapour content.

4.2 A Priori Classification

The retrieval chain is based on four distinct configurations: (1) Cloud Free over Ocean scenes: CF/O; (2) Cloud Free over Land scenes: CF/L; (3) CLoudy over Ocean scenes: CL/O; and (4) CLoudy over Land scenes: CL/L. Only non-precipitation scenes are considered. This means that the chain needs two classification processes: one to determine if a situation is cloudy or clear, and one to eliminate precipitating scenes.

Three options have been studied to identify clear and cloudy scenes in our retrieval scheme: (1) using the cloud cover from the ECMWF operational analysis as an *a priori* flag, (2) using an independent classifier derived from VIS/IR observations onboard geostationary satellites, and (3) developing a dedicated cloud classifier based on the microwave observations. The third option has been investigated in [7]: a microwave cloud classifier has been designed to find whether a scene is clear or has a low-, middle- or high-cloud cover. The training of this statistical method has been performed using a MSG-Seviri (Meteosat Third Generation-Spinning Enhanced Visible and Infrared Imager) cloud classification [18, 19]. In the present study, the cloud cover from the ECMWF operational analysis has been chosen instead, although we are aware of its limitations. The use of the microwave dedicated cloud classifier will be considered in the future. Each atmospheric situation is classified as being clear (cloud cover below 2%) or as being cloudy (cloud cover over 2%). No partial coverage is considered.

Various options are also considered for the precipitation flags: (1) using the precipitation estimates of the ECMWF operational analysis (i.e., short range prediction), (2) using microwave precipitation flags from the literature such as Alishouse et al. [10], Ferraro et al. [22], Weng et al. [55], or Hong and Heygster [25], (3) using the precipitation retrieved by other instruments onboard the same satellite platform. Preliminary tests showed that the quality of the ECMWF *a priori* precipitation analysis is sufficient for this purpose. In the following, the scenes will be eliminated if the ECMWF precipitation estimate (for both large-scale and convective precipitation) is higher than 0.1 mm. This is a conservative estimate as ECMWF reanalysis is down to over-estimate precipitation, especially the occurrence of low values.

4.3 Neural Networks Calibration Scheme

We will only briefly comment on this step since this calibration methodology has been described in [5]. The procedure has been developed for AQUA (AMSR-E/HSB) and MetOp (AMSU-A/MHS) data.

4.3.1 Calibration Model

Neural networks (NN) have been widely used to perform nonlinear transformations from one space to another by following some particular statistical constraints. In this calibration procedure, a Multi-Layered Perceptron (MLP) model [26, 44] is used as a nonlinear mapping model. It is composed of three layers of neurons: the first layer codes the inputs (i.e., observed brightness temperatures), the third layer codes the outputs (i.e., the calibrated brightness temperatures), and the second layer, called the hidden layer, is used to increase the complexity of the NN model. The higher the number of neurons in this hidden layer, the more complex the NN becomes. The NN is trained to reproduce the behaviour described by a database of samples, i.e., the learning dataset composed of the real observations $T_{B}s$ and their associated calibrated $T_{B}s$. The real AMSRE/HSB (respectively the AMSU-A/MHS) observations and the coincident T_B RTTOV simulations presented in Sect. 3 constitute here the learning dataset. The NN is designed to "project" or calibrate the actual real observations into the space of the RTTOV simulations.

Four NNs have been derived corresponding to the four cases previously described: CF/O; CF/L; CL/O; CL/L. Each one of these four learning datasets includes about 100,000 situations, which is largely sufficient to train the four respective NNs. Each one of these calibration cases will be associated to a different retrieval model (see Sect. 4.4). The goal of the calibration procedure is to reduce the RT errors (e.g., bias, range of variability, structure, saturations) but it should be kept in mind that other sources of discrepancies are still present in the learning database and that the calibration procedure cannot suppress them all by building a perfect bridge between real and simulated satellite T_Bs.

4.3.2 Comparison of Real and Simulated Satellite Observations

RMS error differences between simulated $T_{B}s$ and the initial real observations and the calibrated observations are represented in Fig. 3 for the four configurations (CF/O, CF/L, CL/O, and CL/L) and for the AQUA and the MetOp platforms. All the situations where the differences between the simulated and calibrated T_Bs are larger than three standard deviations of these T_B differences are filtered out (corresponding to less than 0.5% of the situations). Most of these situations correspond to precipitating scenes, missed by the precipitation filter. The calibration procedure never degrades the statistics and often reduces the discrepancies between the simulated and observed satellite measurements by a few degrees. This is true for land scenes, for both cloud free and cloudy situations, and for every channel of all instruments. For ocean scenes, the impact is also very interesting for cloud free and cloudy situations, except for lower frequencies ($\leq$10 GHz) on clear scenes, where the impact is close to zero. The calibration procedure never degrades the statistics. Note that a large part of the discrepancy between the simulated and the initial observations is likely due to the RT simulations, not to instrument calibration problems. The simulation errors can arise from errors in the input parameters (e.g., ECMWF analysis, land surface emissivities), as well as from deficiencies of

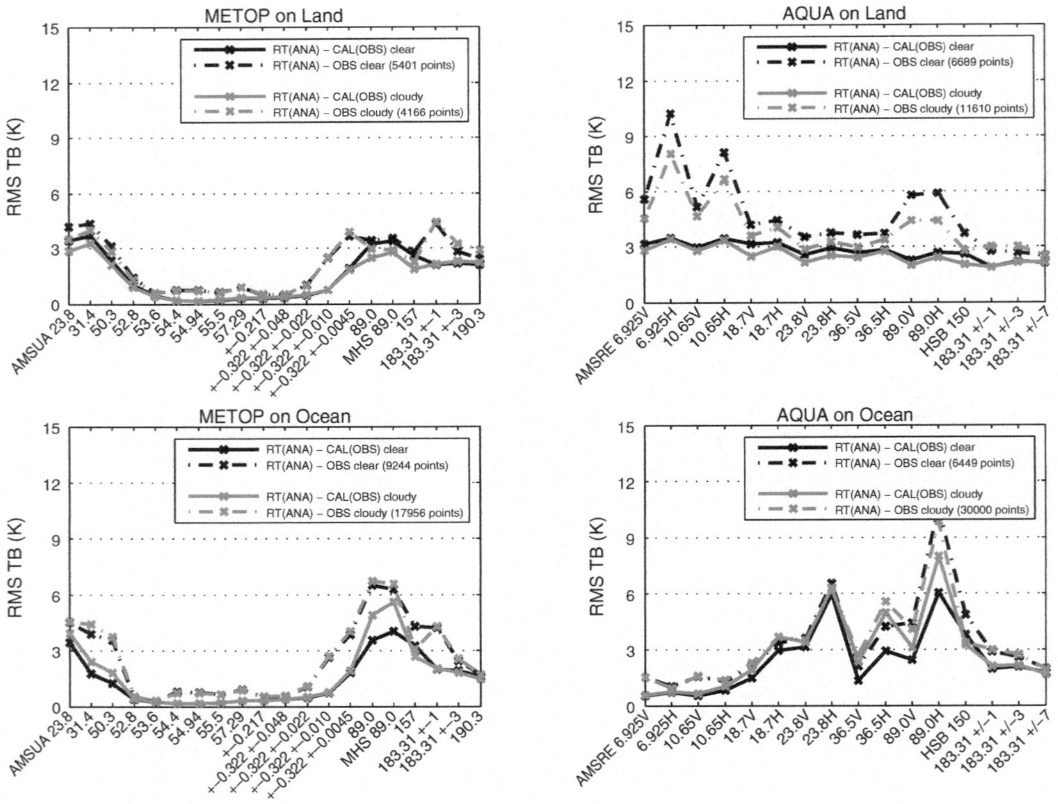

Fig. 3 Root Mean Square differences between the radiative transfer simulations using the ECMWF analysis and the calibrated observations (dashed lines) and the initial observations (solid lines). Different cases are considered: over land (top) and ocean (bottom), for MetOp (left) and AQUA (right), for clear sky (black lines) and cloudy sky (grey lines)

the RT code. Errors in the ECMWF analysis can result from misplacement of clouds. This can be seen best over ocean where strong emission signal in the cloudy case at 89 GHz does not disappear by the use of the calibration (Fig. 3).

For instance, initial errors at low AMSR-E frequencies (at 6 and 10 GHz especially in H polarisation) over land are likely due to extrapolation errors in the emissivities (the land surface emissivities are derived from satellite observations between 19 and 85 GHz). Over the ocean, the errors with AMSR-E in H polarisation, before and after calibration could be related to limitation of the radiative transfer model in its simulation of the sea surface emissivities that is particularly sensitive in H polarisation.

Other statistical results were analysed (not shown) to check the quality of the calibration: (1) the distribution of the errors is narrower when using the calibrated data, (2) the correlation is only slightly increased when the real observations are calibrated, meaning that the calibration transformation is statistically close to linear for most of the samples, and (3) the scatterplots of differences showed that the quality of the calibration procedure is satisfactory for all T_B ranges. The angle dependency of the calibration has been checked and is very limited (not shown): the same calibration procedure is adopted regardless of the scanning angle.

4.4 *Neural Networks Retrieval Scheme*

NN techniques have been very successful in developing computationally efficient algorithms for remote sensing applications. For example, a NN algorithm is applied to retrieve simultaneously the atmospheric temperature and humidity atmospheric profiles from 1000 to 1 hPa over sea using AMSU-A and -B observations [2]. In this section, such a NN retrieval algorithm is used in association with the NN calibration scheme.

4.4.1 Neural Networks Model

The retrieval scheme is composed of a MLP neural network, similar to the one in Sect. 4.3. The NN model is composed of three layers: the first layer codes the inputs of the retrieval scheme (i.e., satellite observations and *a priori* information), the third layer represents the outputs (i.e., the retrieved products), and again, the hidden layer is use to control the complexity of the model. Like in Sect. 4.3, different retrievals need to be considered for the four configurations (CF/O, CF/L, CL/O, and CL/L) and for the Megha-Tropiques, AQUA, and MetOp platforms.

The inputs of the NN models are:

- The calibrated T_B in the RT simulation space from Megha-Tropiques, AQUA, and MetOp microwave instruments (Sect. 3.3);
- An *a priori* information on the atmospheric temperature profile given by the ECMWF operational analysis (Sect. 3.1);

– *A priori* surface emissivities from a microwave emissivity climatology (see Sect. 3.2) and surface skin temperature from the ECMWF analysis, for land surfaces.

No *a priori* information on the Water Vapour (WV) profile is used directly in the retrieval, in order to better assess the information content of the satellite observations, independently of other related information. Nevertheless, the ECMWF operational analysis is used to provide some *a priori* information on the water vapour, by classifying the situation according to their Total Content Water Vapour (TCWV): TCWV $\leq 20\,\mathrm{kg\,m^{-2}}$; $20\,\mathrm{kg\,m^{-2}} <$ TCWV $\leq 35\,\mathrm{kg\,m^{-2}}$; or $35\,\mathrm{kg\,m^{-2}} >$ TCWV. This limits the reduction of the dynamical range of the retrieval as compared to the learning database, which is often observed with statistical mapping. A dedicated NN is used for each of these categories and for each of the NNs considered earlier for the various retrieval configurations. In addition, the retrievals are trained for different scanning angles (5°, 15°, 25°, 35°, 45° and 55°): the actual retrieval of real data at a particular scanning angle will linearly interpolate the result of the two NNs with the closest scanning angles. The total number of NNs is composed of four configurations (clear/cloudy and land/ocean) × three platforms × three humidity ranges × six incidence angles, for a total of 216 NNs.

The intermediate outputs of the retrieval scheme are the 43 water vapour contents of the atmospheric profile from the surface to the top of the atmosphere and the TCWV, in addition to the surface temperature and emissivities over land, and to the surface wind speed over ocean. Note that the surface skin temperature is not retrieved over ocean, as Megha-Tropiques and MetOp do not provide the low frequency observations sensitive to this parameter. We decided to use the relative humidity (between 0 and 100%) because the temperature profile is available from NWP centres, and because this unit allows to easily compare dry and wet atmospheres, and lower and upper atmospheric layers. The architecture of the NNs for the Megha-Tropiques data are: 15+43+15+1=74 (resp. 15+43=58) neurons in the first layer, 50 neurons in the hidden layer, and 43+1+15+1=60 (resp. 43+1+1=45) for the land (resp. ocean) retrieval. For practical reasons, the WV retrieval is performed on the 43 atmospheric layers defined by RTTOV (in particular in order to facilitate the *a posteriori* test described in the following paragraph). However, Madras/Saphir, AMSR-E/HSB, or AMSU-A/MHS instruments cannot provide such high vertical resolution (see Sect. 2.4) and the final retrieval of our inversion scheme is defined on six thicker layers delimited by the surface, 920, 750, 560, 400, 250 hPa, and the top of the atmosphere. The mapping from 43 to 6 layers is an integration of the water vapour content of each finer layers into the thicker layer. The number and the limits of these thicker layers have been optimised to minimise the retrieval uncertainty: All the possible combinations of the 43 original layers into 6 thicker layers have been considered, the retrieval results have been estimated for each one of them, and finally, the six coarse vertical layers associated to the lowest uncertainties have been selected. In the retrieval statistics, all the results will be presented in these six layers.

Following the retrieval, an *a posteriori* test is performed. It uses the retrieval results and the ECMWF *a priori* information as inputs to the RTTOV model. These simulated T_Bs are compared to the observations and when the differences are larger than a fixed threshold (corresponding to three standard deviations of a statistic previously performed) the results are discarded. This is a standard procedure in operational retrieval schemes.

4.4.2 Learning and Testing Datasets

One year of ECMWF operational analysis (Sect. 3.1) is used to build the learning dataset. Using a RT model in all of these surface and atmospheric situations would be too demanding computationally. This would also be inefficient because there would be redundancies in this dataset. Instead, the full year of data is reduced using a uniform sampling procedure. With the data thinning, we keep only 1% of the available points. For the RT simulations, the cloud flag described in Sect. 4.2 is used and all the precipitation scenes are suppressed. The RT simulations are performed for Madras/Saphir, AMSR-E/HSB, AMSU-A/MHS instruments.

The resulting dataset is divided in a learning dataset (representing 80% of the situations) and a generalisation dataset (corresponding to the remaining 20%). Results presented in following section will use the generalisation dataset. The training is performed using a back-propagation algorithm [44].

5 Theoretical Evaluation

In this section, the information content of the microwave instruments on board the Megha-Tropiques, AQUA, and MetOp platforms is evaluated using two approaches: first, measuring directly the performance of the retrieval scheme on simulated observations; second, using the classical information content analysis [41, 42]. The first approach provides uncertainty estimates that are generally too optimistic (reasons will be provided in next section), the information content analysis is also a simplification of the real-world but it is generally used as a tool to compare instrument configurations. The advantage of performing these theoretical evaluations is that no real observations is necessary and it is possible to compare them. In the following sections, the same retrieval scheme will be evaluated using real observations from AQUA and MetOp instruments.

5.1 Theoretical Retrieval Uncertainties Using Simulated Observations

The full NN retrieval scheme has been applied to the testing dataset (not to the learning dataset), for the three platforms (Megha-Tropiques, AQUA, and MetOp), including the four configurations (cloud free/cloudy situation, over land/ocean). Figure 4 represents the RMS errors in % of relative humidity for the retrieval of

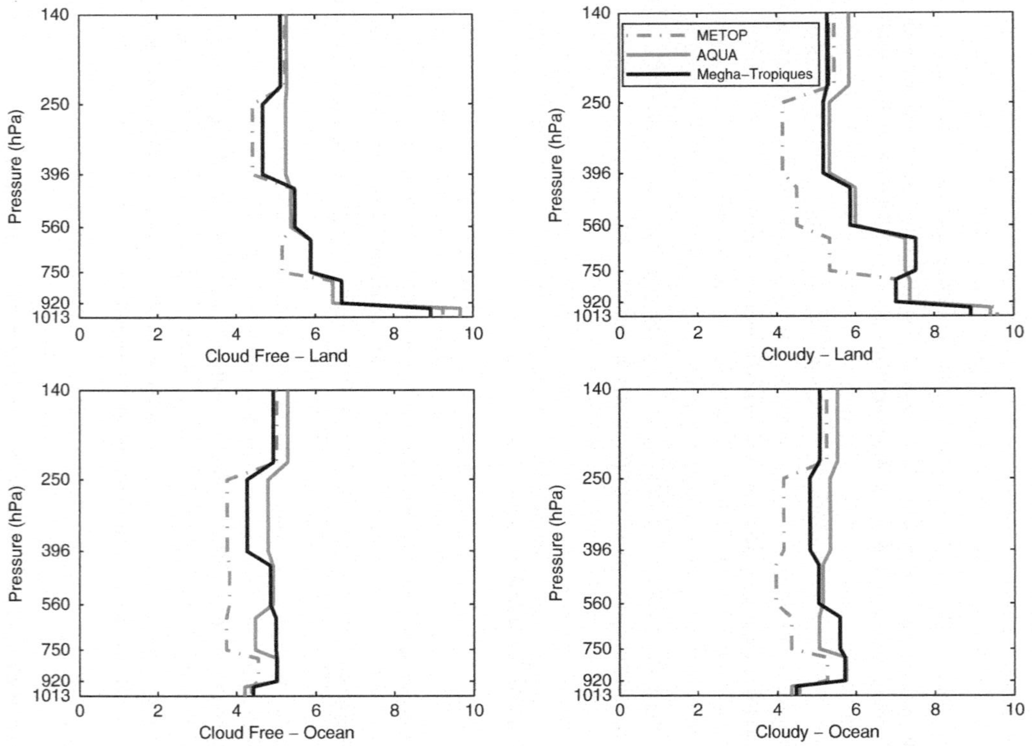

Fig. 4 Theoretical RMS errors for the NN retrieval of the water vapour profile estimated on simulated data. Units are percentage of relative humidity. Statistics are provided for MetOp (grey dot-dashed line), AQUA (grey continuous line), and Megha-Tropiques (black) data, and for four configurations: Cloud Free/Cloudy and over Land/Ocean situations

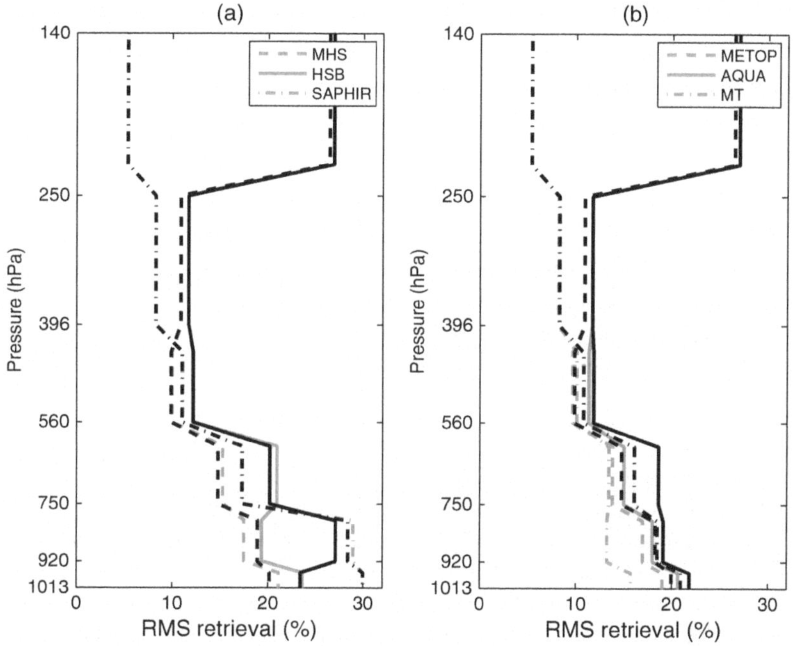

Fig. 5 Information content analysis, for a tropical situation, (**a**) for the three sounding instruments MHS, HSB and SAPHIR, and (**b**) for the instrument combinations for the three platforms MetOp, AQUA and Megha-Tropiques, for an oceanic (grey) and a continental (black) tropical situations

the profile over the six atmospheric layers. There is no bias for any configuration and any platforms (not shown). As expected, land retrievals are more difficult than over ocean, especially for the lower atmospheric layers where the error is increased by a factor two (retrieval errors remain identical for layers higher than 500 hPa). For all cases, results are similar under clear and cloudy conditions. MetOp retrievals are more precise than AQUA and Megha-Tropiques. This can be explained by the lower instrument noise of MHS channels compared to HSB or Saphir (Table 1). Megha-Tropiques and AQUA results are similar. They both include a water vapour sounding instrument and an imager with window channels. The improvement related to additional channels for the Saphir instrument is balanced by an instrument noise that is higher than for HSB. This is in agreement with the results in Fig. 5b.

Table 2 provides the theoretical RMS errors of the retrievals for the TCWV (i.e. the direct retrieval of TCWV, not the integration of the profile) over ocean and land. These statistics are estimated for the three platforms. They are compared, when possible, to the uncertainties of the first guess used as inputs of the retrieval scheme, and specified in the development of the NN. Over ocean, the TCWV is retrieved with 1.6–1.7 kg m^{-2} for both clear and cloudy cases and the three platforms have a similar accuracy. For land surfaces, the AQUA retrievals are slightly better than the Megha-Tropiques and MetOp retrievals that have similar characteristics.

Table 2 Theoretical retrieval errors for Megha-Tropiques, AQUA and MetOp platforms, over ocean and land surfaces, for TCWV

			Retrieval error		
			MT	AQUA	MetOp
Ocean	TCWV	Clr	1.7	1.6	1.6
	(kg m^{-2})	Cld	1.7	1.7	1.7
Land	TCWV	Clr	2.1	1.8	2.0
	(kg m^{-2})	Cld	3.3	2.9	3.3

For all cases, the retrieval errors are better than the first guess errors, which means that the retrieval did bring additional useful information.

Theoretical retrieval errors underestimate the real retrieval uncertainties because all possible sources of uncertainties that could appear in real-world conditions are not considered:

- No coincidence errors between the two considered instruments or with the *a priori* information from the ECMWF operational analysis are introduced;
- No RT errors have been used as this is a difficult quantity to measure;
- The Gaussian nature of the *a priori* information (such as the temperature profile from the ECMWF analysis) is a simplification;
- The statistics are performed on a diverse and large dataset, but these atmospheric situations come from the ECMWF operational analysis, meaning that the profiles are smoother than real profiles or they can be biased if the analysis is not correct;
- Furthermore, the *a priori* is considered, in this study, to have the same statistics for all the situations, which is a simplification. For example, the *a priori* uncertainty on the water vapour should be different for a dry or a humid situation;
- Overall, the dataset used to make these statistics is completely coherent with the dataset used to train the NN, this is less true of course when using real-world observations.

These theoretical uncertainty assessments are representative of perfect inversion conditions, but any additional source of uncertainty will degrade them toward realistic estimates. In particular, the impact of the calibration errors is expected to be very significant. The important point about these theoretical uncertainty estimates is that it is possible to compare the retrieval capacities of the instrument configurations.

5.2 *Information Content Analysis*

In the classical information content analysis [41, 42], the RT is linearised around a first guess solution f_0:

$$(T_B\varepsilon - T_B0) = A \cdot (f - f_0) + \varepsilon, \qquad (2)$$

where f is the geophysical variables to be retrieved (atmospheric WV profile in our case), f_0 is a first guess, $T_B\varepsilon$ is the brightness temperatures observed by the microwave instrument, T_B0 is the brightness temperatures corresponding to the first guess solution f_0, A is the Jacobian of the RT model (Sect. 2.4), and ε the instrument noise (other sources of uncertainties can also be included in this term). If the hypothesis that the variables considered in the problem follow a Gaussian distribution, the Bayesian retrieval can use an iterative procedure:

$$f = f_0 + (A^t \cdot S_\varepsilon - 1 \cdot A + S_f{}^{-1})^{-1} \cdot A^t \cdot S_\varepsilon{}^{-1} \cdot (T_B\varepsilon - T_B0), \tag{3}$$

where S_ε is the covariance of the instrument noise and S_f is the covariance matrix of the first guess error. The iteration starts with the first guess solution f_0 that is updated sequentially using Eq. (3) until convergence is reached. This is a traditional retrieval method used in NWP centres.

It is always very important to have an uncertainty estimate for any retrieval that is made. In the retrieval method of Eq. (3), the uncertainty on the retrieval is given by the retrieval error covariance matrix:

$$Q = (A^t \cdot S_\varepsilon{}^{-1} \cdot A + S_f{}^{-1})^{-1} \tag{4}$$

It is possible to use this expression to measure the quality of retrievals based on the instrument noise information, on the Jacobian of the RT for the considered channels, and the *a priori* information provided by the first guess.

An important information required in the right hand side of Eq. (4) is S_f, the covariance matrix of the first guess error. This represents the *a priori* information on the variables to retrieve, before the inversion. In this experiment, 30% of *a priori* uncertainty is chosen for the water vapour, with no correlation between the atmospheric layers. Given the range of variability of the water vapour in the Tropics, this means that actually no *a priori* information is considered, so the estimated uncertainty will describe the retrieval errors from the satellite observations only.

Another necessary information is the linearisation A of the radiative transfer, which is equal to the Jacobian presented in Sect. 2.4. In Fig. 1, the sensitivities of the water vapour sounding channels are provided for a standard tropical situations. Note that here, only the Jacobians in water vapour are considered, assuming that the other variables are already known.

The term S_ε in Eq. (4) represents the instrument noise provided in Table 1. The RT errors could also be considered and introduced in this covariance matrix, but they are difficult to characterise and they have been neglected in the following. The calibration errors could be introduced also, but in order to compare the various instrument configuration and to assess the intrinsic retrieval scheme capacities, it has been decided not to include them. Since we are interested here in the comparison of the sounding capacities of the instruments, the introduction of these additional terms would not change our conclusions.

Figure 5a (left) presents the theoretical RMS error of the retrieval, in % of RH, for the three sounders MHS, HSB and Saphir, over ocean (grey lines) and land (black lines) surfaces. The land/ocean differences are only present for the lower layers (pressure 600 hPa). As expected from the preliminary analysis of the Jacobians, and from the respective instrument uncertainties, MHS provides the most accurate retrieval of the atmospheric profile in most atmospheric layers. Saphir is extremely efficient for pressures higher than 400 hPa because of its 183.31 GHz ±0.2 channel (see Fig. 1): the uncertainty can reach 5%, which is probably too optimistic and dependent upon over-simplified hypothesis. On the contrary, Saphir does not have surface sensitive channels, and its uncertainties for the lower atmospheric layers is close to 30% which means that it has no information (the *a priori* information has a specified uncertainty of 30% in this experiment). Although Saphir has more channels than the other sounders, the lack of surface sensitive channels and the receiver noises limit the capability of the Saphir instrument alone. However, once combined with the observations from their companion instruments (Fig. 5b), the three missions provide similar information content for water vapour profiling, with an advantage for Megha-Tropiques for higher layers (pressures > 400 hPa) and for the lower atmospheric layers over the ocean. The information provided by the Madras instrument for the lower atmosphere significantly improves the retrieval of the water vapour up to ∼600 hPa and justifies the combination of Saphir/Madras in our retrieval chain.

6 Evaluation of the Water Vapour Retrievals

In this section, the evaluation of the WV retrieved by the inversion chain is presented, as compared to the ECMWF analyses and to radiosonde measurements.

It should be clear, first, that the ECMWF analyses are already a good estimation of the state of the atmosphere, that takes into account a lot of information from satellite data (including AQUA and MetOp observations) and from radiosondes. The analyses can be used as a reference, but the goal of the satellite retrieval is to improve the analysis, in the sense that the retrieved profile, when given as input to the Radiative Transfer (RT) model, should provide a simulation closer to the satellite observations than the simulation using the analysis. Any departure from the analysis can be treated as either error or improvement in the retrieval. These considerations bring the necessity to validate the retrievals in the space of the satellite observations, and this will be developed further in Sect. 7.

Validating satellite WV measurements is very difficult task. An inter-comparison with radiosondes is performed here but only for a short period. This limits the possibility to handle the multiple problems of radiosondes (dry bias, representativity, uncertainties at high altitudes, or different sonde type). We are not using here any GNSS (GPS and the Russian GLONASS) measurements which provide rather accurate total WV measurements on a continuous base and have nowadays become a *de facto* reference [20, 36, 43, 47]. We are not using either the ARM

(Atmospheric Radiation Measurement) program ground-based lidar and microwave radiometer measurements. Such an inter-comparison would represent a study in itself, we decide here to focus on innovative diagnostics to assess the retrieval quality, especially in the T_B-space.

Although the algorithms are globally applicable, they are specifically tested here over the Tropics (±30° in latitude), where the role of water vapour for clouds, circulation and climate sensitivity is of high interest [11]. The results will be presented mostly using the AQUA retrievals. The statistics commented below are processed on a systematic sampling selection from the 2 months of satellite observations described in Sect. 3.4.

6.1 Total Column Water Vapour

The retrieval of the TCWV (Total Column Water Vapour) can be done in two different ways. First, a dedicated NN can be trained to retrieve directly the TCWV. Second, it is possible to integrate vertically the TCWV from the NN retrieved WV profile. The second scheme appears to be more satisfactory in our tests (not shown). This might be surprising, the direct retrieval could be thought to be simpler. A possible explanation could be that the profile retrieval uses more neurons in the inversion process and therefore inversion scheme is able to obtain more information from the observations.

Figure 6 represents a map of the TCWV retrieved from the AQUA observations together with the corresponding ECMWF analysis, for the morning of September

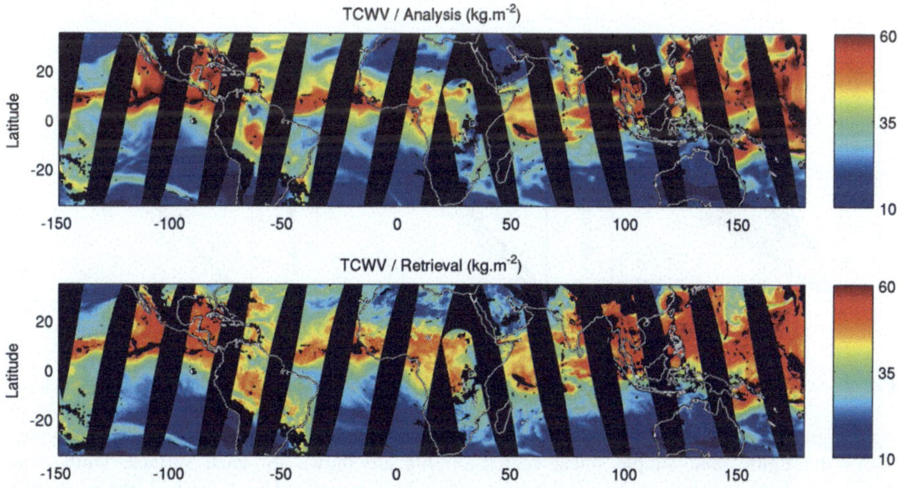

Fig. 6 Example of total column water vapour from the ECMWF analysis (top) and the coincident AQUA retrieval (bottom) on the track of AMSRE and HSB (September 7th, 2002)

7th, 2002, from 00h00 to 11h40 UTC, over the Tropics. The AMSR-E tracks are collocated with the HSB observations. This map (and following maps) shows the retrieval results for the descending and the ascending orbits: the first ones are localised on the west of the map and are observed during local nighttime, the eastern part of the map corresponding to ascending orbits during local daytime. The missing pixels correspond to unprocessed data (precipitating scenes or high elevation).

The spatial structures of the retrieved TCWV are very similar to the analysis (Fig. 6). This is very encouraging considering that no *a priori* information on the WV is used in the retrieval scheme, even if a first guess on the temperature was used by the retrieval. Some differences can be noted however, for instance in the Indian Ocean south of India, or west of Brazil. There is a good continuity in the field between ocean and land surfaces (e.g., west of Africa or Central America), confirming that the land and ocean inversion schemes are robust and coherent with one another: this is a key point as so far the inversion of satellite WV information was limited over land. The retrievals over continents are possible thanks to the surface constraints on the microwave emissivities. The retrieved fields show slightly more spatial noise, but this can be expected as the analyses are smooth by nature.

The TCWVs from the analyses are compared directly to the retrievals in the scatterplots of Fig. 7 (left), over both ocean and land. The bias is limited ($0.39\,\mathrm{kg\,m^{-2}}$), for a standard deviation of the errors equal to $3.6\,\mathrm{kg\,m^{-2}}$. The retrieval slightly underestimates the very high TCWV, especially over land (with a slope of the regression slightly lower than one). This is a general behaviour of statistical retrieval schemes: they tend to dampen the dynamics of the retrieved geophysical variables. Note that this reduction of the dynamics is limited, partly because three neural network inversion models have been trained on three TCWV ranges, as given by the *a priori* ECMWF analysis.

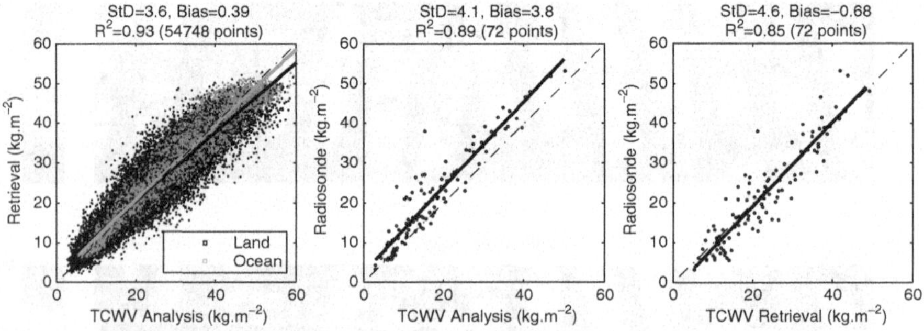

Fig. 7 Scatterplot of total column water vapour: AQUA retrieval versus ECMWF analysis (left), radiosondes versus analysis (middle) and radiosondes versus retrieval (right), for land (black) and ocean (grey). Standard Deviation (StD), bias and coefficient of determination R^2 are provided above the figure, and simple linear regression fits are drawn for each surface. Unlike the radiosondes, analysis and retrieval are performed on both ocean and land. Dashed line is identity. For the left graphic (comparison to analysis), the overall StD=$3.6\,\mathrm{kg\,m^{-2}}$ can be split in 3.1 (RMS=3.12) for oceanic cases, and 4.1 (RMS=4.12) for land cases

In order to check the quality of the retrieval further, September 2002 and January 2003 radiosondes are extracted from the ECMWF operational sounding archive used during the assimilation process in the ERA40/ERA-Interim re-analyses [51]. The temperature and humidity measurements have been quality controlled in order to discard incomplete profiles (threshold of 30 hPa for the temperature and 350 hPa for the humidity), and a vertical extrapolation is applied up to the top of the atmosphere using a climatology [R. Armante, Laboratoire de Méthorologe Dynamique, personal communication, 2008]. The radiosondes must be $\pm 0.3°$ apart at maximum from the closest AMSR-E observation, with a maximum time differences of ± 90 mn. AMSR-E scenes with scattering signatures have been discarded. Seventy-two match-ups fitting the coincidence criteria above have been found, they are almost equally distributed in cloud free and cloudy situations, for night and day observations. The radiosonde measurements only come from continental areas or small islands so the two right panels in Fig. 7 might be biased towards land situations or coastal areas. For each panel, the standard deviation of the errors is provided together with the bias and the R^2 statistics.

These results show that the AQUA retrieval is as good as the analysis compared to the radiosondes: both have StD differences as well as R^2 coefficient in the same range (4.1 and 4.6 kg m^{-2} for the StD, and 85–89% for the R^2). The retrieval is less biased than the analysis, which is surprising considering that the radiosonde observations are assimilated in the analysis. The StD of the difference with the radiosondes is comparable with the departure of the retrieval from the analysis, reinforcing the significance of the comparisons with the analysis. Negative bias of 2.1 kg m^{-2} is observed on the retrieval of night and cloud free situations, while other situations have a bias lower than 0.5 kg m^{-2}. Root Mean Square (RMS) errors show no strong link with the time of observation or the cloud cover. Currently, TCWV is estimated accurately only over ocean regions, for non-precipitating conditions. For comparison purposes, Deblonde and Wagneur [16] obtained a bias of -0.2 kg m^{-2} and a StD of 3.7 kg m^{-2} between SSM/I TCWV estimates and radiosonde observations from small islands. Retrieval results agree better with ECMWF analysis over ocean than over land. Over ocean, our results give a bias of 0.3 kg m^{-2} and a RMS of 2.8 kg m^{-2}. Note that in 2002–2003, RS80 sondes were frequently launched which are known for their humidity bias. Several correction methods have been tested in NWP centres to correct this bias but this is difficult task. Results of our study my change with the use of other sondes in other years.

A previous comparison under similar conditions between SSM/I retrieval and a NWP analysis showed larger differences, with a bias of 1.08 kg m^{-2} and a RMS of 3.25 kg m^{-2} [15]. Aires et al. [3] have developed a similar retrieval scheme using SSM/I observations and obtained 3.8 kg m^{-2} for clear cases and 4.9 kg m^{-2} for cloudy cases, both over land. In [52] a semi-statistical retrieval scheme is developed for TMI observations. The RMS errors compared to radiosondes is equal to 3.5 kg m^{-2} (2.5 when the 5% worst cases are suppressed from the statistics) but this statistics is performed for oceanic cases, and only for the Tropical belt.

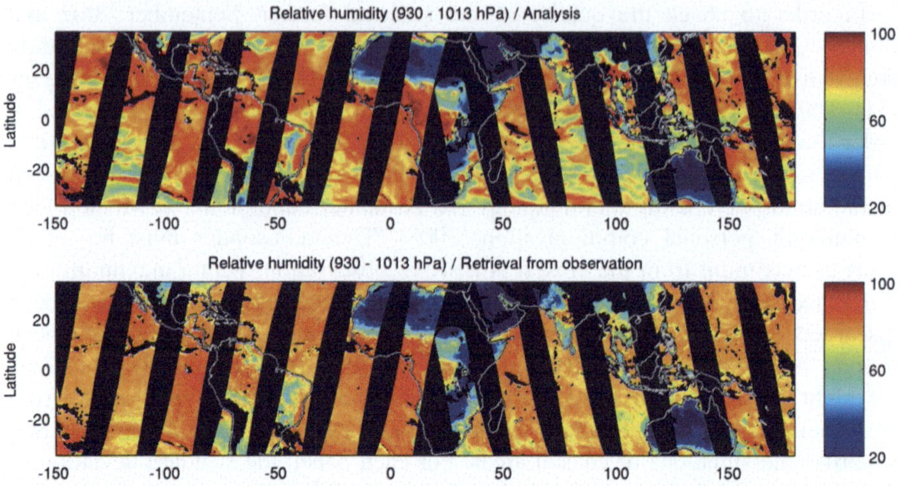

Fig. 8 Example of relative humidity on the lowest layer from the ECMWF analysis (top) and the coincident AQUA retrieval (bottom) on the track of AMSR-E and HSB (September 7th, 2002)

6.2 Atmospheric Water Vapour Profiles

The retrieval of WV profiles with instruments such as AMSU-B, MHS, HSB or Saphir is a true challenge. Although their weighting functions (Fig. 1) sample relatively well the vertical column, they overlap significantly, meaning that the degrees of freedom, or the number of independent pieces of information, are rather limited (about 3–4). When these satellite observations are assimilated in a NWP scheme, the *a priori* information from the atmospheric circulation model helps constrain the vertical structure of the retrieval. It is much more difficult to estimate the vertical structure in a direct retrieval, without *a priori* vertical information.

Similar to Fig. 6, Fig. 8 presents the relative humidity between 920 and 1013 hPa, from the analysis and from the retrieval. Note that this layer is particularly difficult to retrieve, especially over land, due to the surface contribution. Comparison of the maps reveals that the main structures of the WV are captured by the retrieval and the transition from ocean to land is smooth, even for this lower layer. Even meso-scale structures in the WV layer over land are correctly estimated, for instance in Sudan, in Africa, or east of Brazil. This is particularly encouraging because the use of the microwave observations over continents is particularly difficult, especially for the surface-sensitive channels in the lower troposphere [6]. This is to be expected since the surface emissivities have been well constrained in the algorithm. However, some noticeable differences can be observed, even over ocean, for instance in the South Pacific Ocean and in the Indian Ocean, where WV structures present in the analysis are not reproduced by the retrieval.

Figure 9 represents the RMS profile of the difference of the retrieval and the ECMWF analysis. These "departures" are not errors because it cannot be said *a*

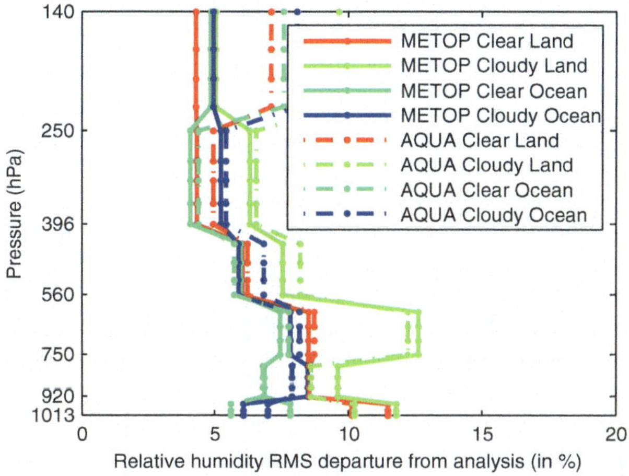

Fig. 9 RMS departure in relative humidity of the ECMWF analysis and the retrieval from AQUA (doted) and MetOp (continuous line) observations: for Cloud Free/Land (red), Cloudy/L (green), Cloud Free/Ocean (cyan) and Cloudy/Ocean (blue) configurations

priori which one is the closest to reality. This statement is part of the challenge in such evaluation. Retrieval seems closer to the analysis over ocean than over land. This can be related with the fact that the analysis assimilates more observations from the microwave sensors over ocean than over land (both in clear and cloudy situations), providing a better agreement. The departures seem to be overall close to the theoretical uncertainty estimates (i.e., between 5 and 10% of RMS error in percent of relative humidity, see Fig. 4). For oceanic cases, the theoretical estimates are higher than the departures. Furthermore the departures have a flat vertical error structure where theoretical errors are stronger in the lower troposphere and smaller in higher atmospheric layers.

For over land and over ocean estimates, the errors are higher for the lower layers over land but the theoretical errors are bigger (about 12%) than the departures (less than 10%). Over land, microwave observation cannot be easily assimilated by the analysis for the lower layers of the atmosphere, due to difficulties in dealing with the surface contribution. It leads to a less accurate description of the RH profile. The situation is worst in cloudy cases where the information from real observations to the analysis is less reliable. In this configuration, analysis is not expected to fit exactly the reality and thus should not be used as a target for our retrieval.

One should also understand that these statistics depend upon the space-time thresholds used in the coincidences. Plus or minus 1.5 h are tolerated in these statistics in order to keep enough data samples. If 10 min were used instead, the RMS profile could be one or two percentage points lower. The RMS profile is represented only up to 250 hPa because the information provided by the satellites beyond this value is low (see weighting functions in Fig. 1). As expected, the retrieval of the lower layer is better over ocean than over land. Overall, MetOp and

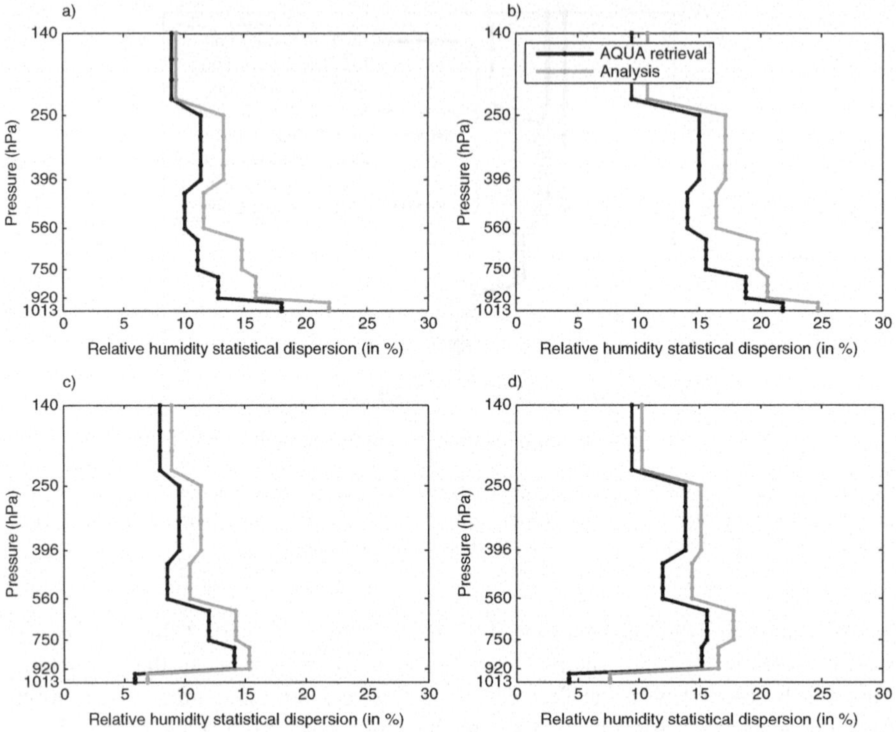

Fig. 10 Statistical "dispersion" of the relative humidity profiles of the ECMWF analysis (grey) and the retrieval from AQUA (continuous line) observations, for the four different configurations (clear/cloudy and land/ocean). These statistics are given from a sample of 2 months (September 2002 and January 2003). (**a**) Clear Land (6689 points). (**b**) Cloudy Land (11610 points). (**c**) Clear Ocean (6449 points). (**d**) Cloudy Ocean (30000 points)

AQUA retrievals have similar statistics. Generally, the AQUA platform performs better close to the surface, due to the larger number of window channels provided by the AMSR-E imager. The contrary usually prevails for the higher layers, likely due to the lower instrument noise of MHS on board MetOp as compared to HSB on board AQUA. The overall statistics are satisfactory with a departure from the analysis that is lower than 10%. The retrieval statistics are stable for angles lower than 49° (the statistics are integrated over all angles up to 49°, and higher angles are excluded).

Another diagnostic is to compare the variability of the resulting WV profiles. Figure 10 represents the statistical variability of the WV profile from the ECMWF analysis and from our AQUA and MetOp retrievals. Over 90% of the variability of the analysis is retrieved by the NN inversion for the two sets of microwave instruments. The variability of the vertical structure is well reproduced, with only a

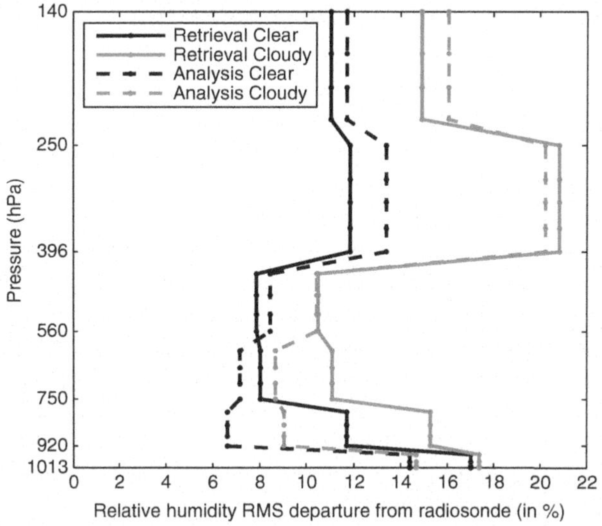

Fig. 11 RMS departure in relative humidity between a collection of radiosonde measurements and ECMWF analysis (dashed line) and coincident AQUA retrieval (continuous line). These statistics are represented over land, for clear (grey) and cloudy (black) scenes

slightly lower variability of the retrieval compared to the analysis. Another reason is that the profiles from the retrieval are in a thinner horizontal resolution than ECMWF analysis that have smoother 1.125° data. How the coarse resolution of the training data set influence the WV variability needs to be investigated in the future (see Kahn et al. [29] for scaling of water vapour variance).

In Fig. 11, the retrieved WV profiles are also compared to the radiosondes, for the same *in situ* measurements as in Fig. 7. The corresponding ECMWF analyses are added to the comparison. For the lower layers below 560 hPa, the analyses are systematically closer to the radiosondes than the retrieval, although the difference between the retrieval and the *in situ* measurements are limited, especially close to the surface. With the radiosondes already assimilated in the analyses, it is expected that they are very close. Nevertheless, this illustrates the difficulty to compare these different kinds of measurements: radiosondes provide a local information that drift up to 100 km away from their launch site, but their spatial average representativity is poor compared to the 20×20 km microwave sensors pixels. WV profiles over land are subject to strong spatial gradients in the low troposphere [48]. The comparison between the retrievals and the radiosondes show differences within 15% in percentage of relative humidity, except for the lower layer [28]. A previous study using radiosondes from RaCCI/LBA field campaign and HSB retrieved water vapour showed similar results for clear sky situations, and more important differences in cloudy cases [33].

7 Brightness Temperature: Space Validation

In a physically-based retrieval scheme, the quality criterion of the inversion process is based on the difference between (1) the actual observations and (2) the RT simulations when using the retrieved atmospheric and surface parameters. The objective of the inversion is to minimise the difference between the observations and these simulations. As a consequence, the improvement of the inversion compared to the a first guess or an *a priori* can be measured by the comparison of the differences between the observations and the RT simulations when using (1) the first guess or the *a priori* information on the state of the atmosphere and (2) the retrieved parameters. If the retrieved variables degrade these statistics, the inversion has not improved the *a priori* information or has mathematically diverged and failed to perform the retrieval. The accuracy in the final retrieval product is largely dependent on the accuracy of radiative transfer calculations. The objective of the inversion is to perform retrieval accurately based on the radiative transfer simulations which is supposed to represent the real world.

Note that the differences between the observations and the simulations are not only related to the quality of the retrieval. The satellite observations include instrument noise and systematic biases and drifts (depending on the sensors), and the RT also induces errors. The objective of the calibration step prior to the inversion process is to limit these two sources of errors, but the calibration is never perfect. In addition, differences in space and time between observations and analysis can also be critical.

In the following, several diagnostics will be presented in order to evaluate the satellite retrievals in the TB-space.

7.1 Brightness Temperature Comparisons

The geophysical variables retrieved by the inversion scheme described in Sect. 4.3 are used as inputs for the RT model (i.e., RTTOV). These include the WV profile, the surface temperature and emissivities over land, and the wind speed at the surface over ocean. The simulations are also performed using the initial ECMWF analysis in order to compare the potential relative improvements provided by the retrievals.

Figure 12 represents the statistics for the MetOp instruments (AMSU-A and MHS, left) and for the AQUA instruments (AMSR-E and HSB, right). Both land (upper part) and ocean (bottom part) are considered. RMS statistics are provided for clear (black) and cloudy (grey) situations. The statistics are presented for RT simulations performed on the retrievals (continuous line) and on the analysis (dashed line). The differences between the simulated T_B result uniquely from the parameters that the inversion scheme retrieves. The simulations with the retrieved products (continuous lines) have always a lower RMS than the simulations with the *a priori* information (dashed ones): this clearly indicates that the retrieved variables represent an improvement in the T_B-space, as compared to the analysis. When the retrievals are used as inputs to the RT code instead of the ECMWF

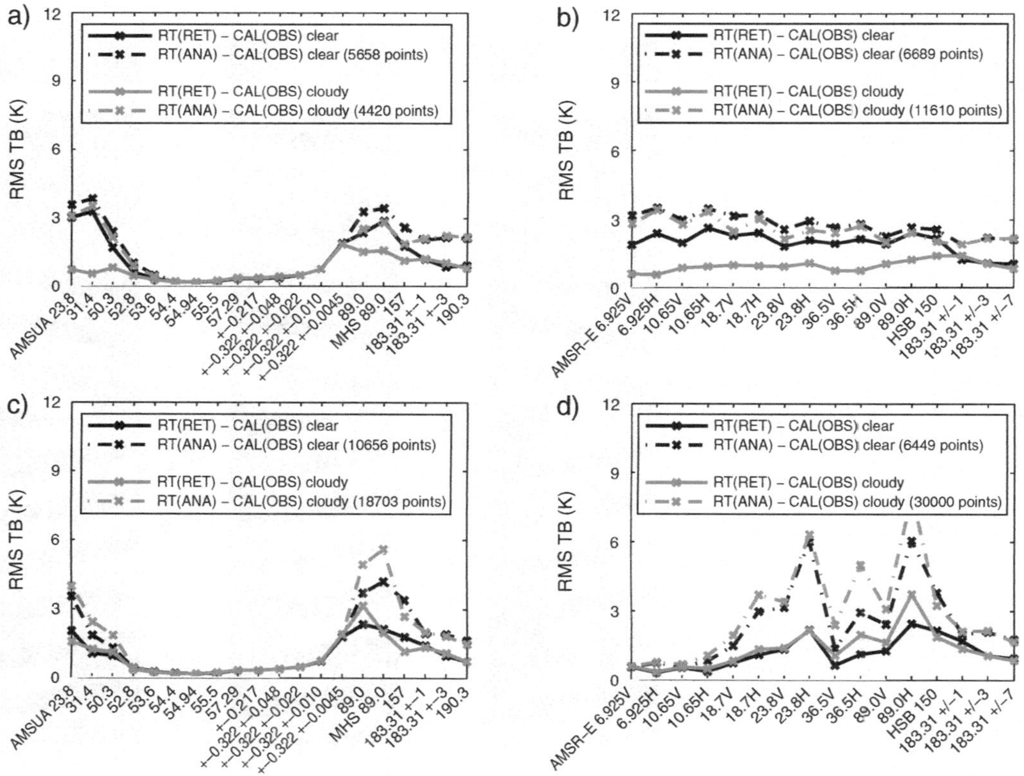

Fig. 12 RMS errors between simulated and observed T_B for AMSU-A and MHS on board MetOp (left), for AMSR-E and HSB on board AQUA (right), over land (top) and ocean (bottom) for clear (black) and cloudy sky (grey), for 2 months (July 2007 and January 2008 for MetOp and September 2002 and January 2003 for AQUA). The continuous line represents the simulations using the retrieval products, the dashed lines are for simulations using the ECMWF analysis and TELSEM surface emissivity. **(a)** METOP on Land. **(b)** AQUA on Land. **(c)** METOP on Ocean. **(d)** AQUA on Ocean

analysis, simulated T_B become closer to the calibrated observations. This is exactly the objective of the retrieval scheme. In particular, the simulations of the 183 GHz channels are improved by the retrieved WV profile for the four configurations, for AQUA and MetOp instruments. The statistics for the WV sounding channels are similar for all cases (land/ocean, clear/cloudy, AQUA/MetOp), meaning that the retrieval is very robust regardless of the conditions. The window channels are improved by better surface temperature and emissivities over land, and surface wind over ocean. Over continental surfaces, better statistics are obtained under cloudy conditions for the surface sensitive channels: this is due to the smaller dynamics of the surface temperature under cloudy cases. Note that part of the improvement can be related to a better space/time coincidence between the retrieval and the satellite data (by construction, the retrieval is performed at the same time and location than the satellite observation).

Figure 13 presents the calibrated observations at 23.8 GHz for vertical polarization from HSB on AQUA (middle), along with the simulations from ECMWF

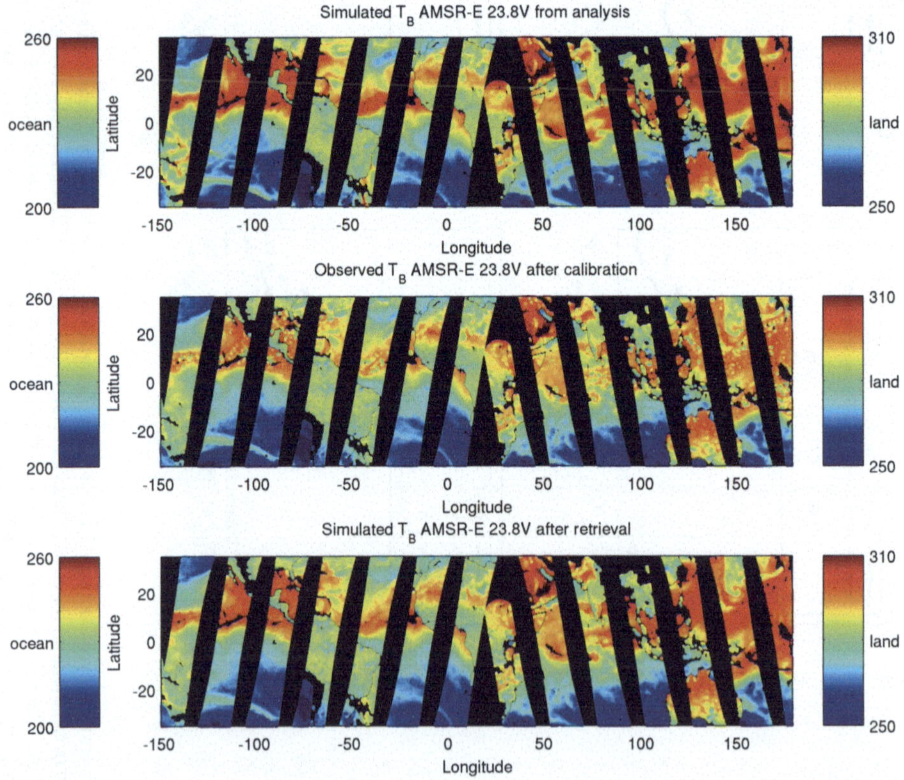

Fig. 13 Example of T_B at 23.8 GHz in vertical polarization of simulation using ECMWF analysis and TELSEM surface emissivity (top), calibrated observations (middle) and simulation using retrieval products (bottom) on the track of AMSR-E and HSB (September 7th, 2002)

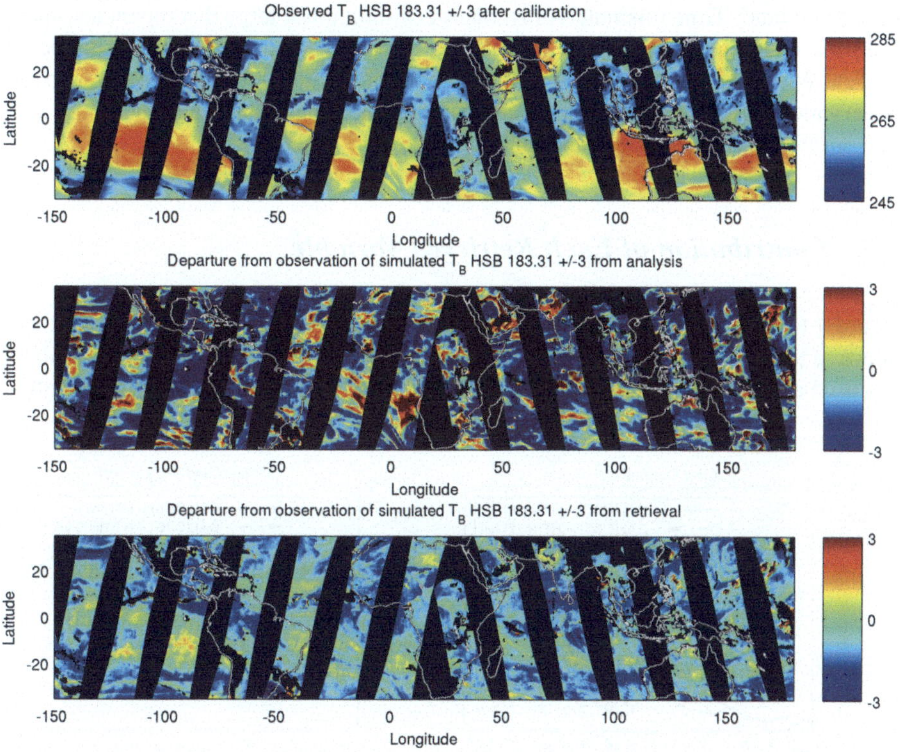

Fig. 14 Example of T_B at 183.81±3 GHz of calibrated observations (top), its departure from simulation using ECMWF analysis and TELSEM surface emissivity (middle), and from simulation using retrieval products (bottom) on the track of AMSR-E and HSB (September 7th, 2002)

analysis (top) and from the retrieval (bottom). This channel is very sensitive to the TCWV, and using the retrieval improves the comparison with the observations. For instance, the observed structures in the Pacific Ocean around 150° East and 20°N or around 140° West and 20°S are better reproduced using the retrieval in the simulation. Over ocean, simulation from analysis seems to present a strong positive bias in very humid atmosphere (up to 5 K from the observation) that is not so reproduced by the retrieval (at 10° of latitude). This figure can be related to the Fig. 6, where retrieval values were way below analysis forecasts over these areas. The possibility of a positive bias from our radiative transfer model is not enough to explain this phenomenon since the RTTOV model is widely used by the NWP centres for forecasts. Furthermore, a bias between observed and simulated T_B would have been reduced by our calibration step.

Figure 14 presents the calibrated observations at 183.81±3 GHz from HSB on AQUA, and the difference between these calibrated observations and the simulation using the retrievals. This channel is strongly correlated to the water vapour in the low troposphere (see Fig. 1). Simulations show significant differences with the observation when using the analysis, with often ≈3 K differences, over both ocean and land. Time mismatch only cannot explain these large discrepancies, since the time difference is less than 30 mn over some areas such as South Atlantic or Africa. When using the retrieved products in the simulations, the agreement with the observations is significantly improved, over both ocean and land, without noticeable discontinuity between ocean and land.

7.2 Contribution of Each Retrieved Variable

In order to assess the impact of each retrieved variable in the improvement in the T_B-space, a similar experiment is conducted when only part of the retrieved variables are used as inputs to the RT code. Similarly to Fig. 12, Fig. 15 shows RMS statistics

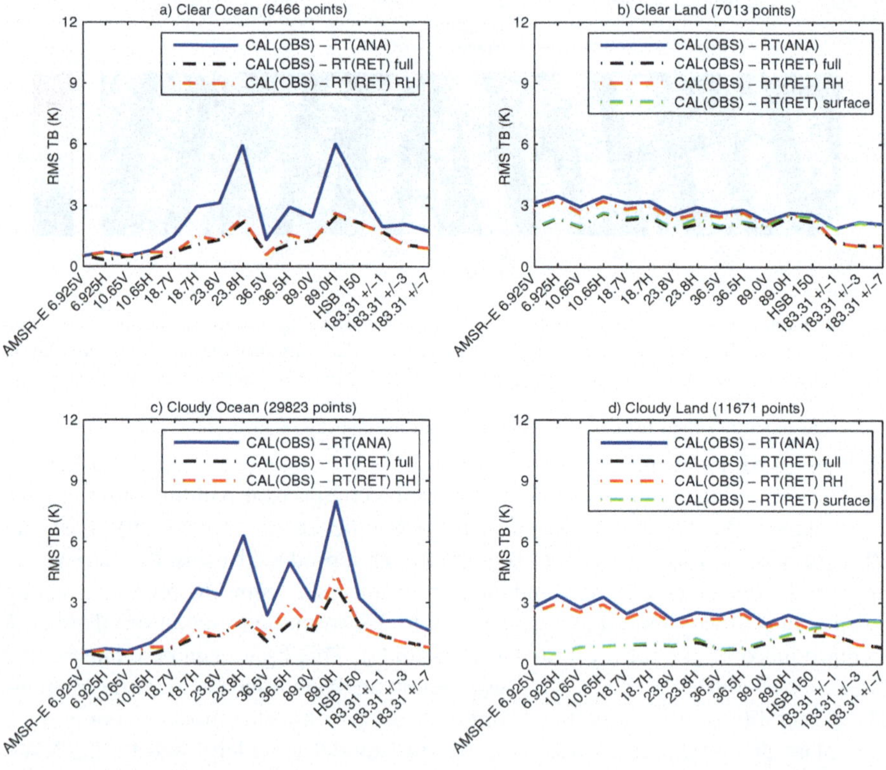

Fig. 15 RMS errors between simulated and observed T_B for AMSR-E and HSB on board AQUA, for Cloud Free/Ocean (**a**), Cloud Free/Land (**b**), Cloudy/Ocean (**c**), and Cloudy/Land (**d**) for 2 months (September 2002 and January 2003). The continuous blue line denotes the simulations using the ECMWF analysis and TELSEM surface emissivity over land, the red dashed line the same simulations introducing only the retrieval of relative humidity profile and the green one only the retrieval of surface emissivities and temperature (land only). The dashed black line denotes the simulations using all the retrieval products

but for AQUA observations (AMSR-E and HSB), for different RT inputs: (1) the ECMWF initial analysis (in blue), (2) all the retrieved variables (black), (3) only the retrieved atmospheric humidity profile (red), and (4) only the retrieved surface parameters (green). For the 183 GHz channels and for both platforms, as expected, the retrieval of the WV is the primary contributor to the reduction of the difference with the observations, the impact of the surface to the signal being very small or null. Even for the window channels, over oceanic situations, the surface retrieval has a limited impact. First, the quality of the oceanic surface parameters in the analysis is good and the departure with the observations when using the analysis is already small. Second, the surface contribution to the radiation is reduced compared to the land case, due the lower emissivities of the ocean. Over land, the retrieval of T_S and the emissivity yields to a significant improvement for all channels up to 100 GHz (with the residual RMS higher for clear cases likely due to the larger variability of T_S under clear condition, as already discussed).

7.3 Evaluation of the Improvement in the Brightness Temperature-Space

Various diagnostics can be designed to automatically monitor the quality of the retrieval: these criteria can help filter out retrievals that have not satisfactorily converged, as well as to assess the degree of confidence in the retrieval.

A measure of the improvement provided by the retrieval with respect to the analysis is given by:

$$|RT(ana) - CAL(obs)| - |RT(ret) - CAL(obs)|$$

where RT(ana) is the RT simulations for the analysis, RT(ret) for the retrieval, and CAL(obs) the calibrated observations. When this quantity is positive, the retrieval reduced the difference with observations, as compared to the analysis.

The histogram of this quantity at 150 and 183.31±3 GHz from HSB (Fig. 16) is biased towards the left, indicating that the retrieval improves the RT simulation as compared to the calibrated observations in more than 50% of the cases. This is true for the four configurations (over land or ocean, and for clear or cloudy scenes), with an improvement up to 10 K. For the 183.31±3 GHz, the distributions are identical for the four configurations, which means that the type of improvement our retrieval brings is similar for cloudy/clear cases, and for land/ocean surfaces, for the relatively page channels (but less for the less page 150 GHz channel). On the contrary, the distributions vary for the 150 GHz channel: they are more symmetrical and slightly biased to the left for the land cases, and more skewed to the left with longer tails for the ocean cases. This means that the improvements can be higher for the oceanic case (they can reach 10 K) and less important for the land case (i.e., up to 5 K). It is important to note that in our convention, the negative cases are improvements. However, positive cases don't have to be necessarily a degradation of the analysis:

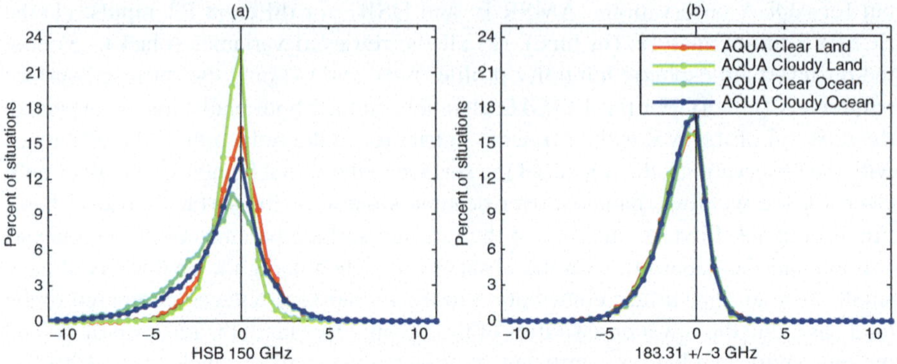

Fig. 16 Normalized distribution of differences between the T_B errors from simulations, using ECMWF analysis and using AQUA retrieval products, as compared to observations, for 150 GHz (**a**) and 183.81±3 GHz (**b**). The statistics were performed for Cloud Free/Land (red), and Cloudy/Land (green), Cloud Free/Ocean (black), Cloudy/Ocean (blue) on 2 months (September 2002 and January 2003). Negative value corresponds to situations where the simulations using the retrieval products are the closest to the actual observations

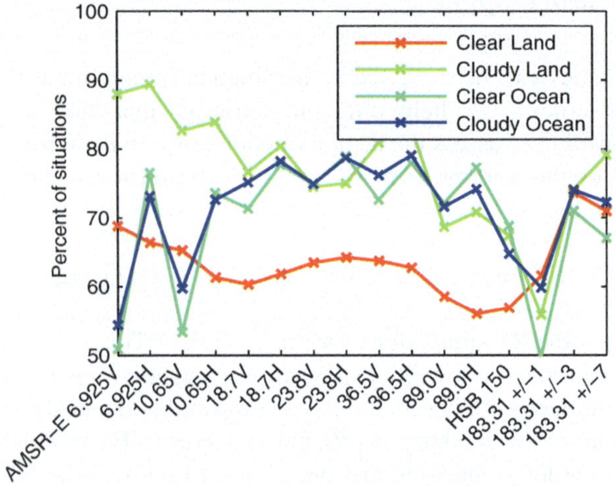

Fig. 17 Proportion of situations improved by the AQUA retrieval versus the ECMWF analysis and TELSEM surface emissivities, considering their departures from the observations in the T_B-space. The statistics were performed for Cloud Free/Land (red), and Cloudy/Land (green), Cloud Free/Ocean (cyan), Cloudy/Ocean (blue) on 2 months (September 2002 and January 2003)

Traditionally, in operational centres, this kind of diagnostics is used to test if the retrieval improves or not the analysis and in the case it degrades it, the retrieval is simply discarded (only a small percentage of the satellite data is actually used in the assimilation system).

The previous figure provides information on the amplitude of the potential improvements of our retrievals. Figure 17 presents the percentage of situations that are improved when the retrieval is used instead of the analysis, for all AMSR-E/HSB channels on board AQUA. Regardless of the situation (land/ocean and clear/cloudy) or the channel, more than 50% of the situations are improved. The percentage of improvements is the highest for window channels over land under cloudy conditions. Note that the percentages of improved situations are of the same order for land and ocean cases, confirming that our retrieval, in most cases, provides similar performances over land and ocean. The higher improvement rates in the 183 ± 3 and ± 7 GHz channels illustrate the ability of our algorithm to retrieve the lower part of the tropospheric humidity, regardless of the cloud cover. This is key information that cannot be obtained from infrared sensors.

The statistics that have been presented showed that the retrievals improve the analysis in a majority of cases, with an impact that is rather significant in T_B-space (few Kelvins). This proves that an increased amount of microwave observations has the potential to be assimilated or used in the retrieval process, in order to improve weather forecast, in particular over the continents.

8 Conclusion

An algorithm has been developed for MW instruments to retrieve water vapour atmospheric profiles over both ocean and land, for clear and cloudy conditions (precipitation cases are not treated). The retrieval is based on a neural network inversion. The algorithm has been trained to the retrieval of water vapour profiles from the microwave instruments on board AQUA, MetOp and Megha-Tropiques. The retrieval chain includes a calibration scheme, as well as a tool to obtain a first guess of the land surface emissivity. The retrieval shows satisfactory theoretical performances over both ocean (STD $= 1.7\,\mathrm{kg\,m^{-2}}$ for clear and cloudy cases) and land (STD $= 2.2$ and $3\,\mathrm{kg\,m^{-2}}$ for the clear and cloudy cases), even for the description of the lower atmosphere. This is very encouraging, given the traditional difficulty to use surface-sensitive channels in assimilation procedures, for the retrieval of the lower atmospheric layers. The uncertainty estimates have been assessed using simulated data and the traditional information content analysis.

The retrieval algorithm performances are tested on observations from AQUA and MetOp platforms. The WV profile and total column content are compared to the ECMWF analysis, as well as to radiosonde measurements. The estimated errors in the total column water vapour content are comparable over land and ocean, and under clear and cloudy cases, with a STD error of $\sim 4.5\,\mathrm{kg\,m^{-2}}$ with respect to the radiosondes that include their own uncertainties. The water vapour profiles are retrieved on six atmospheric layers with a maximum RMS error of 20% in relative humidity in the lower layer over land, without any *a priori* use of water vapour information. The retrieved water vapour fields do not show any obvious discontinuity related to ocean/land or to cloudy/clear transitions. The microwave retrievals are a very promising complement to the infrared estimates, especially when sounding and window channels are used together. Microwave observations are particularly interesting for sounding in the presence of clouds.

The retrieved products have also been evaluated in terms of simulated brightness temperatures. After the inversion process, the retrieved parameters are used as inputs to the RT code and the simulations are compared to the observations. These differences provide a direct test of the performance of the retrieval, and can be used to filter out unsuccessful retrievals. The statistical assessment shows that the retrievals are an improvement compared to the analysis in a majority of cases, and that the improvement in the T_B-space is significant (a few Kelvins). This improvement in the T_B-space is the result of an improvement of the geophysical variables (e.g., humidity profile). However, validation using *in situ* measurements is also necessary for a direct validation of the humidity retrievals.

An important perspective for this work is the combined use of microwave and infrared observations (such as from IASI) for the water vapour retrieval scheme. A good synergy exists between these two types of observations [1, 8] and the neural network approach is a good candidate to exploit it. While this study excluded precipitating cases, and expansion of the retrieval to all sky conditions will be important for water cycle studies in the future.

Acknowledgements We would like to thank the French spatial agency CNES for supporting some of this work during the preparation phase of the Megha-Tropiques mission. We would like also to thank all the scientific and technical team of the Megha-Tropiques mission. This work has been performed over several years with several collaborators: Catherine Prigent, Frédéric Bernardo, Francis Marquisseau, or Hélène Brogniez among others.

References

1. Aires, F. 2011. Measure and exploitation of multi-sensor and multi-wavelength synergy for remote sensing: Part I - Theoretical considerations. *Journal of Geophysical Research* 116: D02301. https://doi.org/10.1029/2010JD014701.
2. Aires, F., and C. Prigent. 2007. Sampling techniques in high-dimensional spaces for satellite remote sensing databases generation. *Journal of Geophysical Research* 112: D20301. https://doi.org/10.1029/2007JD008391.
3. Aires, F., C. Prigent, W.B. Rossow, and M. Rothstein. 2001. A new neural network approach including first-guess for retrieval of atmospheric water vapor, cloud liquid water path, surface temperature and emissivities over land from satellite microwave observations. *Journal of Geophysical Research* 106 (D14): 14887–14907.
4. Aires, F., A. Chedin, N. Scott, and W.B. Rossow. 2002. A regularized neural network approach for retrieval of atmospheric and surface temperatures with the IASI instrument. *Journal of Applied Meteorology* 41 (2): 144–159.
5. Aires, F., F. Bernardo, H. Brogniez, and C. Prigent. 2010. An innovative calibration method for the inversion of satellite observations. *Journal of Applied Meteorology and Climatology* 49 (12): 2458–2473.
6. Aires, F., C. Prigent, F. Bernardo, C. Jiménez, R. Saunders, and P. Brunel. 2011. A tool to estimate land surface emissivities at microwaves (TELSEM) frequencies for use in numerical weather prediction. *Quarterly Journal of the Royal Meteorological Society* 137: 690–699.
7. Aires, F., F. Marquisseau, C. Prigent, and G. Sèze. 2011. A land and ocean microwave cloud classification algorithm derived from AMSU-A and -B, trained using MSG-SEVIRI infrared and visible observations. *Monthly Weather Review* 139 (8): 2347–2366. https://doi.org/10.1175/MWR-D-10-05012.1.

8. Aires, F., M. Paul, C. Prigent, B. Rommen, and M. Bouvet. 2011. Measure and exploitation of multi-sensor and multi-wavelength synergy for remote sensing: Part II - An application for the retrieval of atmospheric temperature and water vapour from METOP. *Journal of Geophysical Research* 116: D02303. https://doi.org/10.1029/2010JD014702.
9. Aires, F., O. Aznay, C. Prigent, M. Paul, and F. Bernardo. 2012. Synergistic multi-wavelength remote sensing versus a posteriori combination of retrieved products: application for the retrieval of atmospheric profiles using MetOp-A. *Journal of Geophysical Research* 117 (D18). https://doi.org/10.1029/2011JD017188.
10. Alishouse, J., S. Snyder, V. Jennifer, and R. Ferraro. 1990. Determination of oceanic total precipitable water from the SSM/I. *IEEE Transactions on Geoscience and Remote Sensing* 28: 811–816.
11. Bony, S., B. Stevens, D.M.W. Frierson, C. Jakob, M. Kageyama, R. Pincus, T.G. Shepherd, S.C. Sherwood, A.P. Siebesma, A.H. Sobel, M. Watanabe, and M.J. Webb. 2015. Clouds, circulation and climate sensitivity. *Nature Geoscience* 8: 261–268.
12. Boukabara, S.-A., et al. 2011. MiRS: An all-weather 1DVAR satellite data assimilation and retrieval system. *IEEE Transactions on Geoscience and Remote Sensing* 49 (9): 3249–3272.
13. Collard, A.D. 2007. Selection of IASI channels for use in numerical weather prediction. *Quarterly Journal of the Royal Meteorological Society* 133 (629): 1977–1991.
14. Deblonde, G., and S.J. English. 2001. Evaluation of the FASTEM-2 fast microwave oceanic surface emissivity model. In *Technical Proceedings of ITSC-XI*, Budapest, 20–26 Sept 2000, 67–78.
15. Deblonde, G., and N. Wagneur. 1997. Evaluation of global numerical weather prediction analyses and forecasts using DMSP special sensor microwave imager retrievals. 1. Satellite retrieval algorithm inter-comparison study. *Journal of Geophysical Research* 102: 1833–1850.
16. Deblonde G., W. Yu, L. Garland, and A.P. Dastoor. 1997. Evaluation of global numerical weather prediction analyses and forecasts using DMSP special sensor microwave imager retrievals, 2, analyses/forecasts inter-comparison with SSM/I retrievals. *Journal of Geophysical Research* 102: 1851–1866.
17. Deeter, M.N. 2007. A new satellite retrieval method for precipitable water vapor over land and ocean. *Geophysical Research Letters* 34 (L02815). https://doi.org/10.1029/2006GL028019.
18. Derrien, M., and H. Le Gléau. 2005. MSG/SEVIRI cloud mask and type from SAFNWC. *International Journal of Remote Sensing* 26: 4707–4732.
19. Derrien, M., and H. Le Gléau. 2010. Improvement of cloud detection near sunrise and sunset by temporal-differencing and region-growing techniques with real-time SEVIRI. *International Journal of Remote Sensing* 31 (7): 1765–1780.
20. Diedrich, H., R. Preusker, R. Lindstrot, and J. Fischer. 2015. Retrieval of daytime total columnar water vapour from MODIS measurements over land surfaces. *Atmospheric Measurement Techniques* 8: 823–836. https://doi.org/10.5194/amt-8-823-2015.
21. Eyre, J.R. 1991. A fast radiative transfer model for satellite sounding systems. *ECMWF Research Department Technical Memorandum*, 176 (Available from the librarian at ECMWF).
22. Ferraro, R., F. Weng, N. Grody, and A. Basist. 1996. An eight-year (1987–1994) time series of rainfall, clouds, water vapor, snow cover, and sea ice derived from SSMI/I measurements. *Bulletin of the American Meteorological Society* 77: 891–905.
23. Gentemann, C.L., F.J. Wentz, C.A. Mears, and D.K. Smith. 2004. In situ validation of tropical rainfall measuring mission microwave sea surface temperatures. *Journal of Geophysical Research* 109: C04021. https://doi.org/10.1029/2003JC002092.
24. Grody, N., J. Zhao, R. Ferraro, F.H. Weng, and R. Boers. 2001. Determination of precipitable water and cloud liquid water over oceans from the NOAA 15 advanced microwave sounding unit. *Journal of Geophysical Research* 106: 2943–2953. https://doi.org/10.1029/2000JD900616.
25. Hong, F., and G. Heygster. 2005. Detection of tropical deep convective clouds from AMSU-B water vapor channels measurements. *Journal of Geophysical Research* 110 (D05205). https://doi.org/10.1029/2004JD004949.

26. Hornik, K., M. Stinchcombe, and H. White. 1989. Multilayer feedforward networks are universal approximators. *Neural Networks* 2: 359–366.
27. Jackson, D.L., and G.L. Stephens. 1995. A study of SSM/I-derived columnar water vapor over the global oceans. *Journal of Climate* 8: 2025–2038. https://doi.org/10.1175/1520-0442(1995)008<2025:ASOSDC>2.0.CO;2.
28. Jethva, H., and J. Srinivasan. 2004. Role of variation in vertical profiles of relative humidity from AMSU-B data. *Geophysical Research Letters* 31 (23). https://doi.org/10.1029/2004GL021098.
29. Kahn, B.H., J. Teixeira, E.J. Fetzer, A. Gettelman, S.M. Hristova-Veleva, X. Huang, A.K. Kochanski, M. Köhler, S.K. Krueger, R. Wood, and M. Zhao. 2011. Temperature and water vapor variance scaling in global models: comparisons to satellite and aircraft data. *Journal of the Atmospheric Sciences* 68 (9): 2156–2168. https://doi.org/10.1175/2011JAS3737.1.
30. Karbou, F., F. Aires, C. Prigent, and L. Eymard. 2005. Potential of AMSU-A and -B measurements for atmospheric temperature and humidity profiling over land. *Journal of Geophysical Research* 110 (D97109). https://doi.org/10.1029/2004JD005318.
31. Kawanishi, T., T. Sezai, Y. Ito, K. Imaoka, T. Takeshima, Y. Ishido, A. Shibata, M. Miura, H. Inahata, and R.W. Spencer. 2003. The advanced microwave scanning radiometer for the Earth Observing System (AMSR-E), NASDA's contribution to the EOS for global energy and water cycle studies. *IEEE Transactions on Geoscience and Remote Sensing* 41: 184–194.
32. Lambrigtsen, B.H., and R.V. Calheiros. 2002. The humidity sounder for Brazil - an international partnership. *IEEE Transactions on Geoscience and Remote Sensing* 41: 352–361.
33. Lima, W.F.A., and L.A.T. Machado. 2006. Analise do sensor HSB na estimativa do conteúdo integrado de vapor d'àgua durante o experimento RaCCI/LBA. *Revista Brasileira de Meteorologia* 21 (2): 1–8.
34. Lindstrot, R., M. Stengel, M. Schröder, J. Fischer, R. Preusker, N. Schneider, T. Steenbergen, and B.R. Bojkov. 2014. A global climatology of total columnar water vapour from SSM/I and MERIS. *Earth System Science Data* 6: 221–233. https://doi.org/10.5194/essd-6-221-2014.
35. Matricardi, M., F. Chevallier, and S. Tjemkes. 2001. An improved general fast radiative transfer model for the assimilation of radiance observations. *ECMWF Research Department Technical Memorandum*, 345. http://www.ecmwf.int/publications.
36. Mears, C.A., J. Wang, D. Smith, and F.J. Wentz. 2015. Inter-comparison of total precipitable water measurements made by satellite-borne microwave radiometers and ground-based GPS. *Journal of Geophysical Research* 120: 2492–2504. https://doi.org/10.1002/2014JD022694.
37. Prigent C., W.B. Rossow, and E. Matthews. 1997. Microwave land surface emissivities estimated from SSM/I observations. *Journal of Geophysical Research* 102: 21867–21890.
38. Prigent, C., F. Aires, and W.B. Rossow. 2006. Land surface microwave emissivities over the globe for a decade. *Bulletin of the American Meteorological Society* 1572–1584. https://doi.org/10.1175/BAMS-87-11-1573.
39. Prigent, C., E. Jaumouille, F. Chevallier, and F. Aires. 2008. A parameterization of the microwave land surface emissivity between 19 and 100 GHz, anchored to satellite-derived estimates. *IEEE Transactions on Geoscience and Remote Sensing* 46: 344–352.
40. Prigent, C., F. Aires, D. Wang, S. Fox, and S. Harlow. 2017. Sea-surface emissivity parameterization from microwaves to millimeter waves. *Quarterly Journal of the Royal Meteorological Society* 143 (702): 596–605. https://doi.org/10.1002/qj.2953.
41. Rodgers, C.D. 1976. Retrieval of atmospheric temperature and composition from remote measurements of thermal radiation. *Reviews of Geophysics* 14 (4): 609–624.
42. Rodgers, C.D. 1990. Characterization and error analysis of profiles retrieved from remote sounding measurements. *Journal of Geophysical Research* 95 (D5): 5587–5595.
43. Roman, J., R. Knuteson, T. August, T. Hultberg, S. Ackerman, and H. Revercomb. 2016. A global assessment of NASA AIRS v6 and EUMETSAT IASI v6 precipitable water vapor using ground-based GPS Suomi Net stations. *Journal of Geophysical Research* 121: 8925–8948. https://doi.org/10.1002/2016JD024806.

44. Rumelhart, D.E., G.E. Hinton, and R.J. Williams. 1986. Learning internal representations by error propagation. In *Parallel Distributed Processing: Explorations in the Microstructure of Cognition*, ed. D.E. Rumelhart, J.L. McClelland, and the PDP Research Group, Vol. I, Foundations, 318–362. Cambridge: MIT Press.
45. Saunders, R.W., M. Matricardi, and P. Brunel. 1999. An improved fast radiative transfer model for assimilation of satellite radiance observations. *Quarterly Journal of the Royal Meteorological Society* 125: 1407–1425.
46. Schlüssel, P., and M. Goldberg. 2002. Retrivela of atmospheric temperature and water vapour from IASI measurements in partly cloudy situations. *Advances in Space Research* 29 (11): 1703–1706.
47. Schröder, L., A. Richter, D.V. Fedorov, L. Eberlein, E.V. Brovkov, S.V. Popov, C. Knöfel, M. Horwath, R. Dietrich, A.Y. Matveev, M. Scheinert, and V.V. Lukin. 2017. Validation of satellite altimetry by kinematic GNSS in central East Antarctica. *The Cryosphere* 11: 1111–1130. https://doi.org/10.5194/tc-11-1111-2017.
48. Steinke, S., S. Eikenberg, U. Löhnert, G. Dick, D. Klocke, P. Di Girolamo, and S. Crewell. 2015. Assessment of small-scale integrated water vapour variability during HOPE. *Atmospheric Chemistry and Physics* 15: 2675–2692. https://doi.org/10.5194/acp-15-2675-2015.
49. Susskind, J., C.D. Barnet, and J.M. Blaisdell. 2003. Retrieval of atmospheric and surface parameters from AIRS/AMSU/HSB data in the presence of clouds. *IEEE Transactions on Geoscience and Remote Sensing* 41 (2): 390–409.
50. Tjemkes, S.A., G.L. Stephens, and D.L. Jackson. 1991. Spaceborne observation of columnar water vapor: SSM/I observations and algorithm. *Journal of Geophysical Research* 96 (10): 10941–10954. https://doi.org/10.1029/91JD00272.
51. Uppala, S.M., P.W. Kallberg, A.J. Simmons, U. Andrae, V. da Costa Bechtold, M. Fiorino, J.K. Gibson, J. Haseler, A. Hernandez, G.A. Kelly, X. Li, K. Onogi, S. Saarinen, N. Sokka, R.P. Allan, E., Andersson, K. Arpe, M.A. Balmaseda, A.C.M. Beljaars, L. van de Berg, J. Bidlot, N. Bormann, S. Caires, F. Chevallier, A. Dethof, M. Dragosavac, M. Fisher, M. Fuentes, S. Hagemann, E. Holm, B.J. Hoskins, L. Isaksen, P.A.E.M. Janssen, R. Jenne, A.P. McNally, J.-F. Mahfouf, J.-J. Morcrette, N.A. Rayner, R.W. Saunders, P. Simon, A. Sterl, K.E. Trenberth, A. Untch, D. Vasiljevic, P. Viterbo, J. Woollen. 2005. The ERA-40 re-analysis. *Quarterly Journal of the Royal Meteorological Society* 131: 2961–3012.
52. Wang, Y., Y. Fu, G. Liu, Q. Liu, and L. Sun. 2009. A new water vapor algorithm for TRMM microwave imager (TMI) measurements based on a log linear relationship. *Journal of Geophysical Research* 114 (D21304). https://doi.org/10.1029/2008JD011057.
53. Wang, J., A. Dai, and C. Mears. 2016. Global water vapor trend from 1988 to 2011 and its diurnal asymmetry based on GPS, radiosonde, and microwave satellite measurements. *Journal of Climate*. https://doi.org/10.1175/JCLI-D-15-0485.1.
54. Wang, D., C. Prigent, L. Killic, S. Fox, C. Jimenez, F. Aires, C. Grassoti, and F. Karbou. 2017. Surface emissivity at microwaves to millimeter waves over polar regions: parameterization and evaluation with aircraft experiments. *Journal of Atmospheric and Oceanic Technology* 34 (5). https://doi.org/10.1175/JTECH-D-16-0188.1.
55. Weng, F., L. Zhao, R. Ferraro, G. Poe, X. Li, and N. Grody. 2003. Advanced microwave sounding unit cloud and precipitation algorithms. *Radio Science* 38: 8086–8096.
56. Wentz, F.J. 2017. A 17-Yr climate record of environmental parameters derived from the tropical rainfall measuring mission (TRMM) microwave imager. *Journal of Climate*. https://doi.org/10.1175/JCLI-D-15-0155.1.

Advances in Clouds and Precipitation Modeling Supported by Remote Sensing Measurements

Takamichi Iguchi and Toshihisa Matsui

1 Introduction

Future prediction is the ultimate goal of running weather and climate models. Good predictions are necessary to satisfy various social needs, not limited to projecting future climate change. Such needs comprise alleviating damages from natural disasters, and advising on agricultural activities. An important concern for short- and long-term predictions is the prediction reliability. Eventually, the correct answer can be obtained by measurements in the future. However, it is then too late to modify errors in the prediction. Therefore, it is essential to prepare the prediction of the future as precise as possible before its implementation. The accumulation of measurement data provides an excellent testbed for those models, through simulations for retrospective periods. So-called hindcast or backtesting is introduced to examine better configurations in the models and analyze how accurate they are. Comparison with measurements is key to validating simulation results and determine the sources of errors and uncertainty.

Surface precipitation is a primary concern of atmospheric models as well as measurements because it is closely related to human activities on the land surface. Liquid water particles associated with precipitation are generated and grown in the atmosphere and subsequently pulled toward the Earth's surface by gravitational force. The coalescence/aggregation of the pre-matured particles can be observed as a cloud. The growth characteristics of cloud particles have a critical role in determining the features of surface precipitation including its location, duration,

T. Iguchi (✉) · T. Matsui
Mesoscale Atmospheric Processes Laboratory, NASA Goddard Space Flight Center, Greenbelt, MD, USA

Earth System Science Interdisciplinary Center, University of Maryland, College Park, MD, USA
e-mail: takamichi.iguchi@nasa.gov; Toshihisa.Matsui-1@nasa.gov

C. Andronache (ed.), *Remote Sensing of Clouds and Precipitation*,
Springer Remote Sensing/Photogrammetry,
https://doi.org/10.1007/978-3-319-72583-3_10

accumulation, and particle phase and types. Better representation of clouds in atmospheric models may lead to better simulation of surface precipitation, latent heating/cooling, and cloud radiative forcing.

Methods to measure clouds are limited compared with those for surface precipitation over land. Ground-based in-situ measurements still can efficiently work by directly observing low-level cloud at high mountains (e.g., [3]) or buildings (e.g., [52]). Aircraft measurements can provide in-site observations even in the high-level clouds (e.g., [14]), although its operation is significantly restricted due to high costs and safety concerns. Balloon-sonde measurements have a reasonable capability to capture the features (e.g., [36, 51]), despite difficulty to place the sonde instruments in the target cloud or precipitation. In addition to these in-situ measurements, various remote sensing measurements are now commonly used. In particular, the developments of the Earth-observing satellites were milestones to substantially extend the measurement capability and the coverage even over ocean far from land.

Productive use of the remote sensing measurement data is essential to further development and improvement of the atmospheric models through hindcast simulations. It is challenging to handle remote sensing measurement data appropriately to evaluate atmospheric simulation results. In general, a reasonable comparison is based on a criterion that both observation and simulation submit the same physical quantities to be compared to each other. In the case of comparison with in-situ measurements, the observed physical quantities are related directly to mass, volume, size, velocity, energy, etc. of the objects. In contrast, remote sensing measurements observe objects indirectly through measuring radiation emitted or scattered by them. Thus, there are overall two options to unify physical quantities calculated from the observations and simulations. The first option is translating primitive signals detected in the measurements into other physical quantities related directly to the composition and structure of the objects. So-called retrieval algorithms need to be developed to estimate such physical quantities through solving inversion problems. The second option is computing measurement-observable signals from the output of atmospheric models through forward radiative transfer models, and subsequently comparing them with the measured direct signals. There are multiple options regarding how deeply the measured direct signals and the model simulation output are processed to unify the quantities to evaluate the atmospheric model simulations.

Which option yields more suitable comparison between the observation and the simulation may be entirely dependent on the specification of numerical models, type of sensor and platform for measurements, availability of other measurement data, target phenomena, etc. Because combinations of measurements and atmospheric models are not limited, merits and demerits of an option are unsettled. Researchers who use such comparisons should pay attention to what option is better and what are its pros and cons [34].

This chapter reviews how to represent clouds and precipitation microphysics in atmospheric models and introduces examples of recent studies based on the comparison between remote sensing measurements and simulations for cloud and precipitation. This chapter is structures in the following way. A short review about clouds and precipitation modeling and parameterization is presented in Sect. 2. Conventional retrieval-based and relatively new signal-based comparisons between measurements and simulations are discussed in Sect. 3. Summary and conclusions are given in Sect. 4.

2 Clouds and Precipitation Parameterizations in Atmospheric Models

2.1 Modeling of Clouds and Precipitation in the Atmosphere

A prototype of clouds and precipitation microphysics parameterization was developed by Kessler ([21]; also [22]). Water condensates in the atmosphere are modeled using two separate categories: clouds and precipitation. The gravitational sedimentation of the cloud type is neglected, whereas the precipitation category has a specific falling velocity so that precipitation can fall toward the surface. Cloud microphysical processes such as auto-conversion from cloud to precipitation through accretion are represented by transfer of mass among the two categories and water vapor. A differential form of the continuity equation for mass densities of the cloud and precipitation categories are presented in Kessler [22, eqs. 8a and 8b]. These equations are numerically solved to calculate the changes of the mass densities.

The design of modeling and parameterization for cloud and precipitation in [21, 22] are primitive compared to those in the mainstream of the present atmospheric models, particularly on the cloud- and eddy-resolving scales. However, the fundamental concept and design are still shared by the most modern models. Atmospheric hydrometeor particles are classified into multiple categories according to the differences of particle characteristics such as densities, phase, and growth processes. Then, the continuity equation for mass density or mixing ratio of each category is numerically solved. Introduction of mass-weighted mean terminal velocity (e.g., [48]) enabled a hydrometeor category to contain particles with different sizes and consequently different velocities on the assumption that the particle size distribution (PSD) is represented by a function of the size. The transfer of mass among hydrometeor categories associated with cloud microphysical processes is generally completed in each spatial-grid point at every timestep. This assumption makes the model programming easier and more suitable to the acceleration of the numerical computation through parallelization.

The categorization of atmospheric hydrometeors can be expanded only from the mass of the categorized particles to their number and also other higher-order moments. On the other hand, a special style of modeling classifies particles with the same particle characteristics but with different sizes through discretization according to its particle size. The combination of categories through various categorizations is unlimited. Increase in the number of the categories degrades the computational speed because more continuity equations are required to be solved. Therefore, the developers need to decide what categories should be chosen from the endless combination to represent clouds and precipitation most efficiently. To date, a significant number of categorization styles has been suggested and developed in the history of modeling clouds and precipitation.

2.2 *Typical Styles of Cloud Microphysics Parameterization*

So-called bulk cloud microphysics parameterizations form the mainstream in recent atmospheric models. Atmospheric hydrometeors are generally classified into two categories of the liquid phase, i.e., cloud and rain droplets, and several categories of the solid phase including cloud ice, snow, graupel, and/or hail. The bulk microphysics parameterizations are further divided into several styles according to the number of moments whose continuity equations are solved: Schemes aiming only at one moment are called one- or single-moment bulk microphysics, which predicts mass concentration only. Two- or double-moment schemes generally consider mass and number of hydrometeor particles.

An example of formulation in the bulk microphysics parameterization can be found in [45, 46]. The k-th moment of a hydrometeor category is formulated as

$$M^k = \int_{x_1}^{x_2} x^k f(x)dx \tag{1}$$

where x is the particle mass, $f(x)$ is the number density size distribution function (number per unit air volume per unit particle size interval); x_1 and x_2 denote the lower and upper limits of the size distribution of the category, respectively. The zeroth moment corresponds to the number density of the categorized hydrometeor and the first moment is the mass density.

The bulk microphysics schemes usually apply special mathematical functions with a few intercept parameters to the PSD. As a result, the moments can be represented in forms suitable for differentiation and integration. This approach is effective to deal with closure problems that possibly appear in the formulation of cloud microphysical processes for the change of the PSD. For example, [45] shows that formulation of particle growth through coagulation and accretion may include a higher-order moment in the equation for the change of a moment.

The selection of the special mathematical functions depends on the design of the schemes. Conventionally, subsets in generalized gamma distribution have been preferred in formulating the hydrometeor PSD. A typical gamma distribution for the number density size distribution is determined by three parameters (e.g., [53]):

$$N(D) = N_0 D^{\mu} \exp(-\lambda D) \tag{2}$$

where D is the particle diameter; N_0, μ, and λ are the intercept parameters of the distribution. If $\mu = 0$, the distribution is identical to the exponential distribution. Any order of moment in Eq. (1) can be represented by the three parameters and the gamma function, so that the moments can be handled easily in analytical calculations. The gamma distribution has been frequently used because this function has been fitted to observed hydrometeor PSD since Marshall and Palmer [30]. In addition, the history of research with gamma distribution fitting to hydrometeor PSD is closely tied to the development of rainfall retrieval techniques using weather radar measurements (e.g., [1]).

To date, many types of bulk microphysics parameterizations have been developed for being used in atmospheric models for various operational and research purposes. A subset of the parameterizations and their references can be found in Khain et al. [25, table 2]. There is a broad range of selectable options of bulk microphysics parameterizations in the Weather Research and Forecasting (WRF) model [47], which is a major regional atmospheric model.

In contrast to the bulk microphysics parameterizations using special mathematical functions, spectral-bin microphysics parameterizations were suggested to circumvent the closure problems by direct discretization of the hydrometeor PSD. Most spectral-bin parameterizations adopt conventional categorization to classify ice-phase hydrometeor particles, which is similar to those used in bulk microphysics parameterizations. However, some spectral-bin parameterizations apply direct discretization also to parameters determining the characteristics of ice-phase particles, e.g., ice particle bulk density, aspect ratio, etc. (e.g., [12]). Also, double-moment spectral-bin microphysics (e.g., [4]) was developed to overcome the discontinuity in the PSD discretized in conventional single-moment spectral-bin parameterizations (e.g., [23]).

A limited number of spectral-bin microphysics parameterizations have been developed, as compared to bulk microphysics. The reason for this small number is that spectral-bin microphysics requires much larger computational costs, whereas it does not yield additional intelligible values, particularly in numerical weather prediction for operational purposes. The information of simulated PSD is usually underutilized because of its high variability and little utility. Although spectral-bin microphysics directly calculates the change of discretized hydrometeor PSD by solving a large number of continuity equations, it does not guarantee better simulations of surface precipitation rates than bulk microphysics.

3 Comparison of Cloud Microphysics Parameterization Output in Atmospheric Models with Remote Sensing Measurements

3.1 Comparison with Retrieved Physical Quantities

Direct radiance or backscatter data obtained from remote sensing measurements are processed into higher-level data products via processes such as variable conversion based on retrieval techniques, spatial and temporal interpolation, and removing noise. The higher-level products are usually composed of useful geophysical parameters. As a result, measurement data is easily handled for the analysis of physical phenomena, particularly by statistical approaches.

The most common approach to comparing hindcast simulation results of atmospheric models with remote sensing measurements is to utilize such higher-level data products. The simulations calculate the same geophysical parameters from the prognostic variables as in the measurement data products. In general, either simulation or observation data in its native spatial resolution are reorganized into the spatial resolution of the other. As a result, the difference between both at the same spatial coordinates can be calculated. This process enables the calculation of statistical parameters and scores such as overall bias, correlation coefficient, root-mean-square errors, and hit rates to compare systematic errors in the simulation results with the observations. Decreasing errors through improving these statistical parameters and scores is an important step towards better simulations. The performance of numerical models or configurations in the same model can be evaluated quantitatively by comparing these statistical parameters and scores. The more comparisons are sampled, the more robust conclusions are expected. This workflow is a long-established pattern in studies to improve cloud and precipitation parameterizations in atmospheric models.

Since higher-level data products are composed of geophysical parameters with high interest in research communities such as surface precipitation rates, matching the simulation results with them is the final goal as long as the data in the products is exactly accurate. Unfortunately, such higher-level products contain errors and uncertainty that are inherent not only in the remote sensing measurements but also in the processes from the calibrated radiance/backscatter data to higher-level products. Thus, the discussion about the comparison results needs close attention to the bias and the uncertainty in each data product in a strict sense. The differences in data products are often significant according to development versions and types of the retrieval algorithms, even if the platforms for measurements are identical. Perfect validation of such data products is impossible because of the limitation of direct measurements, particularly on a global scale covered by Earth-observing satellites. For example, [27] showed the difference in zonally-averaged monthly precipitation rates derived from rainfall retrieval algorithms of different versions and types for the Tropical Rainfall Measuring Mission (TRMM) core satellite measurements. At the version 4 data (initially corrected at-launch algorithm), a range of difference in the precipitation rates according to the selection of the algorithms is about 40% at most in the tropical mean estimates. The range has been narrowed to about 24% at the version 5 after the first substantial improvement. Liu [29] examined the difference between version 6 and 7 of TRMM multi-satellite precipitation analysis (TMPA) products. The version 7 showed higher frequencies of the light rain regime than the version 6 and vice versa in the heavy rain regime.

3.2 Development of Signal Simulators for Comparison with Backscattered Radiance

In the 2000s, particularly after the initial formulation of A-Train satellite constellation (e.g., [50]), the research community realized the significant uncertainties in remote sensing retrievals, including surface precipitation, cloud liquid and ice water contents, and optical depths derived from different sets of satellite instruments and platforms. The reason is that satellite retrieval algorithms must set unique microphysical assumptions to estimate geophysical parameters according to satellite instrument types and available channel frequencies. These microphysical assumptions often differ between algorithms and are also different from corresponding assumptions employed in cloud microphysics parameterization in atmospheric models. To circumvent such inconsistency, an alternative approach was suggested to directly utilize satellite-measured well-calibrated raw signals, such as visible radiance, infrared-microwave brightness temperature, radar reflectivity, and lidar backscatter. So-called satellite signal simulators have been developed to convert model-output variables into quantities that are identical to satellite-observable signals [5, 9, 13, 31, 33, 34].

For precise simulation of satellite signals in the simulators, the distributions of physical quantities describing clouds, precipitation and background environment must be identical between the satellite signal simulators and the atmospheric models. The model-output pressure, temperature, and water vapor and other gaseous species are used to represent molecular extinction and absorption of radiation. Furthermore, the model-output cloud and precipitation must exactly represent emission and scattering signals of remote sensing measurements in terms of hydrometeor particle phase, size, and density in order to represent size-integrated extinction coefficient k , single-scattering albedo $\tilde{\omega}$, and scattering phase function P for radiative transfer process at a particular wavelength of the instrument:

$$k = \frac{\pi}{4} \int D^2 Q_{ext}(D) N(D) dD \tag{3}$$

$$\tilde{\omega} = \frac{\pi}{4k} \int D^2 Q_{scat}(D) N(D) dD \tag{4}$$

$$P = \frac{\int D^2 p_{scat}(D) Q_{scat}(D) N(D) dD}{\int D^2 Q_{scat}(D) N(D) dD} \tag{5}$$

where the particle size distribution $N(D)$ must be identical to the assumption in the cloud microphysics parameterization used in atmospheric models; Q_{ext}, Q_{scat}, and p_{scat} are the extinction efficiency, scattering efficiency, and phase function for the particle of diameter D, respectively. These single-particle elements also must be estimated in consistent assumption of particle phase, density, and possibly shape in the cloud microphysics parameterization of the atmospheric models. Q_{ext}, Q_{scat}

are generally derived from laboratory experiments, whereas p_{scat} can be derived either from analytical or numerical solutions. A non-dimensional size parameter X is often introduced to calculate p_{scat}; it is represented using particle diameter D and normalized wavelength λ:

$$X = \frac{\pi D}{\lambda} \tag{6}$$

When X is negligibly small (<0.002), scatter by an element with the size, e.g., molecular, can be neglected. When X is around 0.002 to 0.2, the scattering regime is categorized in Rayleigh scattering, where its phase function $P(\theta)$ for unpolarized incident radiation is expressed as

$$P_{scat}(\theta) = \frac{3}{4}\left(1 + \cos^2\theta\right) \tag{7}$$

where θ is the scattering angle. The above equation can be applied regardless of any particle shape. A single particle can be treated as a fluffy sphere with effective density ρ_e. In this scattering regime, the dialectic constant can be mixed with different properties, such as air, water, and ice to represent different hydrometeor characteristics through analytic solution of complex medium [2].

When X is over 0.2, forward scattering becomes stronger than Rayleigh scattering, and its scattering phase function becomes more complex. These scattering regimes are categorized into Mie ($0.2 < X < 2000$) and geometric ($2000 < X$) regimes. In these regimes, Lorenz-Mie (e.g., [11]) and T-matrix solutions (e.g., [35]) are widely used as an analytical solution to estimate scattering of spherical and oblate particles. However, the shapes of most ice crystals and their aggregates are totally different from a sphere or oblate. To treat these particles, more computationally intensive solutions are required, such as the Discrete Dipole Approximation [6] and the Finite-Difference Time-Domain method [55]. These numerical solutions are often employed to accurately represent W-band radar backscatter or microwave brightness temperature from ice- or mixed-phase clouds [39].

Once scattering and absorption elements are integrated over size and species, the transfer of radiation is solved to estimate the radiance at the detector. The radiative transfer calculation is based on various mathematical and numerical techniques and assumptions depending on the scattering magnitude (e.g., [49]) and polarization (e.g., [8]) with boundary conditions, e.g., solar constant, surface skin temperature. In the case of active sensors such as radar/lidar, backscattering signals are derived at each target level. Radiance or backscatter is further convolved on the particular satellite footprint size, i.e., the field of view, through a Gaussian weighting function [33].

The consistency between the atmospheric model and the forward signal simulator in terms of the assumption of hydrometeor particle phase, PSD, and other background environment must be identical as possible. It is also important to

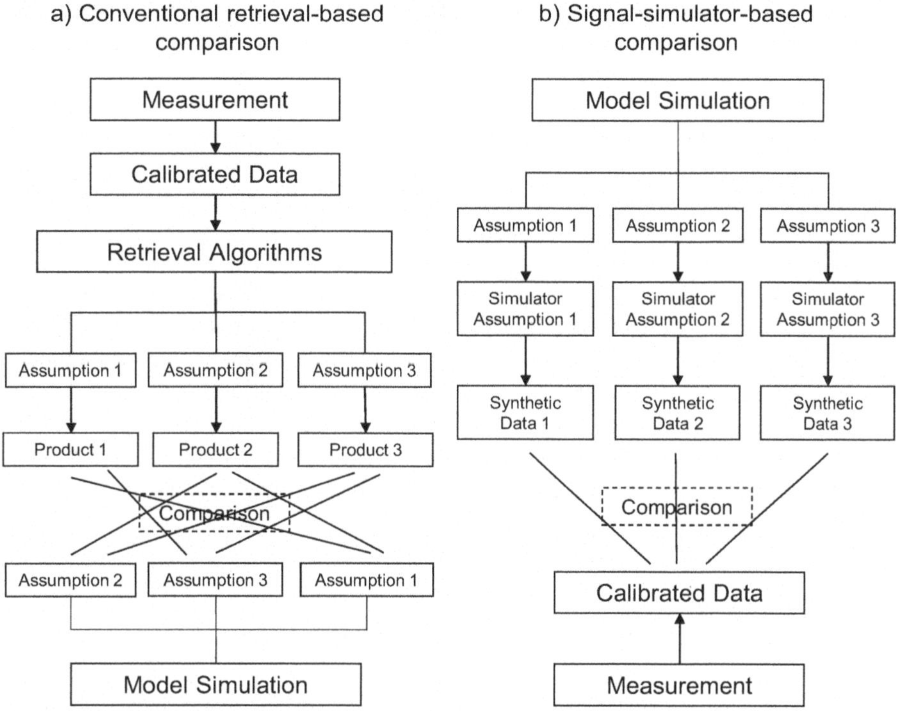

Fig. 1 Schematic chart of workflows of (**a**) retrieval-based comparison and (**b**) signal-simulator-based comparison between remote sensing measurements and model simulations for cloud and precipitation

employ appropriate forward radiative transfer and scattering models that have quality equivalent to those used in the operational retrievals.

Figure 1 shows a schematic chart illustrating the difference between conventional retrieval-based comparison and signal-simulator-based comparison. In the retrieval-based comparison, there could be multiple products for target physical quantities according to the differences in retrieval algorithms. Usually, the assumptions in the retrieval algorithm and the model parameterization are not required to be matched for comparison. Indeed, the difference in the physical quantities derived from the different retrieval algorithms could be a problem in the evaluation of the model simulation results. In contrast, the signal-simulator-based comparison follows a more straightforward workflow to obtain synthetic calibrated radiance data. The difference in the obtained synthetic data is inherent in the difference in the assumptions used in the model parameterizations because the signal simulator adopts the same assumptions as in the model parameterizations. Therefore, the error in the synthetic calibrated data against the observed data is more closely tied to the error of the model simulation coupled with the parameterization [34].

The model evaluation using a signal simulator has these advantages compared to the conventional retrieval-based comparison. However, it requires appropriate knowledge and understanding to interpret the evaluation results using the obtained radiance or backscatter, because they contain various inherent physical quantities. For example, PSD, hydrometeor particle densities, and phase are all relevant parameters to generate microwave emission and scattering signals according to the sensor channel frequencies.

Spectral-bin microphysics parameterizations could inherently outperform bulk microphysics parameterizations in simulating radiance signals through such simulators. The information of hydrometeor PSD is an essential element in the simulation of radiance signals affected by cloud and precipitation. Hydrometeor PSD obtained from model simulations with spectral-bin microphysics can be directly inputed in a signal simulator, so that the simulated signals have significant variability that may be necessary for consistency with observed signals from various PSD in nature. In contrast, the variability in the corresponding simulated signals from bulk microphysics parameterizations are strongly restricted by the assumptions of PSD.

3.3 Review of Previous Case Studies Using Signal Simulators

This sub-section provides a brief review of previous studies based on the comparison between numerical simulations and remote sensing measurements through radiance signal simulators. Since comparison upon retrieved physical quantities is more common and widely used, studies using signal simulators are introduced particularly here.

Iguchi et al. [18] analyzed hindcast simulations of mid-latitude precipitation systems through comparison with shipborne or spaceborne W-band radar measurements. The Japan Meteorological Agency NonHydrostatic Model (JMA-NHM; [44]) coupled with the spectral-bin microphysics of Hebrew University Cloud Model (HUCM; [23]) and the single-moment bulk cloud microphysics [7] was employed for mesoscale hindcast simulations with 3-km horizontal grid spacing. An offline radar-signal simulator [37, 38] was designed to convert the output of the hindcast simulations into synthetic radar reflectivity factor and Doppler velocity that were equivalent to actual shipborne or spaceborne radar measurements. The assumption for the cloud microphysics such as PSD spectra and particle fall velocity in the simulator was consistent with that in the spectral-bin or the bulk microphysics parameterization. The observed radar reflectivity factors without correction of attenuation due to atmospheric hydrometeors and gases were referred to for the comparison so that the attenuation was modeled and included in the signal simulator side to calculate corresponding reflectivity factors from the hindcast simulations.

Figure 2 shows time-height cross sections of radar reflectivity factors derived from the shipborne W-band radar measurements and calculated from the hindcast using JMA-NHM coupled with the spectral bin microphysics through the signal

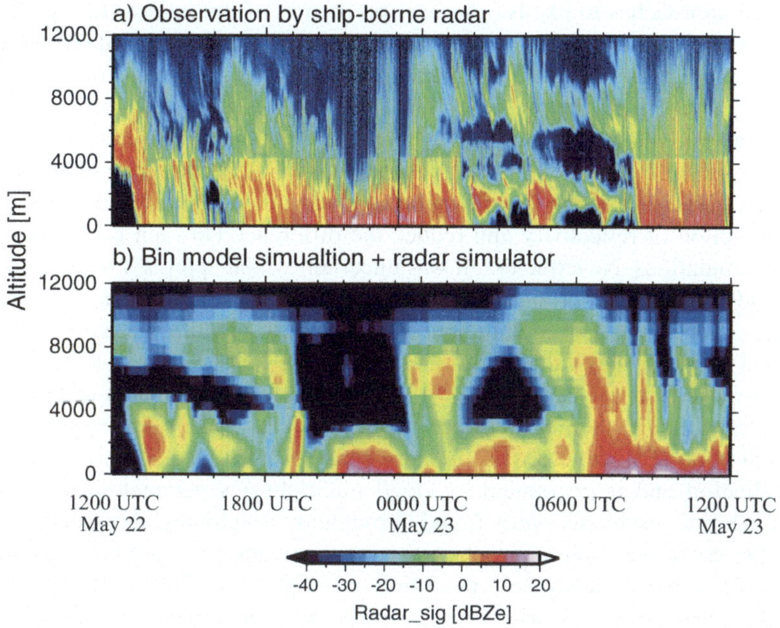

Fig. 2 Time height cross sections (from 1200 UTC 22 May to 1200 UTC 23 May 2001) of the equivalent radar reflectivity factor (dBZe) (**a**) measured by the 95-GHz Doppler radar onboard the research vessel, and (**b**) calculated by the radar-product simulator applied to the outputs from JMA-NHM coupled with the spectral-bin microphysics (from [18] with changes)

simulator. The signal simulator sampled model-output variables in the atmospheric column of the horizontal grid nearest to the coordinates of the research vessel every 1 min. This configuration aimed to make the sampling of the model output as consistent as possible with actual measurements. However, the comparison of the figure plots exhibits some disagreement of cloud occurrence between the measurements and the hindcast simulation, which is not expected to be due to problems in the cloud microphysics parameterization. Such difference is often due to forecast errors under the control of forcing dataset of a deterministic run for the hindcast through the initial and lateral boundary conditions, rather than model physics parameterizations. These errors could obstruct the discussion on the problems in the cloud microphysics parameterization towards its improvement. This problem is common to studies upon comparison with measurements for validation of model physics parameterizations not limited to those using a signal simulator.

The study introduced quantitative discussion such as analysis on the normalized contoured frequency of altitude diagrams (CFADs) to circumvent the effects of such forecast errors on the discussion. The results suggested that the simulation over predicted radar reflectivity factors of ice- and mixed-phase clouds above the freezing level heights. This conclusion was consistent with those from the study using almost the same spectral-bin microphysics parameterization in [28]. However, suggested approaches to resolve the over prediction of the reflectivity factor were different between these publications. Iguchi et al. [18] tested increased fall velocities of snow aggregates and different parameterization for ice nucleation processes from the control run. On the other hand, Li et al. [28] tested changes in bulk density and fall velocity of snow aggregates and functions of ice-particle collection efficiency on temperature and also the introduction of the breakup of large snow aggregates. These changes in both publications worked similarly to diminish the over prediction of reflectivity and reduce the difference between the observations and the simulations. Nevertheless, it was uncertain which approach in these options was valid to solve the problem in the model parameterization. The comparison with the observations lacked sufficient information to identify the error sources in the cloud microphysics calculation. Eventually, the latest version of the spectral-bin microphysics in HUCM (e.g., [24]) introduced breakup process of large snow aggregates, although there is no direct in-situ evidence to justify the update.

Iguchi et al. [18] showed a preliminary study using such a signal simulator for the evaluation and improvement of cloud microphysics parameterization. Many limitations and problems were found throughout developing the discussion. In particular, the spectral-bin microphysics parameterization had a greater number of equations and parameters than typical bulk microphysics. Thus, many parameters could be changed in sensitivity tests to check improvement of the simulated results. The authors, in fact, conducted over 30 types of sensitivity tests, e.g., with modifications of deposition growth speed of ice hydrometeors, snow aggregation efficiency, ice nucleation parameterization, fall speed of hydrometeor particles, etc., and with the composite of these components. Such an approach based on an almost random guess takes unnecessary costs in the simulation and the analysis. Although the publication concluded that increasing the fall velocity of snow aggregates was an effective solution, the conclusion was not based on convincing scientific evidence obtained from direct comparison with in-situ observations in the case study. This problem was a weak point in the scientific discussion, and it could be common to similar studies toward the improvement of model physics parameterization upon the comparison with remote sensing measurements. The research community will need to establish robust and reasonable methods to discuss the probability and likelihood of errors from each component in cloud microphysics parameterizations, beyond conventional approaches just based on sensitivity tests to identify the error sources.

Iguchi et al. [19] attempted validation of a time-dependent melting scheme of ice particles [40] introduced into the HUCM spectral-bin microphysics (e.g., [24]). Melting of ice hydrometeors falling in the atmosphere can cause a bright band, which is identified as a thin horizontally-oriented layer with relatively larger reflectivity below 0 °C temperature level in a radar echo profile. A better understanding of ice particles melting and consequent blight-band formation is a fundamental element in accurate estimation of precipitation rates from radar measurements. Accurate simulation of the bright band structure could be an important subject to promote the understanding of the microphysics process and its interaction with the cloud dynamics. However, most cloud microphysics parameterizations, in typical atmospheric models, require additional arbitrary assumptions in simulating a bright band structure because of simplified modeling of ice particles melting.

Two precipitation events in the fall of 2010 over the southern part of Finland during the field campaign of the Light Precipitation Validation Experiment (LPVEx) were analyzed in [19]. The HUCM spectral-bin microphysics with a time-dependent ice melting scheme was introduced into the Advanced Research WRF (WRF-ARW) model version 3.4 [47]. Two types of WRF hindcast simulation were conducted with or without the time-dependent melting scheme to highlight the effects of ice particles melting on the radar reflectivity and Doppler velocity profiles containing bright band structure. An offline radar simulator module in the Goddard Satellite Data Simulator Unit (G-SDSU) [32, 33] was used to calculate synthetic radar reflectivity and Doppler velocity profiles. Liquid water fractions of melting ice particles, predicted by the WRF simulation, are included in the calculation of the radar simulator for more consistent simulation of a bright band in addition to other standard model output variables.

Figure 3 shows observed and simulated radar reflectivity and Doppler velocity profiles in the form of the normalized contoured frequency of temperature diagrams (CFTDs) in the first precipitation event. The observed radar reflectivity and Doppler velocity were derived from the measurements of vertically-pointing micro rain radar (24.15-GHz transmit frequency) deployed at a field site of the LPVEx campaign. These data were combined with vertical temperature profiles of sounding data observed at a field site to plot the CFTDs. CFTDs are more useful to exhibit the correlation between probability distribution of a target quantity and temperature than conventional CFADs. The correlation is important for a discussion of the ice particles melting and consequent bright-band formation according to the ambient atmospheric temperature because melting is quite sensitive to the change of temperature. In fact, CFADs were plotted in the study for the analysis of bright band structure, but not shown in the publication. The reason was that the vertical temperature profiles of the real atmosphere and the model simulation were not completely identical [19, fig. 3], so that the vertical locations of observed and simulated bright bands could be different due to the error in simulating the temperature profiles. This problem could obscure the focus on the simulation of ice particles melting through the analysis of the bright band structure. The effects of the introduction of the time-dependent melting scheme on the simulation of bright bands are demonstrated in the CFTD plots. The observation CFTD for radar reflectivity in Fig. 3a shows relatively large reflectivity from 0 to 3 °C. The increases of reflectivity, under 0 °C level, seem to correspond to bright-band effects. The WRF-SBM simulation with the time-dependent melting scheme yields similar CFTD structure (Fig. 3b), which shows increases in reflectivity from approximately 0 to 3 °C temperature ranges. In contrast, the model simulation using an outdated instantaneous melting scheme in place of the time-dependent melting scheme is not able to reproduce the CFTD structure (Fig. 3c); increases in reflectivity do not appear at the temperature ranges. The Doppler velocity CFTD from the observations (Fig. 3d) shows gradual increases of the velocity with temperature in the range from 0 to 3 °C. The WRF-SBM simulation with the time-dependent melting scheme

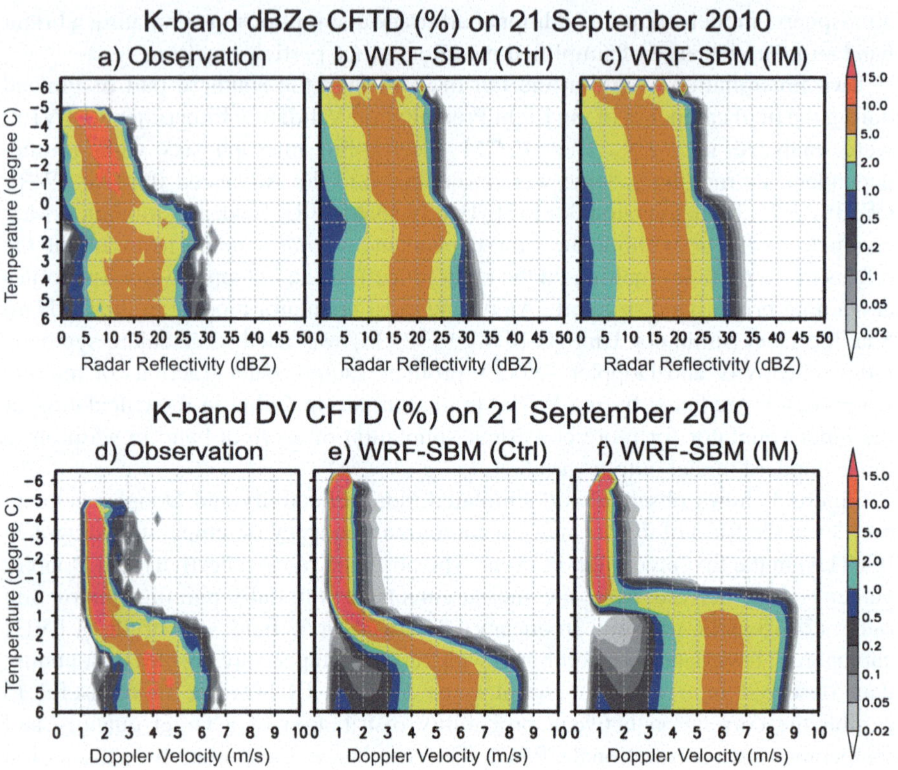

Fig. 3 CFTDs of the water equivalent radar reflectivity factor (dBZ) derived from the micro rain radar measurements at the Jarvenpaa site coupled with the vertical temperature profiles of sounding data at the Jokioinen site and obtained from the WRF coupled with the spectral-bin microphysics simulations in the 21 Sep 2010 case: (**a**) the measurements, (**b**) WRF-SBM simulation using the time-dependent melting scheme, and (**c**) simulation using the instantaneous melting scheme. CFTDs of the mean Doppler velocity ($\mathrm{m\,s^{-1}}$): (**d**) the measurements, (**e**) model simulation using the time-dependent melting scheme, and (**f**) simulation using the instantaneous melting scheme (from [19], with changes)

yields similar CFTD structure (Fig. 3e) again, whereas the CFTD of the simulation using the instantaneous melting scheme (Fig. 3f) does not exhibit such a zone with gradual velocity change.

Iguchi et al. [18] attempted overall validation of the whole structure of the cloud and precipitation, whereas [19] focused on a phenomenon in radar measurements related to a specific cloud microphysics process, i.e., melting of ice particles. The latter was able to suggest a firmer scientific message about the effects of gradual melting of falling ice particles on the formation of bright bands, although details of the disagreement between the observations and the simulations were not discussed and solved sufficiently. Strictly quantitative discussion on the difference between remote sensing measurements and model simulations to obtain robust conclusions is still difficult even by signal simulators, because of multiple factors for uncertainty and errors from both measurements and atmospheric model simulations. One of the solutions is to use multi-sensor multi-frequency approaches to avoid localized tuning of the microphysics [10, 32, 34]. More the observations could provide better constraint for the microphysics processes. Also, emerging polarimetric radars could yield better constraint in the hydrometeor distributions and the storm dynamics [42, 43].

3.4 *Producing a Testbed for Development of Retrieval Algorithms*

Once radiance-based model evaluation or even conventional retrieval-based evaluation has proved that simulated cloud and precipitation are valid, the simulation results can be utilized for testing and developing another retrieval algorithms for remote sensing measurements. The model output contains all grid-scale physical quantities, e.g., pressure, temperature, and water vapor, liquid and ice water contents, etc. First, these physical quantities in the model output are converted through a signal simulator to radiance or backscatter that can be obtained by a target remote sensing instrument. Second, the radiance or backscatter from the simulator is converted to physical quantities through retrieval algorithms that are a target for examination. Third, the retrieved physical quantities are compared with the same quantities in the original model output as a consistency check. A schematic flowchart of the process can be found in [54, fig. 1].

An example of the research was developed in the Global Precipitation Measurement (GPM) mission. GPM is an international mission providing worldwide precipitation measurements by the multiple Earth-observing satellites [15]. The GPM core satellite aboard multiple channels of the radar and radiometer plays a central role in calibrating precipitation algorithms that unifies the measurements of the GPM-constellation satellites. The core satellite started its measurements from space in February 2014 after the platform launch. However, even before the core satellite launch, the first version of GPM precipitation algorithms had to be established and tested upon the existing database, mainly targeting mid- to high-latitude precipitation systems beyond the observation ranges of the pre-existing TRMM core satellite. So-called synthetic GPM simulator had been established on the dataset of the field campaign measurements obtained at various GPM Ground Validation (GV) sites [33]. The GV ground- and aircraft-based observations were used to validate and constrain hindcast simulations using WRF model coupled with a spectral-bin microphysics parameterization [16, 17, 19]. The GV-constrained model output was subsequently inputed into the synthetic GPM simulator to generate GPM microwave bright temperature and dual-frequency radar reflectivity together with simulated geophysical parameters such as precipitation rates and water contents. This GPM simulator consisted of detailed satellite-orbital and sensor-

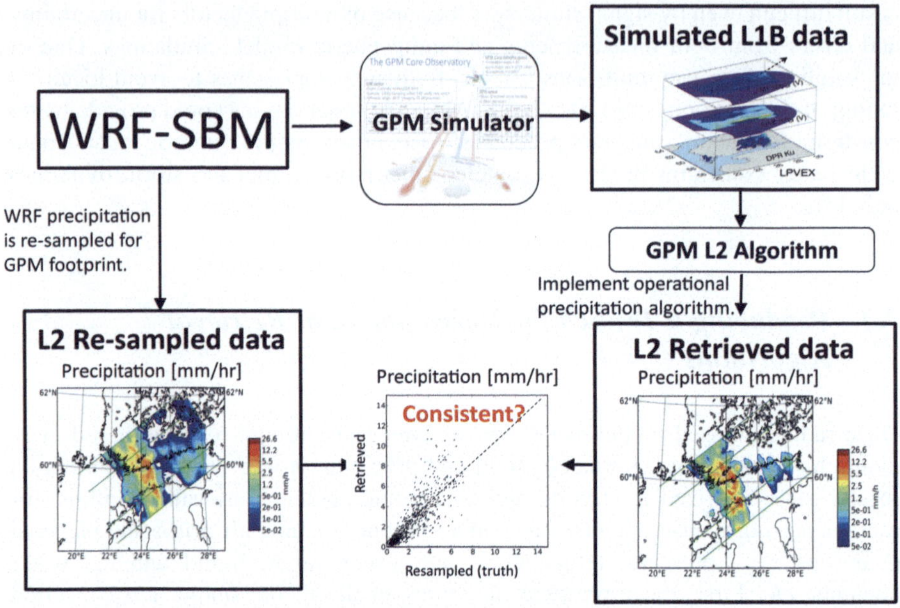

Fig. 4 Flow chart diagram of the GPM synthetic algorithm testing, starting from the WRF coupled with the spectral-bin microphysics (WRF-SBM) simulations to the evaluation of the GPM pre-launch precipitation algorithm using the synthetic-signal data and simulated precipitation rates

scanning modules, unified radiative transfer modules for microwave radiometer and precipitation radars, and realistic antenna gain modules upon the fields of the WRF model simulations. With the database, algorithm developers could test their algorithm even before the launch of the satellite (Fig. 4).

Kidd et al. [26] established a physically-based retrieval scheme to estimate precipitation from cross-track (XT) passive-microwave sensors on board the constellation satellites of the GPM mission. The algorithm is based upon the physically-based Goddard Profiling (GPROF) algorithm using database derived from observation-constrained model simulation. This modified GPROF scheme utilized NASA Multi-scale Modeling Framework (MMF) database to provide unified and comprehensive database, similar to that used in the conically-scanning (CS) retrieval scheme. Approximately one-million columns, i.e., two-dimensional cloud-resolving model profiles, within MMF were randomly chosen from the series of the simulations, and then applied to the GPM simulator [33] to simulate synthetic signals or brightness temperature from the forward radiative transfer by accounting the sensor-specific footprint size and viewing angles at various channel frequencies. The model database was evaluated and improved against the observation through modifying the hydrometeor mixing ratio in the MMF and the specification in the single scattering of the radiative transfer. The database ensures greater consistency across meteorological regimes and surface types by providing a more comprehensive set of global microphysics profiles. The database was corrected for bias against the

CS database to ensure consistency in the final product. Applying the technique to the Microwave Humidity Sounder (MHS) sensor observations, the MMF-GPROF algorithm provided equivalent or superior precipitation estimates to the CS sensors in many regions. The MMF-GPROF algorithm was also applied to the Advanced Technology Microwave Sounder (ATMS) and the Sondeur Atmospherique du Profil d' Humidité Intropicale par Radiométrie (SAPHIR) to provide the Version 3 of constellation XT precipitation estimates.

4 Summary and Conclusions

This chapter introduced approaches to the comparison between remote sensing measurements and hindcast simulations for cloud and precipitation and reviewed several simulation results of atmospheric models toward validation and improvement of cloud microphysics parameterizations. The comparison has been conventionally based on physical quantities estimated from the measurements through retrieval algorithms, against corresponding quantities calculated from the model simulations. This type of approach is still widely accepted as a simple and direct method to identify errors in target quantities with a great interest in the weather and climate research such as surface precipitation rates. However, significant uncertainties in the quantities retrieved from the remote sensing measurements had been recognized gradually, which might degrade the reliability of the validation. Alternatively, a concept of signal simulators was proposed, and simulator modules, matching with various remote sensing instruments, have been developed in the research community. The utilization of such simulators enables signal-based comparison between the measurements and the atmospheric model simulations without inherent uncertainty within the retrieval algorithms. Nevertheless, the agreement of signal radiances derived from measurements and simulations does not guarantee excellent simulation of all prognostic and diagnostic variables including, e.g., surface precipitation rates.

Even though the comparison upon signal simulators is less uncertain, the evaluation of cloud microphysics parameterizations is obstructed by errors and uncertainties due to forcing data driving the simulation, dynamics core, and other physics parameterizations as well as errors in the measurements. Monte Carlo experiments (so-called ensemble forecasting for numerical weather prediction) are a widely-used approach to investigate probability and errors in simulation results through perturbing the initial condition. On the other hand, data assimilation (e.g., [20]) aims at improving accuracy and precision of model forecasting through systematically correcting the simulated fields by incorporating observational data into the assimilation algorithms. Radiance-based assimilation system can utilize remote sensing measurement data even without conversion of quantities obtained from the measurements through retrieval algorithms (e.g., [56]). However, the conventional assimilation cannot be applied directly for the model parameterization improvement.

Some recent studies applied Bayesian analysis directly to exploring the detailed sensitivity of target physics parameters in the model output to specific parameters in physics parameterizations. Posselt [41] perturbed parameters in a single-moment bulk cloud microphysics parameterization to examine their impacts on model output parameters representing cloud structure, dynamics, and latent and radiative heating in idealized cloud-resolving simulations. A huge number of ensemble simulations were generated using a Bayesian Markov Chain Monte Carlo (MCMC) algorithm. The study concluded that assimilation of mean precipitation rates, vertically-integrated liquid and ice water contents, and radiative fluxes could not uniquely constrain a set of parameters investigated in the research. It also suggested that observing systems containing information to uniquely constrain hydrometeor PSD may be necessary to produce unique estimates of parameter values.

The type of analysis based on Bayesian inference has potential to provide a new perspective in the utilization of remote sensing measurements towards advances in cloud and precipitation modeling, beyond approaches upon conventional sensitivity tests based on speculation. It also may yield an important next step in the studies using signal simulators through the discussion on the parameter dependency between the radiance/backscatter signals and target physical quantities.

Acknowledgements The authors acknowledge the support from the NASA Precipitation Measurement Mission (PMM), NASA Modeling Analysis Prediction (MAP), and the Atmospheric System Research (ASR) Program of the U. S. Department of Energy (DOE).

References

1. Atlas, D. 1953. Optical extinction by rainfall. *Journal of Meteorology* 10: 486–488.
2. Bohren, C.F., and L.J. Battan. 1980. Radar backscattering by inhomogeneous precipitation particles. *Journal of the Atmospheric Sciences* 37: 1821–1827.
3. Borys, R.D., and M.A. Wetzel. 1997. Storm Peak Laboratory: a research, teaching, and service facility for the atmospheric sciences. *Bulletin of the American Meteorological Society* 78 (10): 2115–2123.
4. Chen, J.-P., and D. Lamb. 1994. Simulation of cloud microphysical and chemical processes using a multicomponent framework. Part I: Description of the microphysical model. *Journal of the Atmospheric Sciences* 51: 2613–2630.
5. Donovan, D.P., N. Schutgens, J.P.V. Baptista, H. Barker, J.P. Blanchet, A. Belaulne, and J. Testud. 2004. The EarthCARE Simulator. Users Guide and Final Report. KNMI, ESA/ESTEC, MSC, UQAM, GKSS, University of Kiel, CETP-UVSQ and Pennsylvania State University, 28, 48–49.
6. Draine, B.T., and P.J. Flatau. 1994. Discrete-dipole approximation for scattering calculations. *Journal of the Optical Society of America A* 11: 1491–1499.
7. Eito, H., and K. Aonashi. 2009. Verification of hydrometeor properties simulated by a cloud-resolving model using a passive microwave satellite and ground-based radar observations for a rainfall system associated with the Baiu front. *Journal of the Meteorological Society of Japan. Series II* 87: 425–446.
8. Evans, K.F., and G.L. Stephens. 1991. A new polarized atmospheric radiative transfer model. *Journal of Quantitative Spectroscopy and Radiative Transfer* 46: 413–423.

9. Han, Y., P. van Delst, Q. Liu, F. Weng, B. Yan, R. Treadon, and J. Derber. 2006. JCSDA community radiative transfer model (CRTM) - Version 1. NOAA Technical Report NESDIS, 122, 33.
10. Han, M., S.A. Braun, T. Matsui, and C.R. Williams. 2013. Evaluation of cloud microphysics schemes in simulations of a winter storm using radar and radiometer measurements. *Journal of Geophysical Research: Atmospheres* 118: 1401–1419.
11. Hansen, J.E., and L.D. Travis. 1974. Light scattering in planetary atmospheres. *Space Science Reviews* 16: 527–610.
12. Hashino, T., and G.J. Tripoli. 2007. The Spectral Ice Habit Prediction System (SHIPS). Part I: Model description and simulation of the vapor deposition process. *Journal of the Atmospheric Sciences* 64: 2210–2237.
13. Haynes, J.M., Z. Luo, G.L. Stephens, R.T. Marchand, and A. Bodas-Salcedo. 2007. A multipurpose radar simulation package: QuickBeam. *Bulletin of the American Meteorological Society* 88: 1723–1727.
14. Heymsfield, A.J. 2007. On measurements of small ice particles in clouds. *Geophysical Research Letters* 34 (23). https://doi.org/10.1029/2007GL030951.
15. Hou, A.Y., R.K. Kakar, S. Neeck, A.A. Azarbarzin, C.D. Kummerow, M. Kojima, R. Oki, K. Nakamura, and T. Iguchi. 2014. The global precipitation measurement mission. *Bulletin of the American Meteorological Society* 95: 701–722.
16. Iguchi, T., T. Matsui, J.J. Shi, W.K. Tao, A.P. Khain, A. Hou, R. Cifelli, A. Heymsfield, and A. Tokay. 2012. Numerical analysis using WRF-SBM for the cloud microphysical structures in the C3VP field campaign: impacts of supercooled droplets and resultant riming on snow microphysics. *Journal of Geophysical Research-Atmospheres* 117. https://doi.org/10.1029/2012JD018101.
17. Iguchi, T., T. Matsui, A. Tokay, P. Kollias, and W.K. Tao. 2012. Two distinct modes in one-day rainfall event during MC3E field campaign: analyses of disdrometer observations and WRF-SBM simulation. *Geophysical Research Letters*, 39. https://doi.org/10.1029/2012GL053329.
18. Iguchi, T., T. Nakajima, A.P. Khain, K. Saito, T. Takemura, H. Okamoto, T. Nishizawa, and W.K. Tao. 2012. Evaluation of cloud microphysics in JMA-NHM simulations using bin or bulk microphysical schemes through comparison with cloud radar observations. *Journal of the Atmospheric Sciences* 69: 2566–2586.
19. Iguchi, T., T. Matsui, W.K. Tao, A.P. Khain, V.T.J. Phillips, C. Kidd, T. L'Ecuyer, S.A. Braun, and A. Hou. 2014. WRF-SBM simulations of melting-layer structure in mixed-phase precipitation events observed during LPVEx. *Journal of Applied Meteorology and Climatology* 53: 2710–2731.
20. Kalnay, E. 2003. *Atmospheric Modeling, Data Assimilation and Predictability*. Cambridge: Cambridge University Press.
21. Kessler, E. 1969. *On the Distribution and Continuity of Water Substance in Atmospheric Circulations*. Meteorological Monographs. Vol. 32. Berlin: Springer.
22. Kessler, E. 1995. On the continuity and distribution of water substance in atmospheric circulations. *Atmospheric Research* 38: 109–145.
23. Khain, A., M. Ovtchinnikov, M. Pinsky, A. Pokrovsky, and H. Krugliak. 2000. Notes on the state-of-the-art numerical modeling of cloud microphysics. *Atmospheric Research* 55: 159–224.
24. Khain, A.P., V. Phillips, N. Benmoshe, and A. Pokrovsky. 2012. The role of small soluble aerosols in the microphysics of deep maritime clouds. *Journal of the Atmospheric Sciences* 69: 2787–2807.
25. Khain, A.P., K.D. Beheng, A. Heymsfield, A. Korolev, S.O. Krichak, Z. Levin, M. Pinsky, V. Phillips, T. Prabhakaran, and A. Teller. 2015. Representation of microphysical processes in cloud - resolving models: spectral (bin) microphysics versus bulk parameterization. *Reviews of Geophysics* 53: 247–322.
26. Kidd, C., T. Matsui, J. Chern, K. Mohr, C. Kummerow, and D. Randel. 2016. Global precipitation estimates from cross-track passive microwave observations using a physically based retrieval scheme. *Journal of Hydrometeorology* 17: 383–400.

27. Kummerow, C., J. Simpson, O. Thiele, W. Barnes, A.T.C. Chang, E. Stocker, R.F. Adler, A. Hou, R. Kakar, and F. Wentz. 2000. The status of the Tropical Rainfall Measuring Mission (TRMM) after two years in orbit. *Journal of Applied Meteorology* 39: 1965–1982.
28. Li, X., W.-K. Tao, T. Matsui, C. Liu, and H. Masunaga. 2010. Improving a spectral bin microphysical scheme using TRMM satellite observations. *Quarterly Journal of the Royal Meteorological Society* 136: 382–399.
29. Liu, Z. 2015. Comparison of precipitation estimates between Version 7 3-hourly TRMM Multi-Satellite Precipitation Analysis (TMPA) near-real-time and research products. *Atmospheric Research* 153: 119–133.
30. Marshall, J.S., and W.M. Palmer. 1948. The distribution of raindrops with size. *Journal of Meteorology* 5: 165–166.
31. Masunaga, H., T. Matsui, W.-K. Tao, A.Y. Hou, C.D. Kummerow, T. Nakajima, P. Bauer, W.S. Olson, M. Sekiguchi, and T.Y. Nakajima. 2010. Satellite data simulator unit: a multisensor, multispectral satellite simulator package. *Bulletin of the American Meteorological Society* 91: 1625–1632.
32. Matsui, T., X. Zeng, W.-K. Tao, H. Masunaga, W.S. Olson, and S. Lang. 2009. Evaluation of long-term cloud-resolving model simulations using satellite radiance observations and multifrequency satellite simulators. *Journal of Atmospheric and Oceanic Technology* 26: 1261–1274.
33. Matsui, T., T. Iguchi, X. Li, M. Han, W.-K. Tao, W. Petersen, T. L'Ecuyer, R. Meneghini, W. Olson, and C.D. Kummerow. 2013. GPM satellite simulator over ground validation sites. *Bulletin of the American Meteorological Society* 94: 1653–1660.
34. Matsui, T., J. Santanello, J.J. Shi, W.-K. Tao, D. Wu, C. Peters-Lidard, E. Kemp, M. Chin, D. Starr, and M. Sekiguchi. 2014. Introducing multisensor satellite radiance-based evaluation for regional Earth System modeling. *Journal of Geophysical Research: Atmospheres* 119: 8450–8475.
35. Mishchenko, M.I., L.D. Travis, and D.W. Mackowski. 1996. T-matrix computations of light scattering by nonspherical particles: a review. *Journal of Quantitative Spectroscopy and Radiative Transfer* 55: 535–575.
36. Murakami, M., and T. Matsuo. 1990. Development of the hydrometeor videosonde. *Journal of Atmospheric and Oceanic Technology* 7 (5): 613–620.
37. Okamoto, H., T. Nishizawa, T. Takemura, H. Kumagai, H. Kuroiwa, N. Sugimoto, I. Matsui, A. Shimizu, S. Emori, and A. Kamei. 2007. Vertical cloud structure observed from shipborne radar and lidar: midlatitude case study during the MR01/K02 cruise of the research vessel Mirai. *Journal of Geophysical Research: Atmospheres* 112: D08216.
38. Okamoto, H., T. Nishizawa, T. Takemura, K. Sato, H. Kumagai, Y. Ohno, N. Sugimoto, A. Shimizu, I. Matsui, and T. Nakajima. 2008. Vertical cloud properties in the tropical western Pacific Ocean: validation of the CCSR/NIES/FRCGC GCM by shipborne radar and lidar. *Journal of Geophysical Research: Atmospheres* 113. https://doi.org/10.1029/2008JD009812.
39. Olson, W.S., L. Tian, M. Grecu, K.-S. Kuo, B.T. Johnson, A.J. Heymsfield, A. Bansemer, G.M. Heymsfield, J.R. Wang, and R. Meneghini. 2016. The microwave radiative properties of falling snow derived from nonspherical ice particle models. Part II: Initial testing using radar, radiometer and in situ observations. *Journal of Applied Meteorology and Climatology* 55: 709–722.
40. Phillips, V.T.J., A. Pokrovsky, and A. Khain. 2007. The influence of time-dependent melting on the dynamics and precipitation production in maritime and continental storm clouds. *Journal of the Atmospheric Sciences* 64: 338–359.
41. Posselt, D.J. 2016. A Bayesian examination of deep convective squall-line sensitivity to changes in cloud microphysical parameters. *Journal of the Atmospheric Sciences* 73: 637–665.
42. Putnam, B.J., M. Xue, Y. Jung, G. Zhang, and F. Kong. 2017. Simulation of polarimetric radar variables from 2013 CAPS spring experiment storm-scale ensemble forecasts and evaluation of microphysics schemes. *Monthly Weather Review* 145: 49–73.

43. Ryzhkov, A., M. Pinsky, A. Pokrovsky, and A. Khain. 2011. Polarimetric radar observation operator for a cloud model with spectral microphysics. *Journal of Applied Meteorology and Climatology* 50: 873–894.
44. Saito, K., T. Fujita, Y. Yamada, J.-I. Ishida, Y. Kumagai, K. Aranami, S. Ohmori, R. Nagasawa, S. Kumagai, and C. Muroi. 2006. The operational JMA nonhydrostatic mesoscale model. *Monthly Weather Review* 134: 1266–1298.
45. Seifert, A., and K.D. Beheng. 2001. A double-moment parameterization for simulating autoconversion, accretion and selfcollection. *Atmospheric Research* 59: 265–281.
46. Seifert, A., and K.D. Beheng. 2006. A two-moment cloud microphysics parameterization for mixed-phase clouds. Part 1: Model description. *Meteorology and Atmospheric Physics* 92: 45–66.
47. Skamarock, W.C., J.B. Klemp, J. Dudhia, D.O. Gill, D.M. Barker, M.G. Duda, X.Y. Huang, W. Wang, and J.G. Powers. 2008. A description of the Advanced Research WRF Version 3, NCAR technical note, Mesoscale and Microscale Meteorology Division. National Center for Atmospheric Research, Boulder, CO.
48. Srivastava, R.C. 1967. A study of the effect of precipitation on cumulus dynamics. *Journal of the Atmospheric Sciences* 24: 36–45.
49. Stamnes, K., S.-C. Tsay, W. Wiscombe, and K. Jayaweera. 1988. Numerically stable algorithm for discrete-ordinate-method radiative transfer in multiple scattering and emitting layered media. *Applied Optics* 27: 2502–2509.
50. Stephens, G.L., D.G. Vane, R.J. Boain, G.G. Mace, K. Sassen, Z. Wang, A.J. Illingworth, E.J. O'Connor, W.B. Rossow, and S.L. Durden. 2002. The CloudSat mission and the A-Train: a new dimension of space-based observations of clouds and precipitation. *Bulletin of the American Meteorological Society* 83: 1771–1790.
51. Takahashi, T. 2010. The videosonde system and its use in the study of East Asian monsoon rain. *Bulletin of the American Meteorological Society* 91 (9): 1231–1246.
52. Tobo, Y., J. Uetake, Y. Uji, Y. Iwamoto, K. Miura, and R. Misumi. 2017. Routine measurements of atmospheric ice nucleating particles on the Tokyo Skytree, Atmospheric Ice Nucleation Conference - Focus Meeting 9, Leeds, UK. https://aerosol-soc.com/abstracts/routine-measurements-atmospheric-ice-nucleating-particles-tokyo-skytree.
53. Ulbrich, C.W. 1983. Natural variations in the analytical form of the raindrop size distribution. *Journal of Climate and Applied Meteorology* 22: 1764–1775.
54. Woods, C.P., D.E. Waliser, J.-L. Li, R.T. Austin, G.L. Stephens, and D.G. Vane. 2008. Evaluating CloudSat ice water content retrievals using a cloud-Řresolving model: sensitivities to frozen particle properties. *Journal of Geophysical Research: Atmospheres* 113: D00A11.
55. Yang, P., H. Wei, H.-L. Huang, B.A. Baum, Y.X. Hu, G.W. Kattawar, M.I. Mishchenko, and Q. Fu. 2005. Scattering and absorption property database for nonspherical ice particles in the near-through far-infrared spectral region. *Applied Optics* 44: 5512–5523.
56. Zhang, S.Q., T. Matsui, S. Cheung, M. Zupanski, and C. Peters-Lidard. 2017. Impact of assimilated precipitation-sensitive radiances on the NU-WRF simulation of the West African monsoon. *Monthly Weather Review*. https://doi.org/10.1175/MWR-D-16-0389.1.

Index

C. Andronache (ed.), *Remote Sensing of Clouds and Precipitation*,
Springer Remote Sensing/Photogrammetry,
https://doi.org/10.1007/978-3-319-72583-3

T

V

W

X

The manufacturer's authorised representative in the EU is Springer Nature Customer Service Centre GmbH, Europaplatz 3, 69115 Heidelberg, Germany. If you have any concerns regarding our products, please contact ProductSafety@springernature.com

Printed and bound by CPI Group (UK) Ltd, Croydon, CR0 4YY
05/10/2026
02228139-0010